DATA HANDLING IN SCIENCE AND TECHNOLOGY — VOLUME 25

Statistical Design — Chemometrics

DATA HANDLING IN SCIENCE AND TECHNOLOGY
Advisory Editors: S. Rutan and B. Walczak

Other volumes in this series:

Volume 1 Microprocessor Programming and Applications for Scientists and Engineers, by R.R. Smardzewski
Volume 2 Chemometrics: A Textbook, by D.L. Massart, B.G.M. Vandeginste, S.N. Deming, Y. Michotte and L. Kaufman
Volume 3 Experimental Design: A Chemometric Approach, by S.N. Deming and S.L. Morgan
Volume 4 Advanced Scientific Computing in BASIC with Applications in Chemistry, Biology and Pharmacology, by P. Valkó and S. Vajda
Volume 5 PCs for Chemists, edited by J. Zupan
Volume 6 Scientific Computing and Automation (Europe) 1990, *Preceedings of the Scientific Computing and Automation (Europe) Conference,12–15 June, 1990, Maastricht, The Netherlands*, edited by E.J. Karjalainen
Volume 7 Receptor Modeling for Air Quality Management, edited by P.K. Hopke
Volume 8 Design and Optimization in Organic Synthesis, by R. Carlson
Volume 9 Multivariate Pattern Recognition in Chemometrics, illustrated by case studies, edited by R.G. Brereton
Volume 10 Sampling of Heterogeneous and Dynamic Material Systems: Theories of Heterogeneity, Sampling and Homogenizing, by P.M. Gy
Volume 11 Experimental Design: A Chemometric Approach (Second, Revised and Expanded Edition) by S.N. Deming and S.L. Morgan
Volume 12 Methods for Experimental Design: Principles and Applications for Physicists and Chemists, by J.L. Goupy
Volume 13 Intelligent Software for Chemical Analysis, edited by L.M.C. Buydens and P.J. Schoenmakers
Volume 14 The Data Analysis Handbook, by I.E. Frank and R. Todeschini
Volume 15 Adaption of Simulated Annealing to Chemical Optimization Problems, edited by J. Kalivas
Volume 16 Multivariate Analysis of Data in Sensory Science, edited by T. Næs and E. Risvik
Volume 17 Data Analysis for Hyphenated Techniques, by E.J. Karjalainen and U.P. Karjalainen
Volume 18 Signal Treatment and Signal Analysis in NMR, edited by D.N. Rutledge
Volume 19 Robustness of Analytical Chemical Methods and Pharmaceutical Technological Products, edited by M.W.B. Hendriks, J.H. de Boer, and A.K. Smilde
Volume 20A Handbook of Chemometrics and Qualimetrics: Part A, by D.L. Massart, B.G.M. Vandeginste, L.M.C. Buydens, S. de Jong, P.J. Lewi, and J. Smeyers-Verbeke
Volume 20B Handbook of Chemometrics and Qualimetrics: Part B, by B.G.M. Vandeginste, D.L. Massart, L.M.C. Buydens, S. de Jong, P.J. Lewi, and J. Smeyers-Verbeke
Volume 21 Data Analysis and Signal Processing in Chromatography, by A. Felinger
Volume 22 Wavelets in Chemistry, edited by B. Walczak
Volume 23 Nature-inspired Methods in Chemometrics: Genetic Algorithms and Artificial Neural Networks, edited by R. Leardi
Volume 24 Handbook of Chemometrics and Qualimetrics, by D.L. Massart, B.M.G. Vandeginste, L.M.C. Buydens, S. de Jong, P.J. Lewi, and J. Smeyers-Verbeke

DATA HANDLING IN SCIENCE AND TECHNOLOGY—VOLUME 25

Advisory Editors: S. Rutan and B. Walczak

Statistical Design — Chemometrics

R.E. BRUNS

Instituto de Quimica, Universidade Estadual de Campinas, Brazil

I.S. SCARMINIO

Departamento de Quimica, Universidade Estadual de Londrina, Brazil

B. de BARROS NETO

Departamento de Quimica Fundamental, Universidade Federal de Pernambuco, Brazil

ELSEVIER

Amsterdam – Boston – Heidelberg – London – New York – Oxford
Paris – San Diego – San Francisco – Singapore – Sydney – Tokyo

ELSEVIER B.V.
Radarweg 29
P.O. Box 211, 1000 AE
Amsterdam
The Netherlands

ELSEVIER Inc.
525 B Street, Suite 1900
San Diego
CA 92101-4495
USA

ELSEVIER Ltd
The Boulevard,
Langford Lane, Kidlington
Oxford OX5 1GB
UK

ELSEVIER Ltd
84 Theobalds Road
London WC1X 8RR
UK

First edition 2006

Library of Congress Cataloging in Publication Data
A catalog record is available from the Library of Congress.

British Library Cataloguing in Publication Data
A catalogue record is available from the British Library.

ISBN-13: 978-0-444-52181-1
ISBN-10: 0-444-52181-x
ISSN: 0922-3487 (Series)

∞ The paper used in this publication meets the requirements of ANSI/NISO Z39.48-1992 (Permanence of Paper).
Printed in The Netherlands.

Preface

Utility ought to be the principal intention of every publication. Wherever this intention does not plainly appear, neither the books nor their authors have the smallest claim to the approbation of mankind.

Thus wrote William Smellie in the preface to the first edition of *Encyclopaedia Britannica*, published in 1768.

Our book has the modest intention of being useful to readers who wish — or need — to do experiments. The edition you are reading is a translation of a much revised, corrected and expanded version of our original text, *Como Fazer Experimentos*, published in Portuguese. To prepare this edition, every sentence was reconsidered, with the objective of clarifying the text. All the errors that we were able to discover, or the readers were kind enough to point out, have been corrected.

During the last 20 years or so we have spent considerable time teaching chemometrics — the use of statistical, mathematical and graphical techniques to solve chemical problems — to hundreds of students in our own universities, as well as in over 30 different industries. These students came principally from the exact sciences and engineering but other professional categories were also represented, such as management, medicine, biology, pharmacy and food technology. This diversity leads us to believe that the methods described here can be learned and applied, with varying degrees of effort, by any professional who has to do experiments.

Statistics does not perform miracles and in no way can substitute specialized technical knowledge. What we hope to demonstrate is that a professional who combines knowledge of statistical experimental design and data analysis with solid technical and scientific training in his own area of interest will become more competent, and therefore even more competitive.

We are chemists, not statisticians, and perhaps this differentiates our book from most others with similar content. Although we do not believe it is possible to learn the techniques of experimental design and data analysis without some knowledge of basic statistics, in this book we try to keep its discussion at the minimum necessary — and soon go on to what really interests the experimenter — research and development problems. On the other hand, recognizing that statistics is not very dear to the heart of many scientists and engineers, we assume that the reader has no knowledge of it. In spite of this, we arrive earlier at treating experimental problems with many variables than do more traditional texts.

Many people have contributed to making this book a reality. When the first edition came out, the list was already too extensive to cite everyone by name. We have been fortunate that this list has grown considerably since that time and our gratitude to all has increased proportionately. We do, however, wish to thank especially those whose work has allowed us to include so many applications in this edition. These people are cited with specific references when their results are discussed. We are also grateful to Fapesp, CNPq and Faep-Unicamp research granting agencies for partial financial support.

Of course, we remain solely responsible for the defects we have not been able to correct. We count on the readers to help us solving this optimization problem. Our electronic addresses are below. If you know of places where we could have done better, we will be most interested in hearing from you.

Campinas, July 2005.

B. de Barros Neto
Fundamental Chemistry Department
Federal University of Pernambuco
E-mail: bbn@ufpe.br

I. S. Scarminio
Chemistry Department
State University of Londrina
E-mail: ieda@qui.uel.br

R. E. Bruns
Chemistry Institute
State University of Campinas
E-mail: bruns@iqm.unicamp.br

Contents

CHAPTER 1
How statistics can help

For to be possessed of good mental powers is not sufficient; the principal matter is to apply them well. The greatest minds are capable of the greatest vices as well as of the greatest virtues, and those who proceed very slowly may, provided they always follow the straight road, really advance much faster than those who, though they run, forsake it.

Descartes, *Discourse on the Method of Rightly Conducting the Reason and Seeking for Truth in the Sciences*, Part I.

This is a book about good sense. More specifically, about good sense in performing experiments and in analyzing their results. Right at the beginning of his *Discourse on Method*, shortly before the sentence quoted above, Descartes states that "good sense is of all things in the world the most equally distributed, for everybody thinks himself so abundantly provided with it, that even those most difficult to please in all other matters do not commonly desire more of it than they already possess" (Descartes, 1637). If you believe this (Descartes obviously did not), this book is not for you.

Let us assume, however, that you agree with Descartes – after all, you are still reading – and think that not everything that looks obvious is really so obvious. In this case, if you are involved with experimentation, be it in academic life, in industry, or in a research and development laboratory, we believe this book can be very useful to you. With it you can learn to perform experiments and draw the appropriate conclusions from them more economically and efficiently.

In the following chapters we will discuss some relatively simple and easy to use experimental methods. These techniques might even appear obvious after you think a little about them, but this does not detract from their merit or effectiveness. To make this point clearer, let us consider a practical example, very easy to find in real life, especially in industry, where the cost–benefit ratio is always a very important consideration.

Suppose a chemist wishes to obtain the maximum yield from a certain reaction, and that only two variables, the temperature and the concentration of a certain reagent control this reaction. In the nomenclature that we shall adopt in this book, the property of interest, in this case the reaction yield, is the response. The variables that — in principle, at least — affect the response (that is, the temperature and the concentration) are the factors. The function describing this influence is called a response surface. The objective of the research worker is to find out the values — the levels — of the two factors that produce the largest possible response. How would you set about solving this problem?

Here is a suggestion based on common sense. To keep everything under control, we start by fixing one of the factors at a certain level and then vary the other one until we find the level of this second factor that produces the largest yield. By varying only one of the factors at a time, we are making sure that any change in the response is caused only by the changes made in this factor. Next, we set this factor at its optimum level, and then vary the first factor (the one initially held constant), until we also discover which of its possible levels yields the maximum response. Fine. That is all there is to it. The experiment is finished, and we have found the optimum values of the two factors, right?

Wrong! This might be common sense, but certainly it is not *good* sense. Almost everyone we asked agreed that the procedure we have just described is "the logical one". Yet there is another, much more efficient way, to perform the experiment. In fact, with this "common sense" approach the maximum yield would be discovered only in very fortunate circumstances. Contrary to what many people think, it is much better to vary *all factors at the same time*. This is because variables in general can influence each other, and the ideal level for one of them can depend on the levels of the others. This behavior, which is called an interaction between factors, is a phenomenon that happens very frequently. In fact, it is rare to find two factors acting in completely independent ways.

This is just one example of how common sense can be misleading. We will return to it later, for a more detailed treatment. In this chapter, we will just introduce some basic modeling notions and present concise descriptions of the techniques discussed in this book, trying to indicate how they can be useful to the experimenter.

1.1. Statistics can help

Problems are common, especially in industry, for which several properties have to be studied at the same time. These properties, in turn, are affected by a large number of experimental factors. How can we investigate the effects of all these factors on all properties, minimizing our work and reducing the costs of running the experiments? Then, how can we improve the quality of the resulting product? Then again, which experimental factors should be controlled to guarantee the expected quality of the end product?

Research aimed at finding answers to these questions often takes several months of work by scientists, engineers and technical personnel, with quite high costs in terms of salaries, reagents, chemical analyses and physical tests. The main goal of this book is to show that using some statistical concepts can help us answer these questions in a rational and economical way. By using experimental designs based on statistical principles, researchers can extract, from a minimum number of experiments, a maximum of useful information about the system under study.

Some of the most efficient methods to improve or optimize systems, products, and processes are presented in the chapters that follow. These methods are powerful tools, with which several specific objectives can be reached. We can make products with improved properties, shorten their development time, minimize their sensitivities to variations in environmental conditions, and so on.

Returning to our initial example, let us consider some specific questions about how experimental designs can help the researcher reach his objectives faster and in a less costly way. Let us say he[1] already knows that the temperature and concentration, as well as the type of catalyst, affect the yield. How would it be possible to set the temperature and concentration levels to increase productivity? Is it possible to maximize the reaction yield by varying these factors? Would changes in these values produce the same changes in the yield if a different catalyst were used? What experiments should we perform to obtain more information about the system? How can we quantify catalyst efficiency for different combinations of temperature and concentration? How can the levels of the experimental factors be changed to give the largest possible yield, while keeping the mechanical properties of the final product within specification? In the remaining chapters, we discuss statistical techniques for the design and analysis of experiments that will help us find reliable answers to all these questions.

The methods we will discuss do not depend on the nature of the problem under study. They are useful for studying chemical reactions, biological systems and mechanical processes, among many others, and can be applied to all possible scales of interest, from a single laboratory reaction to a full-scale industrial process. The statistical principles involved in all

[1]Or *she*, folks. We are not – definitely not – biased. This is just in order to avoid awkward constructions such as he/she or even the dreadful s(h)e. We promise we shall endeavor to treat both genres with equanimity all over the text.

these cases are exactly the same. Of course, our fondness for statistical methods implies no demeaning of the knowledge the technical expert already has about his system. Far from it. As we have stated in the preface, this is priceless. Statistical tools, no matter how valuable, are only a complement to technical knowledge. In an ideal scenario, these two — basic knowledge of the problem and statistics — should support each other.

1.2. Empirical models

When we attempt to model data obtained from experiments or observations, it is important to distinguish empirical from mechanistic models. We will try to clarify this difference considering two practical examples.

Suppose an astronomer wishes to predict when the next lunar eclipse will occur. As we know, the data accumulated after centuries of speculation and observation led, in the last quarter of the 17th century, to a theory that perfectly explains non-relativistic astronomical phenomena: Newtonian mechanics. From Newton's laws, it is possible to deduce the behavior of heavenly bodies as a logical consequence of their gravitational interactions. This is an example of a mechanistic model: with it we can predict trajectories of planets and stars because we know what causes their movements, that is, we know the mechanism governing their behavior. An astronomer only has to apply Newtonian mechanics to his data and draw the necessary conclusions. Moreover, he need not restrict his calculations to our own solar system: Newton's laws apply universally. In other words, Newtonian mechanics is also a global model.

And now for something completely different, and closer to many of us. A chemical engineer is asked to project a pilot plant based on a reaction recently developed in the research laboratory. She[2] knows that the behavior of this reaction can be influenced by many factors: the initial reagent amounts, the pH value of the reaction medium, the reaction time, the catalyst load, the rates at which the reagents are introduced into the reactor, the presence or absence of light, and so on. Even if a valid kinetic model were available for this reaction, it would be unlikely to account for the influences of all these factors, not to mention others that usually appear during scale-up from laboratory bench top to pilot plant. If we think of a full-scale plant, which is usually the end goal of the entire project, the situation becomes even more complex. Imponderable factors inevitably show up, such as impurity levels of the raw materials, environmental changes (humidity, for one), stability of the whole process and of course equipment aging and deterioration. In a situation so

[2]Or *he*, et cetera.

complicated, only the staunchest optimists among us would dream of discovering a mechanistic model for this process that could match the reliability of Newtonian mechanics in predicting the motion of large bodies. In such — so to speak — dire straits, the researcher is forced to resort to empirical models, that is, models that just try to *describe*, based on the available experimental evidence, the behavior of the process under study. This is totally different from trying to *explain*, from a few very clever laws, what is really taking place, which is what a mechanistic model tries to do.

But even finding a model to describe the behavior of the system may turn out to be a hopeless task. In empirical modeling, we are content if we manage to describe how our process behaves *in the investigated experimental range*. That is, empirical models are also just local models. Using them for making predictions outside the studied experimental domain is done strictly at the risk of the user. Crystal balls might prove just as effective.

If we had to describe this book in a single sentence, we would say that its objective is to teach the most useful techniques for developing empirical models.

1.3. Experimental design and optimization

Most people only think of statistics when faced with a lot of quantitative information to process. From the "common sense" perspective, using statistical methods would be comparable to mining.[3] The statistician would be some powerful miner, capable of exploring and processing mountains of numbers and extracting precious conclusions from them.

As with many things associated with common sense, this is another misconception, or at least an incomplete one. In an experiment, the most important statistical activity is not data analysis, but the actual *design* of the experimental runs that will produce the data. If this is not properly done, the experiment may yield only a sad bunch of meaningless values, where no statistical wizardry will help.

The secret of a good design is to set up the experiment to yield exactly the type of information we are seeking. To do this, first we have to decide clearly what we are looking for. Once again, this might seem obvious, but in fact it is not. Often it is the most difficult part of an experimental project. We would even say that good experimenters are, above all, people who know what they want. Depending on what the researchers want to know, some techniques will be very helpful, whereas others will be

[3]In fact, the expression *data mining* is now commonly used to describe exploratory investigations of huge data banks, usually from a business perspective.

Table 1.1
Evolution of an empirical study. Knowledge of the system increases as we make our way down the table.

Objective	Technique	Chapter
Variable screening	Fractional designs	4
Evaluation of variable influences	Full factorial designs	3
Empirical model-building	Least-squares modeling	5, 7
Optimization	RSM, simplex	6, 8
Mechanistic model-building	Deduction from general principles	...

worthless. If you want to be a good experimental designer, then, start by asking yourself:

- ***What — exactly what — would I like to know once the experiment is finished?***

Yogi Berra, the American baseball legend, was also known for his witticisms, some of them apparently paradoxical. One of them is very apposite here: *"You've got to be careful if you don't know where you're going 'cause you might not get there"*.

Imagine an axis describing the progress of an experimental investigation, starting from a situation of almost no information and proceeding until the development of a global mechanistic model (if the fates so wish). Moving along this axis corresponds to going down the lines in Table 1.1, which summarizes the contents of this book. In the first line, at a stage with very little information, we do not even know what are the most important variables influencing the system we are studying. Our knowledge is perhaps limited to a little practical experience or some bibliographic information. Under these conditions, we should start with a screening study, to discard unimportant variables. Using fractional factorial designs, discussed in Chapter 4, is an efficient way to screen out unimportant factors. Fractional designs are very economical and can be used to study dozens of factors simultaneously.[4]

Once we have identified the really significant factors, the next step should be to quantitatively evaluate their influences on the responses of interest, as well as any possible interactions between them. To do this with a minimum number of experiments, we can employ full factorial designs, treated in Chapter 3. Then, if we want a more detailed

[4]The technical literature reports a study of 127 variables (yes, one hundred and twenty-seven!) based on 128 experimental runs.

description, that is, a more sophisticated model, we can use least-squares modeling, which is the subject of Chapter 5. This is probably the most important chapter of the book, because many of the techniques discussed elsewhere are nothing more than special cases of least-squares modeling. An example is Chapter 7, dedicated to **mixture modeling**. Mixture models have some peculiarities but they also are models fitted by the least-squares procedure.

Sometimes our goal is to optimize our system, that is, maximize or minimize some response. It might be that at the same time we should also meet certain requirements. For example: producing the maximum amount of a certain product, at the lowest possible cost, and without violating its specifications. In this case, an adequate technique is response surface methodology (RSM), presented in Chapter 6 and also based on least-squares fitting. Later, in Chapter 8, we discuss a different optimization technique, the sequential simplex, which just seeks to reach an optimum point, with no regard for model building.

Once we have developed an empirical model, we must check if it is really adequate for the system behavior we are trying to describe. Then, and only then, should we try and draw conclusions from the model. Ill-fitted models belong to science fiction, not to science.

It is impossible to evaluate model fitting without some basic statistical concepts. But do not be alarmed. You will not have to become a master statistician to benefit from the techniques presented in this book. A few notions derived from the (deservedly) famous normal distribution will suffice. These, presented in Chapter 2, are very important if we wish to understand and to correctly apply the methods discussed in the rest of the book. In an effort to lighten the dullness that often plagues the discussion of such concepts, we base our treatment of the normal distribution on solving a practical problem of some relevance to the culinary world.

Applying the methods described in this book would be very tedious without the help of some software to do the calculations and draw the proper graphs. We used to distribute with this book a computer disk containing several programs written for this purpose. Today, the abundance of much more sophisticated programs, not only for Windows but also for Linux, many of them freeware, condemned our disk to obsolescence. An internet search will quickly reveal many interesting programs. A good site to start is www.statistics.com. And if you happen to be interested in our own old-fashioned software (now converted to Windows), you can download it free of charge at www.chemomatrix.-iqm.unicamp.br.

CHAPTER 2

When the situation is normal

A researcher is motivated to do experiments because of a desire to solve certain practical problems (or so we think). We wrote this book to show how, by applying statistical techniques, the efficiency of finding solutions to experimental problems can be improved. We would like to teach the reader to take full advantage of statistical techniques, not only for analyzing experimental results, but also — and principally — for systematically planning the experiments prior to making any measurements.

Deservedly or not, statistics is a discipline that enjoys little popularity in the chemical community and among scientists and engineers in general. Mention of the term immediately brings to mind overwhelming amounts of data inserted into huge tables. Therein lies, buried somewhere, the useful information we seek, and which we hope statistical methods will help us discover.

In fact, data analysis is just part of the contribution that statistics can bring to experimentation. Another part, just as important — perhaps even more important[5] — is helping to design the experiments from which the data will come. Many a researcher has been faced with the sad discovery that lack of proper planning can lead to useless results, no matter how noble the intentions behind the experiment. Even the most sophisticated analytical methods would be unable to draw any conclusions from such data. On second thoughts, almost none. Sir Ronald Aylmer Fisher, who invented many of the techniques that we will discuss, left a memorable reminder: "To call in the statistician after the experiment has been performed is like asking the coroner to do a postmortem. Maybe he can tell what the experiment died of."

[5]We think it is *much more* so.

Fortunately, we can easily avoid this distressing situation by carefully planning the experimental runs, taking all relevant details into account, and then using the appropriate analytical tools. Besides minimizing operational costs, we will thus ensure that our experiments will yield the information we need to correctly approach the original problem. With well-planned experiments it is much easier to draw valid conclusions. In fact, analyzing the results becomes an almost trivial step.

The reverse is also true. Researchers who ignore statistical design methods do so at the peril of arriving at doubtful conclusions. Worse still, their poorly designed experiments may not lead to any conclusions at all, their only practical result being a waste of time and money.

In this book, we present various techniques of experimental design and analysis. With a little study, any researcher can learn to apply them in his daily work. To discuss these techniques correctly, however, we will need a working knowledge of some statistical concepts, almost all ultimately based on the normal distribution. This is the rationale for the title chosen for this chapter.

Several excellent statistics textbooks are available, from the most elementary to the very advanced. Typically, they concentrate on specific areas — social sciences, humanities, health sciences and of course physical sciences and engineering. In general they treat many subjects that are undoubtedly important from a strictly statistical point of view, but not all are relevant to our study of experimental design and analysis. Since we wish to arrive as quickly as possible at practical applications without losing statistical rigor, we present only those statistical concepts most essential to the work of an experimenter in this chapter. As boring as statistics can often appear, it is fundamental for planning and performing experiments. To take advantage of the full potential of the techniques presented in this book, you are strongly advised to master the contents of this chapter. If statistics is not among your best-loved subjects in the syllabus, please bear with us and make a sincere effort to learn the few really vital statistical concepts. You will see it is time well spent.

2.1. Errors

To obtain reliable data we need to carry out well-defined procedures, whose operational details depend on the goal of the experiment. Imagine, for example, that our problem is to determine the concentration of acetic acid in a vinegar sample. Traditionally, this is done with an acid–base titration. Following the usual method, we need to

(a) prepare the primary standard solution;
(b) use it to standardize a sodium hydroxide solution of appropriate concentration; and
(c) do the actual titration.

Each of these steps involves a certain number of basic operations, such as weighings, dilutions and volume readings.

Such determinations may be performed in government regulatory laboratories to certify that the vinegar complies with official standards (at least 4% acetic acid, usually).

Suppose an analyst titrates two samples from different manufacturers, and finds 3.80% of acetic acid for one sample and 4.20% for the other. Does this mean the former sample should be rejected because it does not meet the legal minimum specification?

The truth is we do not know, as yet. We cannot provide a fair answer without an estimate of the uncertainty associated with these values. Each laboratory operation involved in the titrations is subjected to errors. The type and magnitude of these errors, the extents of which we have not yet ascertained, will influence the final results — and therefore our conclusions. The apparently unsatisfactory result might not be due to the sample itself but to inherent variations in the analytical procedure. The same might be said of the result that seems to fall within specification.

2.1.1. Types of error

We all know that any measurement is affected by errors. If the errors are insignificant, fine. If not, we run the risk of making incorrect inferences based on our experimental results, and maybe arriving at a false solution to our problem. To avoid this unhappy ending, we need to know how to account for the experimental errors. This is important, not only in the analysis of the final result, but also — and principally — in the actual planning of the experiments, as we have already stated. No statistical analysis can salvage a badly designed experimental plan.

Suppose that during the titration of the vinegar sample our chemist is distracted and forgets to add the proper indicator to the vinegar solution (phenolphthalein, since we know the equivalence point occurs at a basic pH). The consequence is that the end point will never be reached, no matter how much base is added. This clearly would be a serious error, which statisticians charitably label as a gross error. The person responsible for the experiment often uses a different terminology, not fit to print here.

Statistics is not concerned with gross errors. In fact, the science to treat such mistakes has yet to appear. Little can be done, other than learn the lesson and pay more attention next time. Everyone makes mistakes. The conscientious researcher should strive to do everything possible to avoid committing them.

Imagine now that the stock of phenolphthalein is depleted and the chemist decides to use another indicator that happens to be available, say, methyl red. Since the pH range for the turning point of methyl red is below 7, the apparent end point of the titration will occur before all of the acetic acid is neutralized. Therefore, the vinegar will appear to have a lower concentration of acetic acid than the true one. If several samples were titrated with methyl red, all would appear to have concentrations lower than the actual concentrations. Now our chemist would be committing systematic errors. This type of error always distorts the result in the same direction, deviating either positively or negatively from the true value. In the absence of other types of error, the use methyl red instead of phenolphthalein will always result in an acid concentration lower than the true value.

It is easy to imagine other sources of systematic error: the primary standard might be out of specification, an analytical balance or a pipette might be erroneously calibrated, the chemist performing the titration might read the meniscus from an incorrect angle and so on. Each of these factors will individually influence the final result, always in a characteristic direction.

With care and hard work systematic errors can be minimized. Once the measuring instruments have been ascertained to be working properly, we simply follow the stipulated experimental procedure. For example, if we are supposed to use phenolphthalein, we use phenolphthalein, not methyl red.

Our indefatigable chemist finally satisfies himself that everything has been done to eliminate systematic errors. Then, with strict adherence to the analytical protocol, he proceeds to titrate two samples taken from the same lot of vinegar. Confident that the entire analytical process is now under control, our chemist naturally expects that the two titrations will give the same result. After all, the samples come from the same source. Upon comparing the two values obtained in the titrations, however, he finds that, though similar, they *are not exactly the same*. This can only mean that some error source, fortunately small, is still affecting the results.

To further investigate these errors, the chemist decides to perform several more titrations for other samples taken from the same lot. The results for 20 titrations are given in Table 2.1 and also plotted in Fig. 2.1.[6]

Examining the results of the 20 titrations we see that:

- The values fluctuate, but tend to cluster around a certain intermediate value.
- The fluctuation about the central value seems to be random. Knowing that the result of a specific titration falls below the average value, for example, will not help us predict whether the next titration will result in a value above or below average, nor the extent of the deviation.
- Since most of the concentrations determined are less than 4%, it seems as though the sample is indeed out of specification.

[6]Fellow chemists will undoubtedly notice that this is an absurdly low precision for a volumetric procedure. We are doing some exaggeration for didactical purposes.

Table 2.1
Titration results for 20 samples taken from the same lot of vinegar

Titration no.	Concentration (%)	Titration no.	Concentration (%)
1	3.91	11	3.96
2	4.01	12	3.85
3	3.61	13	3.67
4	3.83	14	3.83
5	3.75	15	3.77
6	3.91	16	3.51
7	3.82	17	3.85
8	3.70	18	4.04
9	3.50	19	3.74
10	3.77	20	3.97

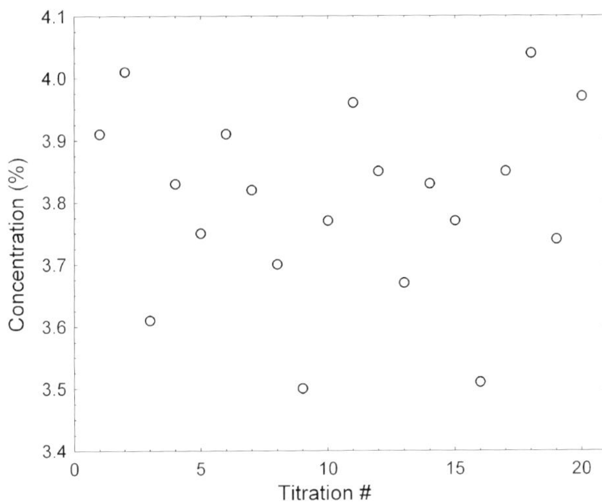

Fig. 2.1. Titration results for 20 samples taken from the same batch of vinegar.

Situations like this are very common in experimental determinations. No matter how hard we try to control all the factors assumed to influence our results, some sources of error always remain. These errors are generally small and tend to occur at random, as stated in the second point above. Sometimes they push the result up, sometimes down, but their effect appears to be due to chance.

Consider the titration. Even if the experimental procedure is strictly followed and all operations are made with utmost care, unpredictable fluctuations in the concentrations will always arise. A small change in the viewing angle while reading the burette, a last droplet that remains in the pipette, and a different shade of the turning point, can all influence the titration result. Since we cannot control these factors, we cannot predict the effects of their variability on the result. These variations lead to errors that seem to be due to chance, and for this reason they are called random errors.

A little reflection should convince us that it is impossible to fully control *all* the variable factors involved in an experiment, however simple the experiment. This implies that any experimental determination will be, to a larger or smaller extent, affected by random errors. If we want to draw valid conclusions, these errors must be taken into account; this is one of the reasons for using statistics.[7]

Exercise 2.1. Think of a simple experiment and identify some factors that would prevent us from obtaining its final result strictly without error.

2.2. Populations, samples and distributions

The first step in treating random errors is to adopt some hypothesis about their distribution. The most common starting point, when treating continuous measurements, is to assume that their error distribution is Gaussian or, as it is usually called, normal. In this section we discuss this hypothesis and its practical consequences by considering the following problem:

- How many beans are needed to make a pot of chili?

Evidently the answer depends, among other things, on the size of the pot. We will assume that our recipe calls for 1 kg of beans, so our question will be recast into determining the number of beans that add up to 1 kg.

One possible solution would be to count all the beans one by one. Since we are interested in a statistical treatment of the question, we can discard this option immediately.[8] We will take an alternative approach. First we will determine the mass of a single bean. Then we will divide 1000 g by this value. The result would be the number of beans in 1 kg, were it not for a complication we will discuss shortly.

Exercise 2.2. Try to guess the number of black beans making up 1 kg. Of course this is not the recommended way to solve this problem (unless you happen to be a psychic), but your guess can be used later in a statistical test.

[7]Note that *error*, in this last sense, should not be viewed as a disparaging term; it is rather a feature of all experimental data with which we must live, like it or not.

[8]Not to mention that life is too short.

Using an analytical balance, our inestimable coworker M.G. Barros (MGB) weighed one bean taken at random from a package of black beans and obtained a value of 0.1188 g. Weighing a second bean, again picked at random, he obtained 0.2673 g. If all the beans had the same mass as the first one, there would be 1000 g/0.1188 g, or about 8418 beans in a kilogram. On the other hand, if each bean has the same mass as the second, this number would fall to 3741. Which of these values is the one we seek?

Neither, of course. Since the mass varies from bean to bean, we should not use individual mass values in our calculations but rather the average taken over all the beans. To obtain this average, we have only to divide the total mass of the package of beans (1 kg) by the number of beans it contains. Unfortunately this bring us back to the starting point — to know how many beans there are in 1 kg, first we need to know … how many beans are there in 1 kg.

If all beans were identical, the average mass would equal the mass of any one of them. To find the answer to our question we could simply weigh any single bean. The problem, of course, is that the mass varies from bean to bean. Worse than that, it varies unpredictably. Who could have guessed that after drawing a bean weighing 0.1188 g from the package, MGB would pick out one weighing exactly 0.2673 g?

We cannot predict the exact mass of a bean drawn from the package, but we can use common sense to set some limits. For example, the mass cannot be less than zero and evidently it must be much less than 1 kg.There are large and small beans, of course, but just by looking at the beans we will see that most of them are about the same size. In other words, we have a situation like that of the vinegar titrations. The individual values vary, but do so around a certain central value. Now, however, the variation is not caused by problems of measurement or instrumentation but by the random element present in our sampling procedure.[9]

In statistics, the set of all possible values in a given situation is called a population. The target of any experimental investigation is always a population. Our objective in collecting and analyzing data is to conclude something about that population.

In any problem it is essential to clearly define the population in which we are interested. Incredible as it may seem, this seemingly banal detail is often not clear to the researcher, who then risks extrapolating his conclusions to systems that fall outside the experimental range she

[9]We are, of course, ignoring errors arising from the weighing process itself. This is of little importance in this example, because, unless the balance is severely malfunctioning, such errors are several orders of magnitude less than the variation due to sampling.

studied. For example, in our gravimetric study of beans the population is the set of individual masses of all the beans in that package. The answer we seek refers to the package as a whole, even if the beans are not all investigated one by one. And unless we introduce another hypothesis (that the package is representative of an entire harvest, for example), our results will refer only to this particular package and to no other.

By individually weighing all beans in the package we would obtain the exact distribution of the weights in the population. We could then calculate the true population mean, that is, the correct average mass of the beans in the package. However, having already rejected the idea of counting all the beans, why would we now weigh them all, one by one? Evidently this is not the solution.

Instead of worrying about the true average (the population mean), which could only be determined by examining all the beans, we will try to be content with an estimate calculated from only some of the beans, that is, from a sample taken from the population. If this sample is sufficiently representative of the population, the sample average should be a good approximation to the population mean, and we might use it to draw conclusions about the population as a whole.

If the sample is to be a realistic and unbiased representation of the entire population, its elements must be chosen in a rigorously random way. In our bean problem, this means the chance of weighing a given bean must be exactly the same for all other beans. After randomly choosing one bean and weighing it, we should return it to the package and mix it thoroughly with the others, so that it has the same chance to be chosen again.[10] Without this precaution, the population will be increasingly modified as we remove more beans and the sample will no longer faithfully represent the original population. This condition is very important in practice, because statistical inferences always assume that the samples are **representative** of the population. When we run an experiment we should always be careful to collect the data so that they are representative of the population we wish to study.

- Population: A collection of individuals or values, finite or infinite.
- Sample: A part of the population, usually selected with the objective of making inferences about the population.
- Representative sample: A sample containing the relevant characteristics of the population in the same proportion that they occur in the population.
- Random sample: A sample of n values or individuals selected in such a way that all possible sets of n values from the population have the same chance of being selected.

[10]This procedure is known as sampling with replacement. If we had to subject the sample to destructive assays, as is sometimes the case with routine inspection of factory production lines, obviously there would be no replacement.

Exercise 2.3. In the bean example the population is finite: the total number of beans can be large but it is limited. The set of all concentrations that in principle can be obtained in the titration of a certain sample constitutes a finite or infinite population? (Note the expression "in principle". Imagine that it is possible to make as many titrations as you wish without running the risk of using up all the stocks of material and reagents.)

2.2.1. How to describe the characteristics of the sample

Table 2.2 contains the individual masses of 140 beans randomly drawn from a package containing 1 kg of black beans, courtesy of the tireless MGB. Examining these data carefully, we can confirm our expectation of a more or less restricted variation. The largest value is 0.3043 g (fifth value in the next-to-last column). The smallest is 0.1188 g, coincidentally the first one. Most of the beans indeed appear to weigh about 0.20 g.

Interpreting the data is easier if we divide the total mass range into small intervals and count the number of beans falling into each interval. In view of the extreme values observed, the 0.10–0.32 g range is wide enough to accommodate all values in Table 2.2. Dividing it into intervals of 0.02 g width and assigning each mass to its proper interval, we obtain the results given in Table 2.3. The center column shows at once that the intervals close to 0.20 g are indeed the ones containing the largest numbers of beans.

Dividing the number of beans in a given interval by the total number of beans in the sample, we obtain the relative frequency corresponding to that interval. For example, the 0.26–0.28 g interval contains 7 beans out of the total of 140. The relative frequency is thus 7 ÷ 140, or 0.050. This means that 5% of the beans weighed between 0.26 and 0.28 g.

The frequencies calculated for all 11 intervals appear in the last column of Table 2.3. It is better to analyze the data in terms of frequencies rather than absolute number of observations, since the theoretical statistical distributions are frequency distributions. Knowing these frequencies we can determine the probabilities of observing certain values of interest. With these probabilities we can in turn test hypotheses about a population, as we shall presently see.

Any data set is more easily analyzed when its values are represented graphically. In the traditional plot for a frequency distribution, each interval is represented by a rectangle whose base coincides with the width of that interval and whose area is identical, or proportional, to its frequency. The geometrical figure thus obtained is called a histogram. Since the sum of all the frequencies must be equal to unity (that is, the percentages must add up to 100%), the total area of the histogram is also equal to unity, if the area of each rectangle is made equal to the frequency of the corresponding interval. Fig. 2.2 shows the histogram of the frequencies of Table 2.3. To facilitate comparison with the data in the

Table 2.2
Masses of 140 black beans drawn at random from a 1 kg package (in g)

0.1188	0.2673	0.1795	0.2369	0.1826	0.1860	0.2045
0.1795	0.1910	0.1409	0.1733	0.2146	0.1965	0.2326
0.2382	0.2091	0.2660	0.2126	0.2048	0.2058	0.1666
0.2505	0.1823	0.1590	0.1722	0.1462	0.1985	0.1769
0.1810	0.2126	0.1596	0.2504	0.2285	0.3043	0.1683
0.2833	0.2380	0.1930	0.1980	0.1402	0.2060	0.2097
0.2309	0.2458	0.1496	0.1865	0.2087	0.2335	0.2173
0.1746	0.1677	0.2456	0.1828	0.1663	0.1971	0.2341
0.2327	0.2137	0.1793	0.2423	0.2012	0.1968	0.2433
0.2311	0.1902	0.1970	0.1644	0.1935	0.1421	0.1202
0.2459	0.2098	0.1817	0.1736	0.2296	0.2200	0.2025
0.1996	0.1995	0.1732	0.1987	0.2482	0.1708	0.2465
0.2096	0.2054	0.1561	0.1766	0.2620	0.1642	0.2507
0.1814	0.1340	0.2051	0.2455	0.2008	0.1740	0.2089
0.2595	0.1470	0.2674	0.1701	0.2055	0.2215	0.2080
0.1848	0.2184	0.2254	0.1573	0.1696	0.2262	0.1950
0.1965	0.1773	0.1340	0.2237	0.1996	0.1463	0.1917
0.2593	0.1799	0.2585	0.2153	0.2365	0.1629	0.1875
0.2657	0.2666	0.2535	0.1874	0.1869	0.2266	0.2143
0.1399	0.2790	0.1988	0.1904	0.1911	0.2186	0.1606

Table 2.3
Frequency distribution of the masses of 140 beans randomly drawn from a package of 1 kg of beans (in g)

Interval (g)	Number of beans	Frequency[a]
0.10–0.12	1	0.007
0.12–0.14	4	0.029
0.14–0.16	11	0.079
0.16–0.18	24	0.171
0.18–0.20	32	0.229
0.20–0.22	27	0.193
0.22–0.24	17	0.121
0.24–0.26	15	0.107
0.26–0.28	7	0.050
0.28–0.30	1	0.007
0.30–0.32	1	0.007
Total	140	1.000

[a] Number of beans in the interval divided by the total number of beans, 140.

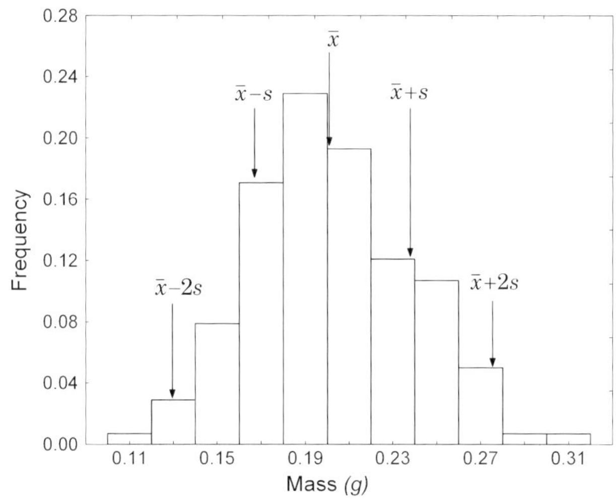

Fig. 2.2. Histogram of the masses of 140 beans drawn at random from a 1-kg package of black beans (in g). The meanings of the symbols are explained in the text.

table, we have made the height of each rectangle, rather than its area, equal to the frequency of the corresponding interval. Since the bases of the rectangles are all the same, this does not change the shape of the histogram.

The advantages of the graphical representation are obvious. The concentration of the masses around the 0.20 g value is noticed immediately, as is the progressive decrease in the number of beans as the masses become further removed from this central value, in both directions. The symmetry of the distribution also stands out: the part to the right of the central region is more or less the mirror image of the part on the left. This feature would be very hard to perceive straight from the values in Table 2.2.

So here is our advice: *if you wish to analyze a data set, one of the first things you should think of doing is plotting the data*. This is the statistical counterpart to the old saying that a picture is worth a thousand words.

Exercise 2.4. Use the data in Table 2.3 to confirm that 54.3% of the beans have masses between 0.18 and 0.24 g.

Exercise 2.5. Draw a histogram of the data in Table 2.1. The literature recommends that the number of rectangles be approximately equal to the square root of the total number of observations. Since the table has 20 values, your histogram should have 4 or 5 rectangles. Five is preferable, because using an odd number of rectangles allows easier visualization of possible symmetries.

The histogram of Fig. 2.2 is a graphical representation of all 140 numerical values of our sample. Its basic characteristics are

- The location of the set of observations at a certain region on the horizontal axis
- Their scattering, or dispersion, about this region

These characteristics can be represented numerically, in an abbreviated form, by several statistical quantities. In the physical sciences, where the variables usually assume values within a continuous interval, the most commonly used quantities are the arithmetic average and the standard deviation, respectively.

The arithmetic average of a data set is a measure of its location, or *central tendency*. This average is given simply by the sum of all the values divided by the total number of elements in the data set. Since this is the only concept of average used in this book, from this page on we will refer to this quantity using only the word "average", with the qualifier "arithmetic" implicitly understood.

A bar placed above the symbol denoting the sample elements usually indicates the average value of a sample. If we use the symbol x to represent the mass of a single bean, the average mass of all beans in the data set is denoted by $\bar{x}$, and given by

$$\bar{x} = \frac{1}{140}(0.1188 + 0.2673 + \cdots + 0.1606) = 0.2024 \text{ g}.$$

With this value[11] we can estimate that the kilogram of beans contains about $1000\,\text{g} \div 0.2024\,\text{g bean}^{-1} = 4940$ beans. This estimate, however, was obtained from the observation of only 140 beans, or 3% of the total assuming the 1-kg package contains about 5000 beans. We cannot expect this estimate to be equal to the exact value, which remains unknown. Our calculation yields the sample average, not the population mean. We will see later how to estimate the uncertainty in this result.

Sample average:

$$\bar{x} = \frac{1}{n}\sum_{i=1}^{n} x_i \tag{2.1}$$

$x_i = i^{\text{th}}$ value; n = number of values in the sample.

Our measure of dispersion will be the standard deviation. To obtain it, we first calculate the deviation, or residual, of each value relative to the

[11]The average is usually expressed with one decimal digit more than the original data. In our example, where the data have four significant digits, this is of no practical importance.

sample average:

$$d_i = x_i - \bar{x}.$$

Then we sum the squares of all these deviations and divide by *n–1*. This gives the variance of the set of observations, represented by the symbol s^2 (Eq. (2.2)).

Note that the variance is the average of the squared deviations except that the denominator is $n - 1$, instead of the total number of observations. To understand this, we must remember that the original observations, obtained by random sampling, were all independent. Even if we knew the masses of the first 139 beans, we would have no way of predicting the mass of the next bean, the 140th. In statistical parlance we say that this set has 140 degrees of freedom. Any single value is independent of the other members of the set.

Sample variance:

$$V(x) = s^2 = \frac{1}{n-1}\sum_{i=1}^{n} d_i^2 = \frac{1}{n-1}\sum_{i=1}^{n} (x_i - \bar{x})^2 \qquad (2.2)$$

x_i = ith value; n = number of values in the sample; $\bar{x}$ = sample average.

With the deviations, a distinction arises. Let us see what happens when we sum the deviations themselves, rather than their squares (all sums are from $i = 1$ to $i = n$):

$$\sum_i d_i = \sum_i (x_i - \bar{x}) = \sum_i x_i - \sum_i \bar{x} = \sum_i x_i - n\bar{x}.$$

Recalling that the average is defined by $\bar{x} = 1/n \sum_i x_i$, we can substitute the $n\,\bar{x}$ term by the sum $\sum_i x_i$. This gives

$$\sum_i d_i = \sum_i x_i - n\bar{x} = \sum_i x_i - \sum_i x_i = 0. \qquad (2.3)$$

In other words, the 140 deviations from the sample average are not all independent. If we already know the deviations of 139 beans, the deviation of the missing one is automatically determined: it is the value needed to make the whole sum equal to zero. The restriction expressed by Eq. (2.3), which comes from calculating the average, removes one degree of freedom from the set of deviations. Of the n deviations, only $n - 1$ can vary randomly. It is only natural that the denominator in the definition of the variance be $n - 1$ instead of n.

The degree of freedom concept is very important. Later we will encounter examples where more than one restriction is imposed on a set of values. With p such restrictions, the number of degrees of freedom will be reduced from n, the total number of elements in the set, to $n - p$. It is this last value that should be used as the denominator in a mean square analogous to Eq. (2.2).

For our sample, where $\bar{x} = 0.2024$ g, the variance is, according to Eq. (2.2),

$$s^2 = \frac{1}{139}\left[(0.1188 - 0.2024)^2 + \cdots + (0.1606 - 0.2024)^2\right] \cong 0.00132 \text{ g}^2.$$

The sample average has the same unit as the original observations, but the variance unit is by definition the square of the original unit of measurement. Interpreting the experimental data is easier if the measures of dispersion and location are given in the same units, so we usually substitute the variance by its positive square root, called the standard deviation. In our example the standard deviation is given by

$$s = \sqrt{(0.00132 \text{ g}^2)} = 0.0363 \text{ g}.$$

Sample standard deviation:

$$s = \sqrt{V(x)} = \sqrt{s^2} \tag{2.4}$$

s^2 = sample variance.

Exercise 2.6. Calculate the average and standard deviation of the first 10 values in Table 2.2 (from 0.1188 to 0.1409 g).

The standard deviation is generally used to define intervals about the average.[12] In our sample of 140 beans, for example, the limits of the interval defined by ± 1 standard deviation are given by 0.2024 ± 0.0363 g or 0.1661 and 0.2387 g. The region spanned by these two values (Fig. 2.2) corresponds to 66.6% of the total area of the histogram. This means that two-thirds of all the observed masses fall between these limits. The region defined by two standard deviations about the average goes from 0.1298 to 0.2750 g, and contains 96.8% of the total area. Subject to certain restrictions to be discussed later, these sample intervals can be used to test hypotheses concerning the underlying population.

These longhand arithmetic manipulations were done solely for didactic reasons. You need not worry about the prospect of calculating endless sums just to obtain averages, standard deviations, or many other statistical quantities. Any scientific electronic calculator comes already pre-programmed for these operations. Moreover, several easily accessible computer programs, some of them freeware, perform not only these calculations but much more complex ones as well. The sooner you learn to use one of these programs the better. If for no other reason, statistics will appear less of a burden.

The $\bar{x} = 0.2024$ g and $s = 0.0363$ g values were obtained from the individual weighing of 140 beans and, therefore, faithfully represent the

[12]The standard deviation is usually calculated with *two* more decimal places than the original data. Again, we are disregarding this detail in our example.

sample: they are sample estimates (or statistics). The values that ultimately interest us, however, are the population parameters. Our primary interest is the number of beans in the entire 1-kg package, not just in the small sample of 140 beans (where, for that matter, the number of course is 140).

Statisticians normally use Latin symbols to represent sample values, reserving the Greek alphabet for population parameters. Following this convention, we represent the population mean and standard deviation of our example by the Greek letters μ and σ, respectively. What can we infer about the values of these parameters, knowing the sample values $\bar{x}$ and s?

2.3. The normal distribution

Suppose that the beans whose masses appear in Table 2.2 are separated from the rest of the package and treated as a small population of only 140 elements. We have already seen in Table 2.3 that 5% of these elements have masses between 0.26 and 0.28 g. Since we know the exact frequency distribution of the masses in this small population, we can say that the probability of randomly drawing a bean from the package with a mass in the 0.26–0.28 g range is 5%. We could do the same to any bean randomly drawn from the package if we knew exactly the frequency distribution in the whole package (that is, in the population). To do that, however, we would have to weigh every single bean.

Imagine now that we knew of a model that adequately represented the distribution of the masses of all the beans in the package. Then we would not need to weigh each bean to make inferences about the population. We could base our conclusions entirely on that model, without having to do any additional experimental work.

This concept — using a model to represent a certain population — is the central theme of this book. It will be present, implicitly or explicitly, in all of the statistical techniques we shall discuss. Even if in some cases we do not formally state our model, you will recognize it from context. Of course our inferences about the population will only be correct insofar as the chosen model is valid. For any situation, however, we will always follow the same procedure:

- Postulate a model to describe the population of interest
- Check the adequacy of this representation
- If satisfied, draw the appropriate conclusions; otherwise, we change the model and try again.

One of the most important statistical models — arguably the most important — is the normal (or **Gaussian**) distribution that the famous mathematician Karl F. Gauss proposed at the beginning of the 19th century, to calculate the probabilities of occurrence of measurement errors.[13] So many data sets were — and still are — well represented by the normal distribution that it has come to be considered the natural behavior for any type of experimental error: hence the adjective *normal*. If on occasion one encountered an error distribution that failed to conform to a Gaussian function, the data collection procedure was usually viewed with suspicion. Later it became clear that many legitimate experimental situations arise for which the normal distribution does not apply. Nevertheless, it remains one of the fundamental models of statistics.

Many of the results we present later are rigorously valid only for data following a normal distribution. In practice this is not so severe a restriction, because almost all the tests we will study remain efficient in the presence of moderate departures from normality, and because we can use adequate experimental planning to reduce the effects of possible non-normalities.

2.3.1. Calculating probabilities of occurrence

A statistical distribution is a function that describes the behavior of a random variable, that is, a quantity that can assume any permissible value for the system to which it refers, but for which the chance of occurrence is governed by some probability distribution. If we could discover or estimate the nature of this distribution, we could calculate the probability of occurrence of any value of interest. We would, in fact, possess a sort of statistical crystal ball we could use to make predictions. Soon we will see how to do this using the normal distribution.

The normal distribution is a **continuous** distribution, that is, a distribution in which the variable can assume any value within a predefined interval. For a normally distributed variable, this interval is $(-\infty, +\infty)$, which means that the variable can — in principle — assume any real value.

A continuous distribution of the variable x is defined by its probability density function $f(x)$, a mathematical expression containing a certain number of parameters. The normal distribution is fully defined by two parameters, its mean and its variance (Eq. (2.5)).

The normal distribution:

$$f(x)\,\mathrm{d}x = \frac{1}{\sigma\sqrt{2\pi}}\mathrm{e}^{-(x-\mu)^2/2\sigma^2}\,\mathrm{d}x \tag{2.5}$$

$f(x)$ = probability density function of the random variable x; μ = population mean; σ^2 = population variance.

[13]Although Gauss is the mathematician most commonly associated with the normal distribution, the history behind it is more involved. At least Laplace, De Moivre and one of the ubiquitous Bernoullis seem to have worked on it too.

To indicate that a random variable x is normally distributed with mean μ and variance σ^2, we use the notation $x \approx N(\mu, \sigma^2)$, where the notation $\approx$ stands for "is distributed in accordance with". If x has zero mean and unit variance, for example, we write $x \approx N(0,\ 1)$ and say that x follows the standard (or **unit**) normal distribution.

Fig. 2.3 shows the famous bell-shaped curve that is the plot of the probability density of the standard normal distribution,

$$f(x) = \frac{1}{\sqrt{2\pi}} e^{-x^2/2}. \tag{2.5a}$$

Note that the curve is perfectly symmetrical about its central point, the mean μ (here equal to zero). The density reaches its maximum value at the mean, and rapidly decreases on both sides as we move away from it. At three standard deviations from the mean, the probability density becomes almost zero. These features are similar to those observed in the histogram of the masses of the 140 beans shown in Fig. 2.2.

The quantity $f(x)\,dx$ is, by definition, the probability of occurrence of the random variable within the interval of width dx around point x. In practical terms, this means that if we randomly extract an x value, the likelihood that it falls within the infinitesimal interval from x to x+dx is $f(x)dx$. To obtain probabilities corresponding to finite intervals — the only ones that have physical meaning — we must integrate the probability density function between the appropriate limits. The integral is the area below the $f(x)$ curve between these limits, which implies that Fig. 2.3 is also a histogram. Since the random variable is now continuous, the

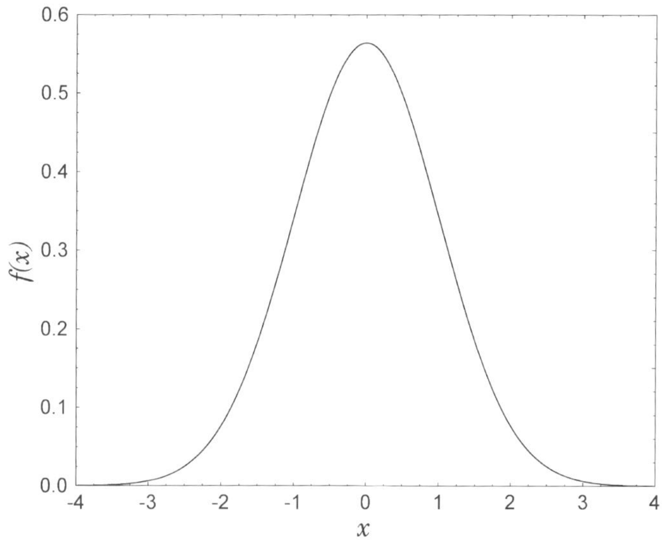

Fig. 2.3. Frequency distribution of a random variable, x $\approx$ N(0, 1). Note that x is the deviation from the mean (zero), measured in standard deviations.

probabilities are calculated by integrals instead of sums. The probability of observing *exactly* (in a strict mathematical sense) any given value becomes zero in this theoretical formulation, because this would be equivalent to making $dx = 0$. For a continuous distribution, therefore, it does not matter if the interval we are referring to is an open or closed one. The probability that $a \leq x \leq b$ is equal to the probability that $a < x < b$:

$$P(a < x < b) = P(a \leq x \leq b) = \int_a^b f(x)\,dx$$

P = probability that the value of variable x of probability density function $f(x)$ will be observed in the $[a, b]$ interval.

As we see in Fig. 2.3, most of the area under the normal curve is within the interval defined by one standard deviation about the mean, and practically all of it is located between $\mu - 3\sigma$ and $\mu + 3\sigma$. To obtain numerical values corresponding to these situations, we integrate, between the appropriate limits, the probability density function given by Eq. (2.5):

$$P(\mu - \sigma < x < \mu + \sigma) = \int_{\mu-\sigma}^{\mu+\sigma} f(x)\,dx = 0.6826 \quad \text{(i.e. 68.26\%)};$$

$$P(\mu - 3\sigma < x < \mu + 3\sigma) = \int_{\mu-3\sigma}^{\mu+3\sigma} f(x)\,dx = 0.9973 \quad (99.73\%).$$

Calculating integrals like these, we can obtain the probabilities corresponding to any limits of interest. In practice, fortunately, we do not have to calculate any integrals at all, because we can read their values from a convenient table, e.g. Table A.1 (see Appendix 2A.1) for intervals of the variable $z \approx N(0, 1)$. These integral values correspond to the standard normal distribution, with zero mean and unit variance, but they can be used to make inferences about any normal distribution, as we shall presently see.[14]

To explain how to use Table A.1 we must introduce the concept of standardization. By definition, to **standardize** a normal variable x with mean μ and variance σ^2 is to transform it by subtracting from each value the population mean and then dividing the result by the population standard deviation:

Standardized normal variable:

$$z = \frac{x - \mu}{\sigma} \qquad (2.6)$$

x = random variable following $N(\mu, \sigma^2)$; z = random variable following $N(0, 1)$.

[14]Many general statistics programs also calculate them for a variety of probability distributions.

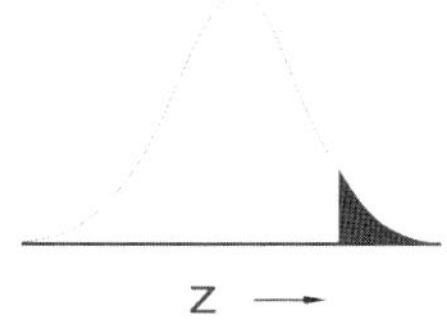

Table A.1
Tail area of standardized normal distribution.

z	0.00	0.01	0.02	0.03	0.04	0.05	0.06	0.07	0.08	0.09
0.0	0.5000	0.4960	0.4920	0.4880	0.4840	0.4801	0.4761	0.4721	0.4681	0.4641
0.1	0.4602	0.4562	0.4522	0.4483	0.4443	0.4404	0.4364	0.4325	0.4286	0.4247
0.2	0.4207	0.4168	0.4129	0.4090	0.4052	0.4013	0.3974	0.3936	0.3897	0.3859
0.3	0.3821	0.3783	0.3745	0.3707	0.3669	0.3632	0.3594	0.3557	0.3520	0.3483
0.4	0.3446	0.3409	0.3372	0.3336	0.3300	0.3264	0.3228	0.3192	0.3156	0.3121
0.5	0.3085	0.3050	0.3015	0.2981	0.2946	0.2912	0.2877	0.2843	0.2810	0.2776
0.6	0.2743	0.2709	0.2676	0.2643	0.2611	0.2578	0.2546	0.2514	0.2483	0.2451
0.7	0.2420	0.2389	0.2358	0.2327	0.2296	0.2266	0.2236	0.2206	0.2177	0.2148
0.8	0.2119	0.2090	0.2061	0.2033	0.2005	0.1977	0.1949	0.1922	0.1894	0.1867
0.9	0.1841	0.1814	0.1788	0.1762	0.1736	0.1711	0.1685	0.1660	0.1635	0.1611
1.0	0.1587	0.1562	0.1539	0.1515	0.1492	0.1469	0.1446	0.1423	0.1401	0.1379
1.1	0.1357	0.1335	0.1314	0.1292	0.1271	0.1251	0.1230	0.1210	0.1190	0.1170
1.2	0.1151	0.1131	0.1112	0.1093	0.1075	0.1056	0.1038	0.1020	0.1003	0.0985
1.3	0.0968	0.0951	0.0934	0.0918	0.0901	0.0885	0.0869	0.0853	0.0838	0.0823
1.4	0.0808	0.0793	0.0778	0.0764	0.0749	0.0735	0.0721	0.0708	0.0694	0.0681
1.5	0.0668	0.0655	0.0643	0.0630	0.0618	0.0606	0.0594	0.0582	0.0571	0.0559
1.6	0.0548	0.0537	0.0526	0.0516	0.0505	0.0495	0.0485	0.0475	0.0465	0.0455
1.7	0.0446	0.0436	0.0427	0.0418	0.0409	0.0401	0.0392	0.0384	0.0375	0.0367
1.8	0.0359	0.0351	0.0344	0.0336	0.0329	0.0322	0.0314	0.0307	0.0301	0.0294
1.9	0.0287	0.0281	0.0274	0.0268	0.0262	0.0256	0.0250	0.0244	0.0239	0.0233
2.0	0.0228	0.0222	0.0217	0.0212	0.0207	0.0202	0.0197	0.0192	0.0188	0.0183
2.1	0.0179	0.0174	0.0170	0.0166	0.0162	0.0158	0.0154	0.0150	0.0146	0.0143
2.2	0.0139	0.0136	0.0132	0.0129	0.0125	0.0122	0.0119	0.0116	0.0113	0.0110
2.3	0.0107	0.0104	0.0102	0.0099	0.0096	0.0094	0.0091	0.0089	0.0087	0.0084
2.4	0.0082	0.0080	0.0078	0.0075	0.0073	0.0071	0.0069	0.0068	0.0066	0.0064
2.5	0.0062	0.0060	0.0059	0.0057	0.0055	0.0054	0.0052	0.0051	0.0049	0.0048
2.6	0.0047	0.0045	0.0044	0.0043	0.0041	0.0040	0.0039	0.0038	0.0037	0.0036
2.7	0.0035	0.0034	0.0033	0.0032	0.0031	0.0030	0.0029	0.0028	0.0027	0.0026
2.8	0.0026	0.0025	0.0024	0.0023	0.0023	0.0022	0.0021	0.0021	0.0020	0.0019
2.9	0.0019	0.0018	0.0018	0.0017	0.0016	0.0016	0.0015	0.0015	0.0014	0.0014
3.0	0.0013	0.0013	0.0013	0.0012	0.0012	0.0011	0.0011	0.0011	0.0010	0.0010

Adapted from Statistical Tables for Biological, Agricultural and Medical Research, 6th edition, by R. A. Fisher and F. Yates, Oliver and Boyd, London, 1963.

As an example, let us suppose that the masses of a population of beans are normally distributed, with $\mu = 0.2024$ g and $\sigma = 0.0363$ g. We are now making two assumptions that we will need to scrutinize later:

- That the masses follow a normal distribution
- That the population parameters are the same as those calculated for the sample

This is our first attempt to use a model to describe experimental data. For the moment we will assume it is adequate.

The standardized mass is, according to Eq. (2.6),

$$z = \frac{x - 0.2024\ \text{g}}{0.0363\ \text{g}},$$

where x is the mass of a bean. Since the numerator and the denominator have the same units, z is dimensionless.

The numerical value of z represents the deviation of x from the population mean μ, expressed in number of standard deviations. This becomes clearer when we rewrite Eq. (2.6) as $x = \mu + z\sigma$. Taking $z = -2$, for example, we have $x = \mu - 2\sigma$, that is, x is two standard deviations below the average. In our example, the actual mass of the bean corresponding to $z = -2$ would be $x = 0.2024\ \text{g} - 2 \times 0.0363\ \text{g} = 0.1298\ \text{g}$.

Exercise 2.7. The 20 titration results in Table 2.1 have average and standard deviation of 3.80 and 0.1509, respectively. Use these values to standardize (in the statistical sense we have just seen) the result of a titration. What concentration would be obtained in a titration whose result was 2.5 standard deviations above the average?

The effect of standardization becomes evident when we employ the definition of the standardized variable to substitute x by z in the general expression for the normal distribution. From Eq. (2.6) we have $x = \mu + z\sigma$ and, therefore, $\mathrm{d}x = \sigma\ \mathrm{d}z$. Substituting these two expressions into Eq. (2.5) we have

$$f(x)\ \mathrm{d}x = \frac{1}{\sigma\sqrt{2\pi}} \mathrm{e}^{-(\mu+z\sigma-\mu)^2/2\sigma^2}\sigma\ \mathrm{d}z,$$

which reduces to

$$f(z)\ \mathrm{d}z = \frac{1}{\sqrt{2\pi}} \mathrm{e}^{-z^2/2}\ \mathrm{d}z.$$

Note that the left-hand side is expressed in terms of z because the expression is now a function of z not x. This equation is identical to Eq. (2.5a). Standardization changes the scale and shifts the origin of the random variable axis. Thus, the original variable x, which is distributed as $N(\mu, \sigma^2)$, is transformed into a new variable z that follows the standard normal distribution, $z \approx N(0, 1)$. Since this transformation does not depend on the values of μ and σ, we can always use the standard normal distribution to describe the behavior of *any* normal distribution.

2.3.2. Using the tails of the standard normal distribution

Table A.1 contains values of the **right-tail areas** of the standard normal distribution, from $z = 0.00$ to 3.99. The first column contains the value of z to the first decimal place, while the top line in the table gives the second decimal. To find the value of the tail area for a given value of z we look in the table at the appropriate intersection of line and column. The value corresponding to $z = 1.96$, for example, is at the intersection of the line corresponding to $z = 1.9$ and the column headed by 0.06. This value, 0.0250, is the fraction of the total area located to the right of $z = 1.96$. Since the curve is symmetrical about the mean, an identical area is located *to the left* of $z = -1.96$, in the other half of the Gaussian (Fig. 2.4). The sum of these two tail areas, right and left, equals 5% of the total area. From this we can conclude that the remaining 95% of the area lies between $z = -1.96$ and 1.96. If we randomly extract a value of z, there is 1 chance in 20 (5%) that this value will lie below -1.96 or above 1.96. The other 19 times, chances are that it will fall inside the $[-1.96, 1.96]$ interval.

If we accept that the normal model adequately represents the population distribution of the masses of the beans, we can use Table A.1, together with the values of the sample statistics, to answer questions about the probability of occurrence of values of interest. For example:

- What is the probability that a bean picked at random weighs between 0.18 and 0.25 g?

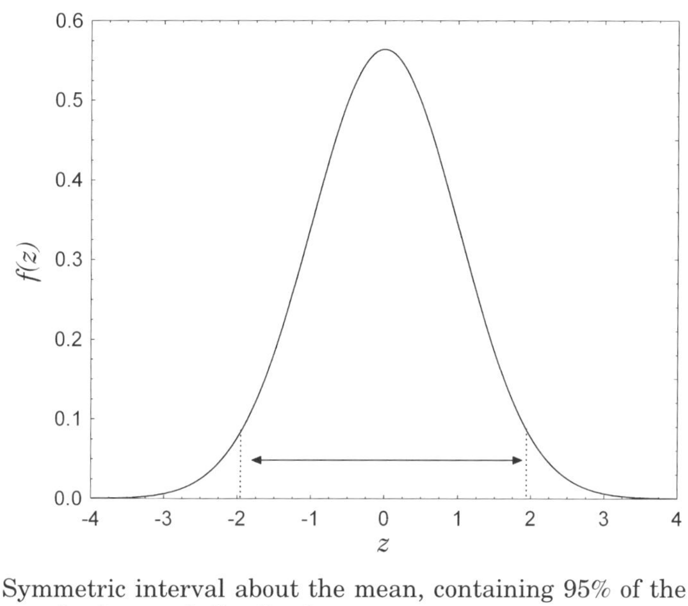

Fig. 2.4. Symmetric interval about the mean, containing 95% of the total area under the standard normal distribution curve.

First we need to standardize the masses:

$$z_1 = \frac{0.18\ \text{g} - 0.2024\ \text{g}}{0.0363\ \text{g}} = -0.62,$$

$$z_2 = \frac{0.25\ \text{g} - 0.2024\ \text{g}}{0.0363\ \text{g}} = 1.31.$$

With this transformation, the question no longer refers directly to the masses but to the value of z. Now, the question is "What is the probability that z falls in the $[-0.62, 1.31]$ range?"

This probability corresponds to the area between the limits indicated by the arrow in Fig. 2.5. This is the total area, which is one, minus the areas of the two tails, the one on the right, above $z = 1.31$, and the other on the left, below $z = -0.62$. The area on the right can be read directly from Table A.1, by looking up the value for $z = 1.31$, which is 0.0951. The area of the tail on the left cannot be read directly from the table, because it does not have negative values. For reasons of symmetry, however, the area below $z = -0.62$ must be equal the area to the right of $z = 0.62$, which is 0.2676.

Subtracting the areas of the two tails from the total area, we obtain the desired probability: $(1.0 - 0.0951 - 0.2676) = 0.6373$. The answer to our original question, then, is that 63.73% of the beans (about two-thirds) should weigh between 0.18 and 0.25 g. Remember, however, that this answer is based on the validity of our two assumptions: that the

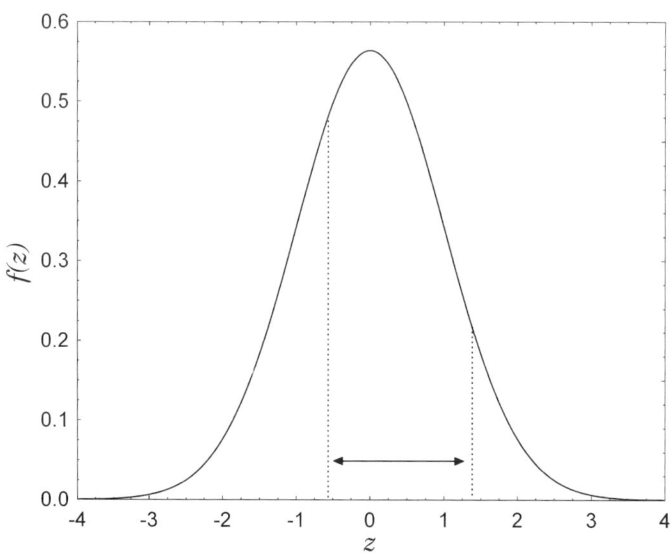

Fig. 2.5. Area corresponding to P $(-0.62 < z < 1.31)$.

distribution of the masses of the beans is normal and that the numerical values of the population parameters are equal to the sample statistics.

Exercise 2.8.

(a) What is the probability that a bean will weigh more than 0.18 g?
(b) Define the limiting masses for an interval containing 95% of the beans.
(c) Your answer to Exercise 2.2 can be transformed into an estimate for the average mass of a bean. Based on what we have seen so far, what are the chances of encountering a bean with a mass larger than or equal to this one?

Exercise 2.9. The area of the tail to the left of a point in a probability density function is called the cumulative probability at that point, and represents the probability that the observed value of the variable will be equal to or less than the value defined by the point. Cumulative probabilities will be used in the next chapter, where we will perform analyses based on normal plots. Use Table A.1 to determine the cumulative probabilities of the following values in a normal population:

$$(a) - \infty; \quad (b)\mu; \quad (c)\mu - \sigma; \quad (d)\mu + \sigma; \quad \text{and} \ (e) + \infty.$$

What do you think the cumulative probability curve for a normal distribution would look like?

The histogram in Fig. 2.2 appears quite symmetrical. Nothing about its appearance seems to contradict the hypothesis that the sample came from a normal population.

For a quantitative test of the adequacy of the normal hypothesis, we may compare the observed frequencies with those predicted by theory. Since our sample consists of a reasonably large number of observations (140), we can expect it to be a fair approximation of the population distribution of the masses of the beans. If the population — the masses of all of the beans in the 1-kg package — deviates drastically from normality, we should be able to discover some evidence of this behavior in the sample occurrence frequencies. In other words, the sample frequencies should differ considerably from those expected from the normal distribution.

Using the sample values $\bar{x} = 0.2024$ g and $s = 0.0363$ g to determine the limits of the intervals $[\bar{x} - s, \ \bar{x} + s]$,$[\bar{x} - 2s, \ \bar{x} + 2s]$and $[\bar{x} - 3s, \ \bar{x} + 3s]$, we find that these correspond to, respectively, 66.6, 96.8 and 100% of the total area of the sample histogram. For a truly normal variable, the corresponding population intervals, $[\mu - \sigma, \ \mu + \sigma]$, $[\mu - 2\sigma, \ \mu + 2\sigma]$ and $[\mu - 3\sigma, \ \mu + 3\sigma]$, would contain 68.3, 95.4 and 99.7% of all the observations. These values are in excellent agreement with the sample values. The worst difference is less than 2%. Judging by this result, we may rightfully use the normal distribution to describe our sample. The comparison, however, is subjective. We did not specify how large a difference would have to be, to compel us to reject the normal hypothesis.

The discussion of more objective criteria is beyond the scope of this book. In Chapter 3 we present an alternative test for normality, in which we must decide if a certain graph appears sufficiently linear. Since we must base our decision on how we judge the appearance of that plot, this test also has a subjective component and so is not much of an improvement on the one just described.

2.3.3. Why is the normal distribution so important

Fortunately we have a very good reason for not worrying too much about the absence (in this book) of a rigorous test for evaluating the normality of a distribution. The statistical techniques that we present are robust with respect to departures from normality. Even if the population of interest is not normal, these techniques still can be used, because they continue to be approximately valid.

This robustness arises from the central limit theorem, one of the fundamental theorems of statistics, which might be stated as follows:

- If the total dispersion of a certain random variable results from a sum of random dispersions of several independent variables of similar importance, its distribution will tend toward normality, whatever the natures of the distributions of the individual variables themselves.

The classic illustration of the central limit theorem comes from throwing dice. The probability that we observe a certain number of points throwing a single die is shown in Fig. 2.6(a). The possible outcomes are the integers from 1 to 6, and if the die is not biased all of them have the same chance to occur, leading to a distribution that is far from normal.

Suppose now that five dice are thrown instead of one, or the same die is thrown five times in a row, and the average of the five observed values is calculated. This average is a function of five random variables, each one with a distribution independent of those of the others, since the value observed on any given die or throw does not affect the values of the others. Moreover, each throw contributes with the same weight to the final result — none of the five observations is more important than any of the others. The two premises of the central limit theorem, (1) independent variations and (2) of equal importance, are therefore satisfied, and the result appears in Fig. 2.6(b): the distribution of the averages already begins to look like a normal distribution. As more observations are used to calculate the average, the tendency toward normality becomes more pronounced, as can be seen for the distribution of the averages of the outcomes for 10 dice (Fig. 2.6(c)).

Often the overall error of an experimental result is the aggregate of several essentially independent individual errors, with no individual error predominating. In the titration, for example, we pointed out the

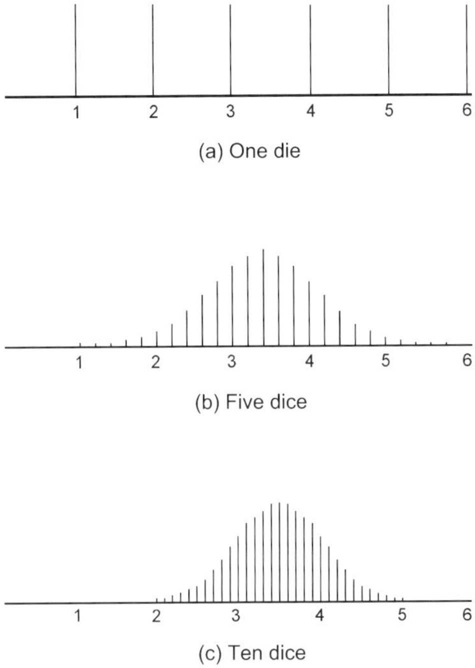

Fig. 2.6. Frequency distributions of average values for throwing straight dice.

reading error of the burette, the error caused by the drop left in the pipette, the error owing to differently perceived shades of the end point and so on. It is the same with the beans: the mass of each bean depends on its degree of dehydration, the action of diseases, genetic factors, etc. A priori, we have no reason to imagine that these errors — either for the beans or for the titrations — follow normal distributions. But at the same time we have no reason to suppose that their individual contributions are dependent on each other or that one of them is much more important than the others. The central limit theorem then tells us that the final error should be distributed in an approximately normal way, and should become more so as the number of independent sources of individual error increases. In practice, situations like this are very common. In most cases we can rely on the normal hypothesis and do not worry about more sophisticated tests until overwhelming evidence to the contrary arises.

Perhaps the central limit theorem justifies the rapturous remark of Sir Francis Galton, the inventor of linear regression: "Hardly will there be something as impressive for the imagination as the admirable form of cosmic order expressed by the Law of the Frequency of Errors (*that is, the normal distribution*). Had the Greeks known of it, they would certainly have personified and deified it".

2.3.4. Calculating confidence intervals for the mean

The main reason we want a model is the prospect of using it to make inferences about population parameters. Let us forget for a while that we weighed 140 beans. Suppose that only one bean had been randomly selected and that its mass was 0.1188 g, the first value in Table 2.2. What does this value allow us to say about the mean population mass, μ.

If the masses of the beans follow a normal distribution, the interval $[\mu - 1.96\sigma,\ \mu + 1.96\sigma]$ contains 95% of all the possible observations. This means that the single measurement 0.1188 g has a 95% probability of being within this interval. Of course this also implies that there is a 5% chance that it falls outside. If we assume the normal model holds, we can say that we have 95% confidence in the double inequality

$$\mu - 1.96\sigma < 0.1188\ \text{g} < \mu + 1.96\sigma.$$

Taking the inequality on the left and adding 1.96σ to both sides, we have

$$\mu < 0.1188\ \text{g} + 1.96\sigma.$$

Subtracting 1.96σ from the inequality on the right, we also have

$$0.1188\ \text{g} - 1.96\sigma < \mu.$$

Combining the two, we arrive at a 95% confidence interval for the population mean:

$$0.1188\ \text{g} - 1.96\sigma < \mu < 0.1188\ \text{g} + 1.96\sigma.$$

To find the numerical limits of the interval, all we need is the value of the population standard deviation. Assuming, for example, that $\sigma = 0.0363\ \text{g}$ (even though we know this is only a sample statistic), we have

$$0.0477\ \text{g} < \mu < 0.1899\ \text{g}.$$

From these values, and the assumptions we made, we can say that the total number of beans in the 1-kg package lies between 5266 and 20,964. Even so, there is still a 5% chance we are wrong! This result is hardly encouraging, but we should not be too surprised. Basing our estimate on the mass of just one bean, we could not expect to draw exceedingly impressive conclusions. Soon we will make estimates based on average values, and will see that they are much more precise, which is to be expected, since average values are more adequate representations of the entire population.

Confidence interval for the population mean, based on one observation:

$$x_i - z\sigma < \mu < x_i + z\sigma \tag{2.7}$$

$x_i =$ individual value; $\mu =$ population mean; $\sigma =$ population standard deviation; $z =$ point of the $N(0, 1)$ distribution corresponding to the desired confidence level.

2.3.5. Interpreting confidence intervals

Students often have the impression that when we calculate confidence limits we are estimating the value of the population mean with a certain probability. Unfortunately, this is not true. While the mass of a bean is a random variable, the population mean μ is not. It is instead a perfectly determined value, which we could find out exactly by examining the entire population. If we weighed all the beans in the package, we could calculate the population mean of the beans in our example without any uncertainty. Therefore, assigning a probability to it makes no sense.

The formal interpretation of a confidence interval is this: if we construct *all* possible intervals corresponding to a certain confidence level, say α, then α percent of them will contain the population mean, while the other $(100 - \alpha)$ percent intervals will not. This means that if we determined all the 95% confidence intervals corresponding to the individual masses of the beans in the package, the population mean should lie within 95% of them, and outside the other 5%. The bad news is that we do not know how to distinguish the correct intervals from the wrong ones, nor how to assign probabilities to the values within the intervals. The average point of the interval, in particular, has no special meaning. To say that the total number of beans lies between 5266 and 20,964, for example, does not imply in any way that the most probable value is the average of these extremes, 13,115. *None of the values within a confidence interval is more probable than any of the others*. This is illustrated in Fig. 2.7(a–c), which shows 95% confidence intervals calculated from a computer simulation in which five-element samples were randomly drawn from an exactly normal distribution with mean $\mu = 0.2024$ and standard deviation $\sigma = 0.0363$. Each of the three figures contains 100 confidence intervals.[15] The squares represent the sample averages, which are scattered around the population mean, represented by the horizontal line. The whiskers mark the limits of the intervals, which vary widely from one sample to another. The gray circles indicate those intervals that failed to include the population mean. Sometimes the mean is above, some times below their limits. Note that there are five such intervals in Fig. 2.7(a), but eight in Fig. 2.7(b) and just three in Fig. 2.7(c). Yet they all are 95% intervals. The 95% figure, which was obtained by chance in Fig. 2.7(a), is in fact the limit for an infinite number of 95% confidence intervals. Of course this is a simulation, in which the population mean is known right from the start.

[15] Details of this calculation will be given in Section 2.6.

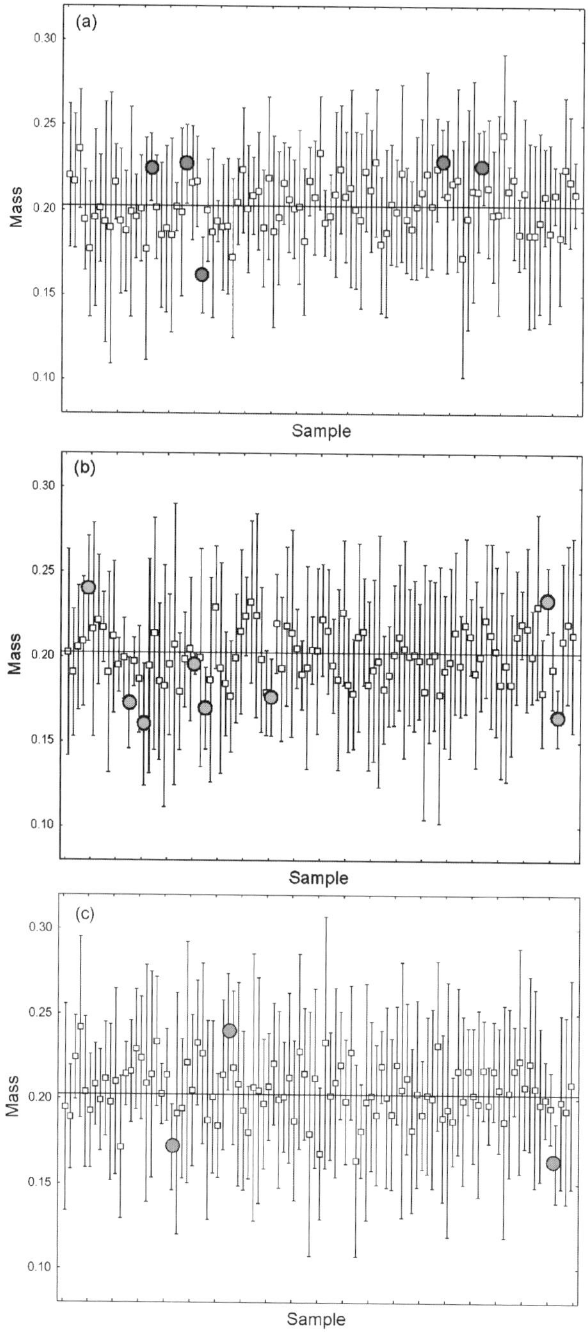

Fig. 2.7. 95% confidence intervals for the mean, calculated from random 5-element samples drawn from a normal population with $\mu = 0.2024$ and $\sigma = 0.0363$.

In a real experiment, all we will have is a single confidence interval, and we will not even know if it is one of the "right" ones or not. This is what we mean when we say it is just a *confidence* interval.

Exercise 2.10. Calculate, from the mass of the second bean in Table 2.2, a 95% confidence interval for the total number of beans in 1 kg.

2.4. Covariance and correlation

To illustrate the concept of random variable, we have used the mass of a bean randomly drawn from its package. We could just as well have chosen the volume of the bean instead of its mass, were it not for the fact that measuring the volume of small irregular objects is much more difficult than weighing them. Nonetheless, the volume of the bean — call it y — is, like its mass, also a random variable. Knowing a given bean's volume does not enable us to predict the exact volume of another bean in the package. The values of the variable y are independent of one another, in the same way that the values of the masses are mutually independent. We expect, therefore, that a histogram of all y values will be like that of the masses: an approximately normal distribution about a central value.

But now there is a difference. The mass and volume of *a given bean* are not independent of each other; a relation exists between them — the bean's density. If we pick a bean with mass above average we expect its volume to be also above average, and vice versa. For any given bean, therefore, the deviations of these two variables with respect to their averages will tend to have the same algebraic sign. We say *tend* because it is unlikely that the density of every bean is exactly the same. If it were so, we would not speak of a tendency but rather of certainty. Then, we could measure one variable and use the linear relation *volume* = *mass* / *density* to unequivocally determine the value of the other variable. A plot of volume against mass would give a line with slope equal to the inverse of the density, as shown in Fig. 2.8(a). In practice, since there are several reasons for the density to vary from one bean to another, e.g., the degree of dehydration, the effect of diseases and the genetic makeup, we should expect the plot to look more like the one of Fig. 2.8(b), where the linearity of the set of points is disturbed by a certain amount of scatter.

We can see in Fig. 2.8(b) that high values of y tend to occur together with high values of x, and vice versa. In these cases, we say that the two random variables present a certain covariance, that is, a tendency to deviate concertedly from their respective averages. We can obtain a numerical measure of the covariance from the products of the deviations $(x_i - \bar{x})$ and $(y_i - \bar{y})$ for each member of the sample. Since in this example the two deviations tend to have the same sign, be it positive or

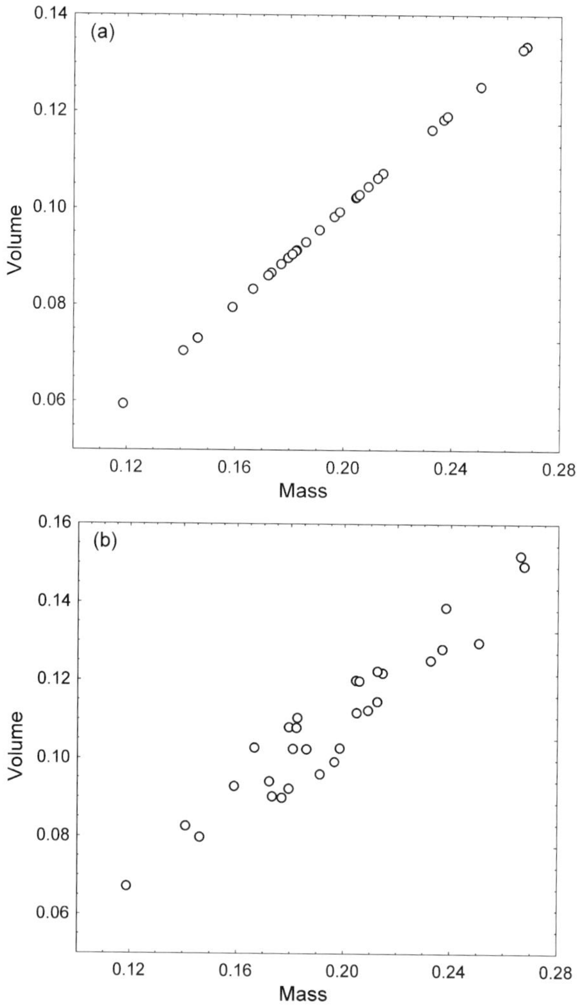

Fig. 2.8. (a) Exact linear relation between volume and mass. (b) Linear relation between the two variables, perturbed by a random error.

negative, the products $(x_i - \bar{x})(y_i - \bar{y})$ will be positive much more frequently than negative. The larger the predominance of positive products, the greater will be the (positive) covariance between x and y. By definition, the covariance is the average of the products of the deviations (Eq. 2.8).

Note the analogy with the definition of the variance, Eq. (2.2). The denominator is again $n - 1$, because only $n - 1$ of the n products of deviations are independent. Note also that $\mathrm{Cov}(x,x)$ is the same as the variance of x.

Sample covariance of the variables x and y:

$$\mathrm{Cov}(x,y) = \frac{1}{n-1}\sum(x_i - \bar{x})(y_i - \bar{y}) \tag{2.8}$$

(x_i, y_i) = individual values for the ith sample element; $(\bar{x}, \bar{y})$ = sample averages; n = number of elements in the sample.

Since the covariance depends on the units of measurement of x and y, it is difficult to use it as a standard to compare the degree of statistical association of different pairs of variables. If, for example, we were investigating oranges instead of beans, the numerical value of the covariance between mass and volume, measured in the same units used for the beans, would be much larger, but it would still mean the same thing: a more or less constant density.

Instead of trying to use the covariance itself as a standard for comparing the degree of statistical association of different pairs of variables, we apply a scale factor to it, dividing each individual deviation from the average by the standard deviation of the corresponding variable. This results in a sort of normalized covariance, which is called the correlation coefficient of the two variables (Eq. (2.9)). This definition forces the correlation coefficient of any pair of random variables to always be restricted to the $[-1,+1]$ interval. The correlations of different pairs of variables are then measured on the same scale (which is dimensionless, as can be deduced from Eq. (2.9)) and can be compared directly.

Sample correlation coefficient of the variables x and y:

$$r(x,y) = \frac{1}{n-1}\sum\left(\frac{x_i - \bar{x}}{s_x}\right)\left(\frac{y_i - \bar{y}}{s_y}\right) \tag{2.9}$$

(x_i, y_i) = individual values for the ith sample element; $(\bar{x}, \bar{y})$ = sample averages; (s_x, s_y) = sample standard deviations; n = number of elements in the sample.

Statistically independent variables have a zero correlation coefficient. The converse, however, is not true, because the correlation coefficient measures only the *linear* association between two variables. A zero correlation coefficient only indicates that a linear relation is not present. Other types of dependence may be present and not be reflected in the numerical value of the correlation coefficient. Exercise 2.11 illustrates one such possibility.

Exercise 2.11. Let y and x be two variables obeying the equation $y = x^2$, in the $[-a, +a]$ interval. (a) What is the value of the correlation coefficient between y and x? (Do not perform any calculations; make a graph of the function and try to use geometric arguments.) (b) Can you think of other functions that would give the same result?

Variables related in a perfectly linear way have a correlation coefficient equal to +1, if increasing one results in the other increasing proportionally,

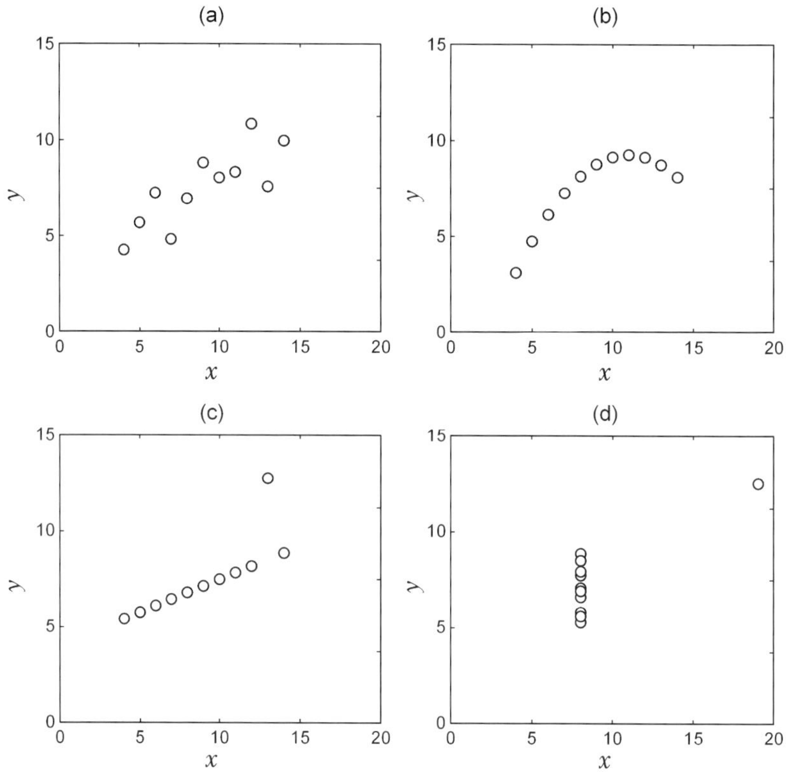

Fig. 2.9. Four data sets with the same correlation coefficient, r = 0.82, but representing very different realities. (Data from Anscombe, 1973).

or equal to −1, if increasing one results in the other decreasing proportionally. Intermediate correlation values indicate that the relation between the two variables can be regarded at most as partially linear.

The numerical value of the correlation coefficient is often used in scientific work as an argument to support the existence of a relation between two variables. Extreme care is necessary with such arguments, for the numerical values can be very misleading. Sometimes a quick look at a graph of the variables is all it takes to discard conclusions based only on the value of the correlation coefficient. We will return to this point in Chapter 5, where we discuss empirical model-building. For the time being, as an example — and also as a warning — we show in Fig. 2.9 the graphs of four sets of pairs of values of variables x and y, all with exactly the same correlation coefficient, 0.82 (Anscombe, 1973). If we were to judge the data only by the value of r, we would conclude that the relation between the variables is the same in each case, which is evidently far from the truth.

By the way, what conclusions would you draw from the four graphs in Fig. 2.9? In spite of appearing quite trivial to science students, two-dimensional plots like those of Figs. 2.8 and 2.9 are extremely important for quality management, because of their efficiency in revealing patterns of association between variables (as is clearly demonstrated by Fig. 2.9). In this context, they are labeled scatter plots, one of the "seven indispensable tools for quality improvement" recommended by the quality guru Ishikawa (1985).[16] In Ishikawa's opinion, all seven tools, which are very simple, should be learned and used by anyone working in industry. We concur. Treating these seven tools in any depth is beyond the scope of this book, but we think it is worth the effort to learn to use them. Good references for this are Montgomery (1997) and Oakland and Followell (1990).

The averages and standard deviations used in Eqs. (2.8) and (2.9) are sample values. Sometimes we have to measure the deviations with respect to *population* values, that is, use μ_x instead of $\bar{x}$ and μ_y instead of $\bar{y}$. When this is the case, we also use n instead of $n-1$ in the denominator, because the means in relation to which the deviations are measured are no longer obtained from the sample values. The deviations are not subjected to any restriction, and therefore retain all n degrees of freedom of the original observations. Note, however, that the covariance and the correlation coefficient remain *sample* values. The difference is now they are calculated relative to the population means.

Exercise 2.12. The values below are volumes, in milliliters, of the beans whose masses appear in the first line of Table 2.2. Calculate the covariance and the correlation coefficient between the masses and volumes of these seven beans.

0.108 0.214 0.143 0.195 0.148 0.144 0.174

2.5. Linear combinations of random variables

Suppose that x_1 and x_2 are random variables with population parameters (μ_1, σ_1^2) and (μ_2, σ_2^2), respectively. The expression

$$y = a_1 x_1 + a_2 x_2,$$

where a_1 and a_2 are constants, is a linear combination of x_1 and x_2. It defines a new random variable, y, whose distribution depends on the individual distributions of x_1 and x_2. In this section we will see how the parameters of the distribution of y are related to those of the distributions of x_1 and x_2.

Starting from n pairs of values (x_1, x_2) extracted from their respective populations we can calculate a set of n values of y. The average value of y

[16]The other six are the check sheet, the flowchart, the histogram, the Pareto chart, the cause and effect (or fishbone) diagram and stratification.

in this set is

$$\bar{y} = \frac{1}{n}\sum y = \frac{1}{n}\sum (a_1x_1 + a_2x_2).$$

The sum is carried out over all the n pairs of values; the indices are omitted to simplify the expressions. Taking the individual sums of x_1 and x_2, we have

$$\bar{y} = a_1\left(\frac{1}{n}\sum x_1\right) + a_2\left(\frac{1}{n}\sum x_2\right),$$

or

$$\bar{y} = a_1\bar{x}_1 + a_2\bar{x}_2,$$

which means that

- The average of the linear combination is the linear combination of the averages of each variable.

Analogously, the variance of the linear combination is given by

$$\begin{aligned} s_y^2 &= \frac{1}{n-1}\sum (y-\bar{y})^2 \\ &= \frac{1}{n-1}\sum (a_1x_1 + a_2x_2 - a_1\bar{x}_1 - a_2\bar{x}_2)^2 \\ &= \frac{1}{n-1}\sum [a_1(x_1-\bar{x}_1) + a_2(x_2-\bar{x}_2)]^2 \\ &= \frac{1}{n-1}\sum \left[a_1^2(x_1-\bar{x}_1)^2 + a_2^2(x_2-\bar{x}_2)^2 + 2a_1a_2(x_1-\bar{x}_1)(x_2-\bar{x}_2)\right] \\ &= a_1^2\left[\frac{1}{n-1}\sum (x_1-\bar{x}_1)^2\right] + a_2^2\left[\frac{1}{n-1}\sum (x_2-\bar{x}_2)^2\right] \\ &\quad + 2a_1a_2\left[\frac{1}{n-1}\sum (x_1-\bar{x}_1)(x_2-\bar{x}_2)\right]. \end{aligned}$$

Using Eqs. (2.2) and (2.9) we can rewrite this last expression as

$$s_y^2 = a_1^2s_1^2 + a_2^2s_2^2 + 2a_1a_2s_1s_2r(x_1, x_2),$$

where s_1^2 and s_2^2 are the variances and $r(x_1, x_2)$ is the correlation coefficient between x_1 and x_2. This result is not as simple as that for the average and, because of the last term, depends on the degree of correlation between the variables.

These results are easily extended to the general case of a linear combination of p variables,

$$y = a_1x_1 + a_2x_2 + \cdots + a_px_p = \sum_i a_ix_i. \tag{2.10}$$

Then

$$\bar{y} = \sum_i a_i \bar{x}_i, \tag{2.11}$$

$$s_y^2 = \sum_i a_i^2 s_i^2 + 2\sum_i \sum_{j>i} a_i a_j s_i s_j r(x_i, x_j). \tag{2.12}$$

Note that these sums are carried out over the number of variables included in the linear combination, *p*, and not over the number of elements in the sample, *n*. The sums over *n* are already implied in the calculation of s_i, s_j and $r\ (x_i,\ x_j)$.

The use of the Latin letters in Eqs. (2.11) and (2.12) indicates that these expressions refer to sample values. Applying the same approach to population values, we obtain analogous expressions, Eqs. (2.13) and (2.14), in which, following convention, Greek letters are used.

Population parameters for a linear combination of random variables:

$$\mu_y = \sum_i a_i \mu_i \tag{2.13}$$

$$\sigma_y^2 = \sum_i a_i^2 \sigma_i^2 + 2\sum_i \sum_{j>i} a_i a_j \sigma_i \sigma_j \rho(x_i, x_j) \tag{2.14}$$

$y = \sum_i a_i x_i$; (μ_i, σ_i^2) = population mean and variance of variable x_i .

Eqs. (2.13) and (2.14) are absolutely general and can be applied to any linear combination of random variables.

The case for which the variables are independently distributed is of special interest. By definition, the correlations between all pairs of variables are zero, and the second sum in Eq. (2.14) vanishes. The variance of the linear combination then becomes simply

$$\sigma_y^2 = \sum_i a_i^2 \sigma_i^2. \tag{2.15}$$

The average value of a sample of *n* elements extracted at random from a certain population is a case of even greater interest. Consider, for example, the average mass of a bean in a sample of 10 beans taken from a 1-kg package. By sampling repeatedly we will have a large set of average values. What will be the distribution of these average values?

The average of *n* observations is, in fact, a special case of linear combination, in which all the coefficients in Eq. (2.10) are equal to 1/*n*:

$$\bar{x} = \frac{1}{n}\sum_i x_i = \frac{1}{n}x_1 + \frac{1}{n}x_2 + \cdots + \frac{1}{n}x_n.$$

If the selection of the sample elements is strictly random there will be no correlation between the observations, and we can use Eq. (2.15) to calculate the variance. Moreover, since we are assuming that the observations are always made on the same population, all observations are individually distributed with the same population variance, σ^2. Therefore, by substituting $1/n$ for a_i and σ^2 for σ_i^2 in Eq. (2.15) we can express the variance of sample average values as

$$\sigma_{\bar{x}}^2 = \sum_i \left(\frac{1}{n}\right)^2 \sigma^2 = \left(\frac{1}{n}\right)^2 \sum_i \sigma^2 = \left(\frac{1}{n}\right)^2 n\sigma^2 = \frac{\sigma^2}{n}. \tag{2.15a}$$

This is a population value, and refers to the statistical distribution of the averages of *all* random samples with n elements that might be extracted from the population. The variance of this distribution is smaller than the variance of the distribution of individual observations, σ^2, by a factor inversely proportional to sample size. The distribution of the averages is therefore narrower than the distribution of the individual values, and the more so the larger the sample taken to calculate the average.

To obtain the mean of a distribution of average sample values, $\bar{x}$, we use Eq. (2.13), substituting $1/n$ for a_i and μ for μ_i

$$\mu_{\bar{x}} = \sum_i \frac{1}{n}\mu = \frac{1}{n}\sum_i \mu = \frac{1}{n}n\mu = \mu. \tag{2.13a}$$

This means that the individual observations and the sample averages are distributed about the *same* mean value μ. The distribution of averages, however, is narrower, as we have just seen: its standard deviation is only $\sigma/\sqrt{n}$. For example, if each sample contains 100 observations, we expect the width of the histogram of the averages to be one-tenth of the width of the distribution of the individual values.

Fig. 2.10(b) shows the distribution of the average masses of 140 random samples of 10 beans taken from the same population that generated the histogram of Fig. 2.2, which appears again in Fig. 2.10(a). (The scale is expanded in relation to the one of Fig. 2.2, to accommodate both histograms.) The narrowing of the distribution of average masses relative to the distribution of the individual masses is clear.

We have seen that the individual values represented in Fig. 2.10(a) have an average mass of 0.2024 g and a standard deviation of 0.0363 g. If these values were identical to the population parameters, the mean of the average masses of the samples of 10 beans would be the same, but the standard deviation would be less: $0.0363/\sqrt{10} = 0.0115$ g. The values for the distribution of averages in Fig. 2.10(b) are 0.1929 and 0.0128 g, respectively. The agreement is very good, especially if we recall that this is just a sample estimate, not a population parameter.

Exercise 2.13. What is the variance of the distribution of the difference $x_1 - x_2$, where x_1 and x_2 are two standardized normal variables that are perfectly positively

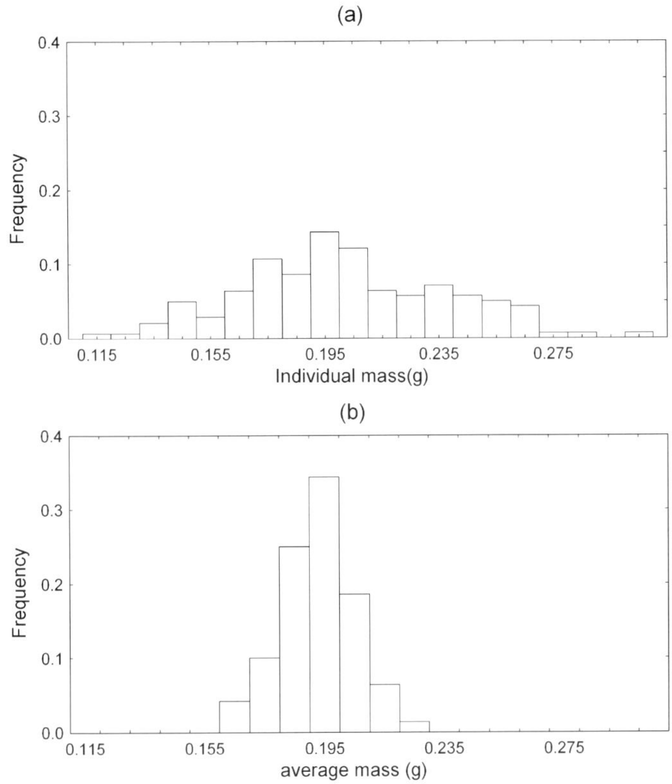

Fig. 2.10. (a) Distribution of the individual masses of 140 beans. (b) Distribution of the average masses of 140 10-bean samples.

correlated, that is, have unit correlation coefficient? What if the correlation coefficient were zero?

2.6. Random sampling in normal populations

In the last section, we drew several important conclusions without restricting the nature of the distribution of the observations in any way. We only had to assume that the samples were random. In this section we shall add another premise: that the samples also come from *normal* populations. As we have already commented, this hypothesis is perfectly acceptable in many real situations, because of the central limit theorem. Under these two restrictions — random samples taken from normal populations — the sample values follow specific distributions that can be used to calculate confidence intervals. Before showing how this can be done, we shall state, without proof, some conclusions we shall need. If you are interested, you can find the proofs in a book on advanced statistics, such as Dudewicz and Mishra (1985).

Let us consider samples of n elements, randomly drawn from a normal population of mean μ and variance σ^2. It can be shown that the sample values $\bar{x}$ and s^2 obey the following:

- The sample averages $\bar{x}$ are also normally distributed with the same mean μ, but with a variance equal to σ^2/n. (In the last section we arrived at a similar conclusion, but we said nothing about the nature of the distribution.)
- The random variable t defined by $t = (\bar{x} - \mu)/(s/\sqrt{n})$ follows a t distribution with $n - 1$ degrees of freedom.
- The random variable χ^2 defined by $\chi^2 = (n - 1)s^2/\sigma^2$ follows a χ^2 distribution, also with $n - 1$ degrees of freedom.

Distribution of sample estimates from normal populations:

$$\bar{x} \approx N\left(\mu, \frac{\sigma^2}{n}\right) \tag{2.16}$$

$$\frac{\bar{x} - \mu}{s/\sqrt{n}} \approx t_{n-1} \tag{2.17}$$

$$(n - 1)\frac{s^2}{\sigma^2} \approx \chi^2_{n-1} \tag{2.18}$$

x = random variable distributed as $N(\mu, \sigma^2)$; $(\bar{x},\ s^2)$ = estimates of (μ, σ^2) obtained from random samples of n elements; $t_{n-1} = t$ distribution with $n-1$ degrees of freedom; $\chi^2_{n-1} = \chi^2$ distribution with $n-1$ degrees of freedom.

From these relations we can obtain confidence intervals, using arguments similar to those of Section 2.3. To illustrate, we shall again use a sample of 10 beans drawn from the 1-kg package.

Suppose that the masses of these beans are the first 10 values in Table 2.2. For this sample, we already know (if you did Exercise 2.6) that $\bar{x} = 0.1887$ gand $s = 0.0423$ g. If the masses follow a normal distribution, the sample average $\bar{x}$ is also normally distributed according to Eq. (2.16). Subtracting the population mean μ and dividing the result by its standard deviation, $\sigma/\sqrt{n}$, yields a standard normal variable:

$$\frac{\bar{x} - \mu}{\sigma/\sqrt{n}} = z \approx N(0, 1) \tag{2.19}$$

For a 95% confidence interval z is 1.96, as we saw in Section 2.3(b). This means that there are 95 chances out of 100 that $-1.96 < z < 1.96$, or that

$$-1.96 < \frac{\bar{x} - \mu}{\sigma/\sqrt{n}} < 1.96.$$

Rearranging the expression to isolate the population mean, μ, we arrive at two inequalities,

$$\mu < \bar{x} + 1.96\frac{\sigma}{\sqrt{n}} \quad \text{and} \quad \bar{x} - 1.96\frac{\sigma}{\sqrt{n}} < \mu,$$

that can be recombined into a single expression:

$$\bar{x} - 1.96\frac{\sigma}{\sqrt{n}} < \mu < \bar{x} + 1.96\frac{\sigma}{\sqrt{n}}. \tag{2.20}$$

To find numerical values for the limits of this interval we need the value of the population standard deviation. Let us assume one more time that the estimate calculated for the 140 beans is approximately correct. We then have $\sigma/\sqrt{n} = 0.0363/\sqrt{10} = 0.0115$ g. Finally, recalling that $\bar{x} = 0.1887$ g for our sample, we can write

$$0.1887\ \text{g} - 1.96 \times 0.0115\ \text{g} < \mu < 0.1887\ \text{g} + 1.96 \times 0.0115\ \text{g},$$

or

$$0.1662\ \text{g} < \mu < 0.2112\ \text{g}.$$

From this expression we can say, with 95% confidence,[17] that 1 kg of beans contains between 4735 and 6017 beans. Recall that using the mass of only one bean (the first value of Table 2.2), we estimated a range of 5266–20,964. Clearly, the new interval is much more precise. We can further improve the precision by using a sample with more elements.

Confidence interval for the population mean, using the normal distribution:

$$\bar{x} - z\frac{\sigma}{\sqrt{n}} < \mu < \bar{x} + z\frac{\sigma}{\sqrt{n}} \tag{2.21}$$

$\bar{x}$ = sample average; μ = population mean; σ = population standard deviation; n = number of elements in the sample; z = point of the $N(0,1)$distribution corresponding to the desired confidence level.

So far we have assumed that the population standard deviation was known. Yet we only knew a sample estimate. It is true it had been obtained from a quite large sample, but seldom is an experiment replicated 140 times. We shall now see how to break free of this unsatisfactory restriction and obtain confidence intervals without relying on population values or performing an inordinate number of replicate experiments.

Exercise 2.14. Consider the 140 values in Table 2.2 as a single random sample from a normal population. Assume that the standard deviation calculated from these values is identical to the population value. Now, how many beans are there in a 1-kg package?

In 1908, W.S. Gosset, a chemist working for the Guinness brewery who wrote his scientific papers under the pseudonym *Student*, published a manuscript on the deduction of the "curve representing the frequency distribution of a quantity z, which is obtained by dividing the distance

[17]In the sense discussed in Section 2.3.5.

between the mean of a sample and the mean of the population by the standard deviation of the sample" (Student, 1908).

The words are Student's own, who is obviously referring to the distribution of the random variable $(\bar{x} - \mu)/s$. Today we prefer to include the factor $\sqrt{n}$ and refer instead to the distribution of the variable $(\bar{x} - \mu)/(s/\sqrt{n})$. This expression is identical to that of Eq. (2.19), except for the substitution of the population standard deviation, σ, by the sample standard deviation, s. Because of this change, the variable no longer obeys the standard normal distribution. Rather, it obeys the one Student deduced, which — it must be stressed — is rigorously valid for random samples drawn from a normal population. With Student's distribution we can compare the deviation of the sample average from the population mean, $(\bar{x} - \mu)$, with a standard deviation $s/\sqrt{n}$ calculated from the same values used to obtain the sample average $\bar{x}$. The population value $\sigma/\sqrt{n}$ is no longer needed. We will see later that this feature of Student's distribution is very important for assessing the fit of empirical models.

Equation 2.17 is the mathematical expression of Student's conclusions. The new random variable is represented by the symbol t_{n-1}, and its distribution is called the t distribution or Student's distribution. The subscript $n - 1$ reminds us that the form of the distribution depends on the size of the sample. Actually there are several distributions, each one corresponding to a certain number of degrees of freedom involved in determining the value of s.

It is important to stress, at this point, that one must use the number of degrees of freedom for calculating the standard deviation, not the average with Student's distribution. As we will see later, the average and the standard deviation can be obtained from different sets of observations. The value of n in t_{n-1} is not necessarily the same value of n used to calculate the average, and which appears in the denominator of Eq. (2.17).

Table A.2 contains t values for some right-tail areas of Student's distribution. The areas appear at the top of the table, in boldface. Like the standard normal distribution, the t distribution is symmetric about a zero mean, so we only need to know the tail values for one side.

The number of degrees of freedom ν used to estimate s is in the first column, also in boldface. For our 10-bean sample this number is $\nu = n - 1 = 9$, which means that the appropriate values can be found in the ninth line. For example, to find the t value at the 95% confidence level we locate the value at the intersection of the ninth line with the column corresponding to the 0.025 tail area. This gives us $t = 2.262$.

For the normal distribution, at the same confidence level, we used $z = 1.96$, which shows that the t distribution is broader, resulting in wider confidence intervals. This makes sense, because the use of s to estimate σ introduces another source of error into our results, which will become less certain as the size of the sample decreases. If our sample only had two elements, for example, the value of t would increase to no less than 12.706, at the same

Table A.2
Points for the t probability distribution with ν degrees of freedom.

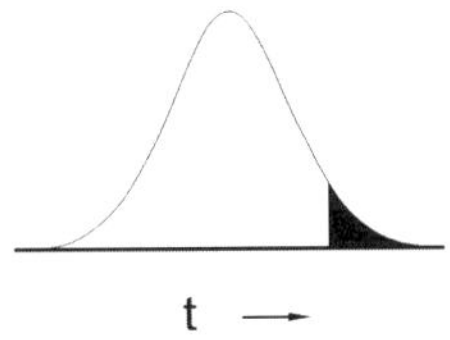

ν	Probability area 0.4	0.25	0.1	0.05	0.025	0.01	0.005	0.0025	0.001	0.0005
1	0.325	1.000	3.078	6.314	12.706	31.821	63.657	127.32	318.31	636.62
2	0.289	0.816	1.886	2.920	4.303	6.965	9.925	14.089	22.326	31.598
3	0.277	0.765	1.638	2.353	3.182	4.541	5.841	7.453	10.213	12.924
4	0.271	0.741	1.533	2.132	2.776	3.747	4.604	5.598	7.173	8.610
5	0.267	0.727	1.476	2.015	2.571	3.365	4.032	4.773	5.893	6.869
6	0.265	0.718	1.440	1.943	2.447	3.143	3.707	4.317	5.208	5.959
7	0.263	0.711	1.415	1.895	2.365	2.998	3.499	4.029	4.785	5.408
8	0.262	0.706	1.397	1.860	2.306	2.896	3.355	3.833	4.501	5.041
9	0.261	0.703	1.383	1.833	2.262	2.821	3.250	3.690	4.297	4.781
10	0.260	0.700	1.372	1.812	2.228	2.764	3.169	3.581	4.144	4.587
11	0.260	0.697	1.363	1.796	2.201	2.718	3.106	3.497	4.025	4.437
12	0.259	0.695	1.356	1.782	2.179	2.681	3.055	3.428	3.930	4.318
13	0.259	0.694	1.350	1.771	2.160	2.650	3.012	3.372	3.852	4.221
14	0.258	0.692	1.345	1.761	2.145	2.624	2.977	3.326	3.787	4.140
15	0.258	0.691	1.341	1.753	2.131	2.602	2.947	3.286	3.733	4.073
16	0.258	0.690	1.337	1.746	2.120	2.583	2.921	3.252	3.686	4.015
17	0.257	0.689	1.333	1.740	2.110	2.567	2.898	3.222	3.646	3.965
18	0.257	0.688	1.330	1.734	2.101	2.552	2.878	3.197	3.610	3.922
19	0.257	0.688	1.328	1.729	2.093	2.539	2.861	3.174	3.579	3.883
20	0.257	0.687	1.325	1.725	2.086	2.528	2.845	3.153	3.552	3.850
25	0.256	0.684	1.316	1.708	2.060	2.485	2.787	3.078	3.450	3.725
30	0.256	0.683	1.310	1.697	2.042	2.457	2.750	3.030	3.385	3.646
40	0.255	0.681	1.303	1.684	2.021	2.423	2.704	2.971	3.307	3.551
60	0.254	0.679	1.296	1.671	2.000	2.390	2.660	2.915	3.232	3.460
120	0.254	0.677	1.289	1.658	1.980	2.358	2.617	2.860	3.160	3.373
∞	0.253	0.674	1.282	1.645	1.960	2.326	2.576	2.807	3.090	3.291

Adapted from Pearson, E. S. and Hartley, H. O. *Biometrika Tables for Statisticians, Vol. 1, 3rd edition*, Cambridge, Cambridge University Press, 1966.

confidence level. That is, if we wish to have the same 95% confidence with so small a sample we must accept an interval about five times larger than the one for the sample of 10. Of course this does not take into account the variation in standard deviations values between the two samples.

The reverse is true: other things being equal, the larger the sample the narrower the interval. As the number of degrees of freedom increases to infinity, the t distribution reduces to the standard normal distribution. You can confirm this by checking the values in the last line of Table A.2.

With the t distribution, therefore, we can calculate a new confidence interval using only sample values (Eq. (2.22)). Note the similarity of this equation with Eq. (2.21), which is based on the population standard deviation.

Confidence interval for the population mean, using the *t* distribution:

$$\bar{x} - t_{n-1}\frac{s}{\sqrt{n}} < \mu < \bar{x} + t_{n-1}\frac{s}{\sqrt{n}} \tag{2.22}$$

$\bar{x}$ = sample average; s = sample standard deviation; μ = population mean; n = number of elements in the sample; t_{n-1} = point of the t distribution (with $n-1$ degrees of freedom) corresponding to the desired confidence level.

In our example, since all information comes from a single sample, the value of n under the square root sign is the same as that appearing in t_{n-1}. As we have said, the value of n in t_{n-1} arises from the number of observations used to estimate the standard deviation whereas the n in $\sqrt{n}$ is the number of observations used to calculate the sample average. These values, however, can be different. Later we will see examples in which we combine information from several different samples to estimate a standard deviation. With this procedure, the value of s – and therefore the value of t_{n-1} – will have more degrees of freedom than that for a single sample. As a consequence, the confidence intervals will become narrower, and our predictions more precise.

For the 95% confidence interval and a sample of 10 elements, Eq. (2.22) becomes

$$\bar{x} - 2.262\frac{s}{\sqrt{10}} < \mu < \bar{x} + 2.262\frac{s}{\sqrt{10}}.$$

Substituting the values for the sample containing 10 beans, $\bar{x} = 0.1887\,\text{g}$ and $s = 0.0423\,\text{g}$, we arrive at the interval $0.1584\,\text{g} < \mu < 0.2190\,\text{g}$, which corresponds to 4566–6313 beans kg^{-1}. As expected, the uncertainty is larger than that of our previous estimate, which was based on the idea that the sample standard deviation could serve as the population value.

Exercise 2.15. Use the seven values in the last line of Table 2.2 and determine, at the 99% confidence level, the number of beans in a kilogram.

Exercise 2.16. Repeat Exercise 2.14, this time using Student's distribution. Assume that the number of degrees of freedom is 120 to obtain the t values from Table A.2. Compare your results with those of Exercise 2.14.

Exercise 2.17. How can we relate the numbers in the last line of Table A.2 to the values in Table A.1?

As the number of degrees of freedom increases the t values converge, rapidly at first and then more slowly, to the z values of the standard normal distribution. As the sample becomes larger, therefore, the difference between the two distributions tends to become negligible. In practice, the t distribution should be used if the number of degrees of freedom of the standard deviation estimate is less than 30. For larger samples, Eq. (2.21) is considered satisfactory.

The different confidence intervals calculated so far for the mean mass of a bean, in the text as well as in the exercises, are graphically compared in Fig. 2.11. We can see a narrowing of the interval as the number of beans in the sample increases. Since this effect varies with the square root of n, increases in sample size become less attractive after a while. For example, to reduce by half the intervals obtained from a 140-bean sample, we would have to weigh 420 more beans (to have a total of $4 \times 140 = 560$ beans). One wonders if so much extra work is warranted by the increase in precision.

Exercise 2.18. The two 95% confidence intervals for the mean mass calculated from the masses of single beans have the same width, as shown in Fig. 2.11. On the other hand, the intervals for the number of beans in 1 kg obtained from the same data have very different widths. For the bean weighing 0.1188 g the 95% confidence interval ranges from 5266 to 20,964 beans. For the one weighing 0.2673 g the limits are 2955 and 5098, giving the impression that this determination is more precise than the other. Is this true? Why?

Starting from Eq. (2.18) and proceeding in exactly the same way as we did for deriving the confidence interval of the mean, we can determine a confidence interval for the population variance. As an example, we will use our sample with 10 beans. The χ^2 values are in Table A.3, which reads

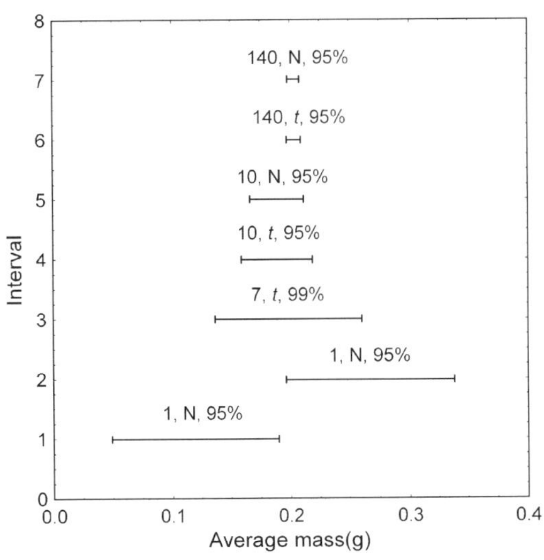

Fig. 2.11. Confidence intervals for the mean mass of a black bean. The number of beans in the sample, the distribution used in the estimate, and the confidence level are indicated above each interval.

like Table A.2, with the difference that the χ^2 distribution has an asymmetric shape, more spread out on the right side. For this reason we must look in two columns to determine the interval limits.

For the 95% confidence interval, we need the points corresponding to 0.025 (right-hand tail) and 0.975 (also for the right-hand tail, and therefore corresponding to 0.025 of the tail to *the left*, which is what we really need). With nine degrees of freedom these values are 19.0 and 2.70, respectively. We can conclude that there is a 2.5% chance that $\chi^2 > 19.0$ and also a 2.5% chance that $\chi^2 < 2.70$. Thus, there is a 95% probability that χ^2 lies between these two limits, that is, that $2.70 < \chi^2 < 19.0$, or, by Eq. (2.18),

$$2.70 < (n-1)\frac{s^2}{\sigma^2} < 19.0.$$

Table A.3
Probability points for the χ^2 distribution with ν degrees of freedom.

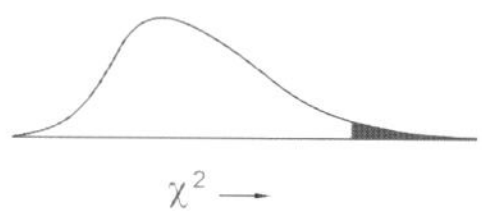

Probability area								
ν	0.995	0.99	0.975	0.95	0.5	0.05	0.01	0.005
1	—	—	—	—	0.455	3.84	6.63	7.88
2	0.010	0.020	0.051	0.103	1.39	5.99	9.21	10.6
3	0.072	0.115	0.216	0.352	2.37	7.81	11.3	12.8
4	0.207	0.297	0.484	0.711	3.36	9.49	13.3	14.9
5	0.412	0.554	0.831	1.15	4.35	11.1	15.1	16.7
6	0.676	0.872	1.24	1.64	5.35	12.6	16.8	18.5
7	0.989	1.24	1.69	2.17	6.35	14.1	18.5	20.3
8	1.34	1.65	2.18	2.73	7.34	15.5	20.1	22.0
9	1.73	2.09	2.70	3.33	8.34	16.9	21.7	23.6
10	2.16	2.56	3.25	3.94	9.34	18.3	23.2	25.2
11	2.60	3.05	3.82	4.57	10.3	19.7	24.7	26.8
12	3.07	3.57	4.40	5.23	11.3	21.0	26.2	28.3
13	3.57	4.11	5.01	5.89	12.3	22.4	27.7	29.8
14	4.07	4.66	5.63	6.57	13.3	23.7	29.1	31.3
15	4.60	5.23	6.26	7.26	14.3	25.0	30.6	32.8
16	5.14	5.81	6.91	7.96	15.3	26.3	32.0	34.3
17	5.70	6.41	7.56	8.67	16.3	27.6	33.4	35.7
18	6.26	7.01	8.23	9.39	17.3	28.9	34.8	37.2
19	6.84	7.63	8.91	10.1	18.3	30.1	36.2	38.6
20	7.43	8.26	9.59	10.9	19.3	31.4	37.6	40.0
25	10.5	11.5	13.1	14.6	24.3	37.7	44.3	46.9
30	13.8	15.0	16.8	18.5	29.3	43.8	50.9	53.7

Adapted from Pearson, E. S. and Hartley, H. O. *Biometrika Tables for Statisticians, Vol. 1, 3rd edition*, Cambridge, Cambridge University Press, 1966.

Rewriting this inequality to isolate σ^2, we have

$$(n-1)\frac{s^2}{19.0} < \sigma^2 < (n-1)\frac{s^2}{2.70}.$$

Finally, substituting $n = 10$ and $s = 0.0423\,\text{g}$, we arrive at the expression $0.0008\ \text{g}^2 < \sigma^2 < 0.0060\ \text{g}^2$, which can be interpreted like the other intervals we have seen. The population variance — and consequently its standard deviation — are very difficult to estimate as precisely as the population mean. In this case, the variance interval spans nearly an order of magnitude. Also, we should note that hypothesis tests directly involving the variance are not robust with respect to deviations from normality. For this reason, they should be used with caution. When possible, it is preferable to use tests involving means.

Exercise 2.19. Calculate, from the seven last values in Table 2.2, a 99% confidence interval for the standard deviation of the mass of one bean.

Comparing the variances of two populations is a very important tool for assessing the fit of statistical models. This subject will be discussed in detail in Chapter 5, but we shall take this opportunity to present the appropriate statistical distribution, which is also derived from the normal distribution.

Consider two random samples, extracted from two possibly distinct normal distributions. Since the variance of each one follows its own χ^2 distribution, we start from Eq. (2.18) and write $s_1^2/\sigma_1^2 \approx \chi^2_{\nu_1}/\nu_1$ for one and $s_2^2/\sigma_2^2 \approx \chi^2_{\nu_2}/\nu_2$ for the other, where ν_1 and ν_2 are their respective degrees of freedom. It can be demonstrated that the ratio $(\chi^2_{\nu_1}/\nu_1)/(\chi^2_{\nu_2}/\nu_2)$ follows an F distribution with ν_1 and ν_2 degrees of freedom, which allows us to write the expression as

$$\frac{s_1^2/\sigma_1^2}{s_2^2/\sigma_2^2} \approx F_{\nu_1,\nu_2},$$

and therefore

$$\frac{s_1^2}{s_2^2} \approx \frac{\sigma_1^2}{\sigma_2^2} F_{\nu_1,\nu_2}. \qquad (2.23)$$

This last expression can be used to test hypotheses about the relation between population variances. In particular, we can test the possibility that they are identical, that is, that $(\sigma_1^2/\sigma_2^2) = 1$. For this we will need Table A.4, which contains the points corresponding to some areas of the right-hand tail of the F distribution. We discuss this subject further in Section 2.7.4.

2.7. Applying the normal distribution

For the different reasons we have already discussed, the normal distribution describes a large number of real processes quite well, and this permits us to use it as a model to solve various problems of practical interest. In this last section we present some common applications.

2.7.1. Making comparisons with a reference value

Suppose we are given the task of deciding whether the batch of vinegar we discussed in Section 2.1 obeys the regulation requiring it to be 4% acetic acid.[18] To accomplish this, we perform the first three titrations of Table 2.1, whose results are 3.91, 4.01 and 3.61%. Two of these values lie below what they should be, but this might be due to natural fluctuations in the titration process, in which case nothing is really wrong with this batch of vinegar.

Ruling out crass and systematic errors, the titration values can vary only because of random errors. If so, the central limit theorem implies that the average values of sets of titrations made on the same batch should follow Student's distribution. The average of the three titrations,

$$\bar{x} = \frac{1}{3}(3.91 + 4.01 + 3.61) = 3.843\%,$$

is expected to be a point on the t distribution for the averages of three elements.

To arrive at a conclusion about the vinegar, we need to decide if the evidence furnished by the three results is incompatible with the hypothesis that they came from a population with mean $\mu = 4\%$. From Eq. (2.21) we have

$$3.843\% - t_2 \frac{s}{\sqrt{3}} < \mu < 3.843\% + t_2 \frac{s}{\sqrt{3}}.$$

Substituting $s = 0.2082\%$, that is, the standard deviation of the three titration results, and $t_2 = 4.303$ (from Table A.2, with a 95% confidence level), we have

$$3.32\% < \mu < 4.36\%.$$

Conclusion: since the confidence interval contains the 4% reference value, we cannot affirm that this value is not the true value of the average for this batch of vinegar. (The specialist would say: we cannot reject the null hypothesis.) Even though the average of the three samples is below

[18]We shall initially assume that this is the average value required, not the minimum value. This restriction will be removed in Exercise 2.20.

Table A.4
F distribution at different confidence levels.

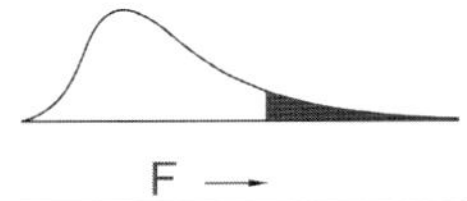

		Degrees of freedom for the numerator (v_1)																	
	v_1	1	2	3	4	5	6	7	8	9	10	12	15	20	30	40	60	120	∞
Degrees of freedom for the denominator (v_2)	v_2	*At 10% confidence level*																	
	1	39.86	49.50	53.59	55.83	57.24	58.20	58.91	59.44	59.86	60.19	60.71	61.22	61.74	62.26	62.53	62.79	63.06	63.33
	2	8.53	9.00	9.16	9.24	9.29	9.33	9.35	9.37	9.38	9.39	9.41	9.24	9.44	9.46	9.47	9.47	9.48	9.49
	3	5.54	5.46	5.39	5.34	5.31	5.28	5.27	5.25	5.24	5.23	5.22	5.20	5.18	5.17	5.16	5.15	5.14	5.13
	4	4.54	4.32	4.19	4.11	4.05	4.01	3.98	3.95	3.94	3.92	3.90	3.87	3.84	3.82	3.80	3.79	3.78	3.76
	5	4.06	3.78	3.62	3.52	3.45	3.40	3.37	3.34	3.32	3.30	3.27	3.24	3.21	3.17	3.16	3.14	3.12	3.10
	6	3.78	3.46	3.29	3.18	3.11	3.05	3.01	2.98	2.96	2.94	2.90	2.87	2.84	2.80	2.78	2.76	2.74	2.72
	7	3.59	3.26	3.07	2.96	2.88	2.83	2.78	2.75	2.72	2.70	2.67	2.63	2.59	2.56	2.54	2.51	2.49	2.47
	8	3.46	3.11	2.92	2.81	2.73	2.67	2.62	2.59	2.56	2.54	2.50	2.46	2.42	2.38	2.36	2.34	2.32	2.29
	9	3.36	3.01	2.81	2.69	2.61	2.55	2.51	2.47	2.44	2.42	2.38	2.34	2.30	2.25	2.23	2.21	2.18	2.16
	10	3.29	2.92	2.73	2.61	2.52	2.46	2.41	2.38	2.35	2.32	2.28	2.24	2.20	2.16	2.13	2.11	2.08	2.06
	11	3.23	2.86	2.66	2.54	2.45	2.39	2.34	2.30	2.27	2.25	2.21	2.17	2.12	2.08	2.05	2.03	2.00	1.97
	12	3.18	2.81	2.61	2.48	2.39	2.33	2.28	2.24	2.21	2.19	2.15	2.10	2.06	2.01	1.99	1.96	1.93	1.90
	13	3.14	2.76	2.56	2.43	2.35	2.28	2.23	2.20	2.16	2.14	2.10	2.05	2.01	1.96	1.93	1.90	1.88	1.85
	14	3.10	2.73	2.52	2.39	2.31	2.24	2.19	2.15	2.12	2.10	2.05	2.01	1.96	1.91	1.89	1.86	1.83	1.80
	15	3.07	2.70	2.49	2.36	2.27	2.21	2.16	2.12	2.09	2.06	2.02	1.97	1.92	1.87	1.85	1.82	1.79	1.76
	16	3.05	2.67	2.46	2.33	2.24	2.18	2.13	2.09	2.06	2.03	1.99	1.94	1.89	1.84	1.81	1.78	1.75	1.72
	17	3.03	2.64	2.44	2.31	2.22	2.15	2.10	2.06	2.03	2.00	1.96	1.91	1.86	1.81	1.78	1.75	1.72	1.69
	18	3.01	2.62	2.42	2.29	2.20	2.13	2.08	2.04	2.00	1.98	1.93	1.89	1.84	1.78	1.75	1.72	1.69	1.66
	19	2.99	2.61	2.40	2.27	2.18	2.11	2.06	2.02	1.98	1.96	1.91	1.86	1.81	1.76	1.73	1.70	1.67	1.63
	20	2.97	2.59	2.38	2.25	2.16	2.09	2.04	2.00	1.96	1.94	1.89	1.84	1.79	1.74	1.71	1.68	1.64	1.61
	25	2.92	2.53	2.32	2.18	2.09	2.02	1.97	1.93	1.89	1.87	1.82	1.77	1.72	1.66	1.63	1.59	1.56	1.52
	30	2.88	2.49	2.28	2.14	2.05	1.98	1.93	1.88	1.85	1.82	1.77	1.72	1.67	1.61	1.57	1.54	1.50	1.46
	40	2.84	2.44	2.23	2.09	2.00	1.93	1.87	1.83	1.79	1.76	1.71	1.66	1.61	1.54	1.51	1.47	1.42	1.38
	60	2.79	2.39	2.18	2.04	1.95	1.87	1.82	1.77	1.74	1.71	1.66	1.60	1.54	1.48	1.44	1.40	1.35	1.29
	120	2.75	2.35	2.13	1.99	1.90	1.82	1.77	1.72	1.68	1.65	1.60	1.55	1.48	1.41	1.37	1.32	1.26	1.19
	∞	2.71	2.30	2.08	1.94	1.85	1.77	1.72	1.67	1.63	1.60	1.55	1.49	1.42	1.34	1.30	1.24	1.17	1.00

Table A.4. (*Continued*)

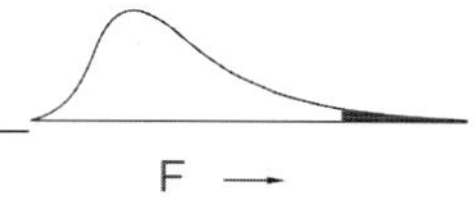

v_1	1	2	3	4	5	6	7	8	9	10	12	15	20	30	40	60	120	∞
	Degrees of freedom for the numerator (v_1)																	
At 5% confidence level																		
1	161.4	199.5	215.7	224.6	230.2	234.0	236.8	238.9	240.5	241.9	243.9	245.9	248.0	250.1	251.1	252.2	253.3	254.3
2	18.51	19.00	19.16	19.25	19.30	19.33	19.35	19.37	19.38	19.40	19.41	19.43	19.45	19.46	19.47	19.48	19.49	19.50
3	10.13	9.55	9.28	9.12	9.01	8.94	8.89	8.85	8.81	8.79	8.74	8.70	8.66	8.62	8.59	8.57	8.55	8.53
4	7.71	6.94	6.59	6.39	6.26	6.16	6.09	6.04	6.00	5.96	5.91	5.86	5.80	5.75	5.72	5.69	5.66	5.63
5	6.61	5.79	5.41	5.19	5.05	4.95	4.88	4.82	4.77	4.74	4.68	4.62	4.56	4.50	4.46	4.43	4.40	4.36
6	5.99	5.14	4.76	4.53	4.39	4.28	4.21	4.15	4.10	4.06	4.00	3.94	3.87	3.81	3.77	3.74	3.70	3.67
7	5.59	4.74	4.35	4.12	3.97	3.87	3.79	3.73	3.68	3.64	3.57	3.51	3.44	3.38	3.34	3.30	3.27	3.23
8	5.32	4.46	4.07	3.84	3.69	3.58	3.50	3.44	3.39	3.35	3.28	3.22	3.15	3.08	3.04	3.01	2.97	2.93
9	5.12	4.26	3.86	3.63	3.48	3.37	3.29	3.23	3.18	3.14	3.07	3.01	2.94	2.86	2.83	2.79	2.75	2.71
10	4.96	4.10	3.71	3.48	3.33	3.22	3.14	3.07	3.02	2.98	2.91	2.85	2.77	2.70	2.66	2.62	2.58	2.54
11	4.84	3.98	3.59	3.36	3.20	3.09	3.01	2.95	2.90	2.85	2.79	2.72	2.65	2.57	2.53	2.49	2.45	2.40
12	4.75	3.89	3.49	3.26	3.11	3.00	2.91	2.85	2.80	2.75	2.69	2.62	2.54	2.47	2.43	2.38	2.34	2.30
13	4.67	3.81	3.41	3.18	3.03	2.92	2.83	2.77	2.71	2.67	2.60	2.53	2.46	2.38	2.34	2.30	2.25	2.21
14	4.60	3.74	3.34	3.11	2.96	2.85	2.76	2.70	2.65	2.60	2.53	2.46	2.39	2.31	2.27	2.22	2.18	2.13
15	4.54	3.68	3.29	3.06	2.90	2.79	2.71	2.64	2.59	2.54	2.48	2.40	2.33	2.25	2.20	2.16	2.11	2.07
16	4.49	3.63	3.24	3.01	2.85	2.74	2.66	2.59	2.54	2.49	2.42	2.35	2.28	2.19	2.15	2.11	2.06	2.01
17	4.45	3.59	3.20	2.96	2.81	2.70	2.61	2.55	2.49	2.45	2.38	2.31	2.23	2.15	2.10	2.06	2.01	1.96
18	4.41	3.55	3.16	2.93	2.77	2.66	2.58	2.51	2.46	2.41	2.34	2.27	2.19	2.11	2.06	2.02	1.97	1.92
19	4.38	3.52	3.13	2.90	2.74	2.63	2.54	2.48	2.42	2.38	2.31	2.23	2.16	2.07	2.03	1.98	1.93	1.88
20	4.35	3.49	3.10	2.87	2.71	2.60	2.51	2.45	2.39	2.35	2.28	2.20	2.12	2.04	1.99	1.95	1.90	1.84
25	4.24	3.39	2.99	2.76	2.60	2.49	2.40	2.34	2.28	2.24	2.16	2.09	2.01	1.92	1.87	1.82	1.77	1.71
30	4.17	3.32	2.92	2.69	2.53	2.42	2.33	2.27	2.21	2.16	2.09	2.01	1.93	1.84	1.79	1.74	1.68	1.62
40	4.08	3.23	2.84	2.61	2.45	2.34	2.25	2.18	2.12	2.08	2.00	1.92	1.84	1.74	1.69	1.64	1.58	1.51
60	4.00	3.15	2.76	2.53	2.37	2.25	2.17	2.10	2.04	1.99	1.92	1.84	1.75	1.65	1.59	1.53	1.47	1.39
120	3.92	3.07	2.68	2.45	2.29	2.17	2.09	2.02	1.96	1.91	1.83	1.75	1.66	1.55	1.50	1.43	1.35	1.25
∞	3.84	3.00	2.60	2.37	2.21	2.10	2.01	1.94	1.88	1.83	1.75	1.67	1.57	1.46	1.39	1.32	1.22	1.00

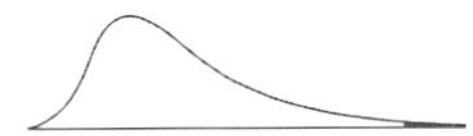
F →

At 1% confidence level

1	4052	4999	5403	5625	5764	5859	5928	5982	6022	6056	6106	6157	6209	6261	6287	6313	6339	6366
2	98.50	99.00	99.17	99.25	99.30	99.33	99.36	99.37	99.39	99.40	99.42	99.43	99.45	99.47	99.47	99.48	99.49	99.50
3	34.12	30.82	29.46	28.71	28.24	27.91	27.67	27.49	27.35	27.23	27.05	26.87	26.69	26.50	26.41	26.32	26.22	26.13
4	21.20	18.00	16.69	15.98	15.52	15.21	14.98	14.80	14.66	14.55	14.37	14.20	14.02	13.84	13.75	13.65	13.56	13.46
5	16.26	13.27	12.06	11.39	10.97	10.67	10.46	10.29	10.16	10.05	9.89	9.72	9.55	9.38	9.29	9.20	9.11	9.02
6	13.75	10.92	9.78	9.15	8.75	8.47	8.26	8.10	7.98	7.87	7.72	7.56	7.40	7.23	7.14	7.06	6.97	6.88
7	12.25	9.55	8.45	7.85	7.46	7.19	6.99	6.84	6.72	6.62	6.47	6.31	6.16	5.99	5.91	5.82	5.74	5.65
8	11.26	8.65	7.59	7.01	6.63	6.37	6.18	6.03	5.91	5.81	5.67	5.52	5.36	5.20	5.12	5.03	4.95	4.86
9	10.56	8.02	6.99	6.42	6.06	5.80	5.61	5.47	5.35	5.26	5.11	4.96	4.81	4.65	4.57	4.48	4.40	4.31
10	10.04	7.56	6.55	5.99	5.64	5.39	5.20	5.06	4.94	4.85	4.71	4.56	4.41	4.25	4.17	4.08	4.00	3.91
11	9.65	7.21	6.22	5.67	5.32	5.07	4.89	4.74	4.63	4.54	4.40	4.25	4.10	3.94	3.86	3.78	3.69	3.60
12	9.33	6.93	5.95	5.41	5.06	4.82	4.64	4.50	4.39	4.30	4.16	4.01	3.86	3.70	3.62	3.54	3.45	3.36
13	9.07	6.70	5.74	5.21	4.86	4.62	4.44	4.30	4.19	4.10	3.96	3.82	3.66	3.51	3.43	3.34	3.25	3.17
14	8.86	6.51	5.56	5.04	4.69	4.46	4.28	4.14	4.03	3.94	3.80	3.66	3.51	3.35	3.27	3.18	3.09	3.00
15	8.68	6.36	5.42	4.89	4.56	4.32	4.14	4.00	3.89	3.80	3.67	3.52	3.37	3.21	3.13	3.05	2.96	2.87
16	8.53	6.23	5.29	4.77	4.44	4.20	4.03	3.89	3.78	3.69	3.55	3.41	3.26	3.10	3.02	2.93	2.84	2.75
17	8.40	6.11	5.18	4.67	4.34	4.10	3.93	3.79	3.68	3.59	3.46	3.31	3.16	3.00	2.92	2.83	2.75	2.65
18	8.29	6.01	5.09	4.58	4.25	4.01	3.84	3.71	3.60	3.51	3.37	3.23	3.08	2.92	2.84	2.75	2.66	2.57
19	8.18	5.93	5.01	4.50	4.17	3.94	3.77	3.63	3.52	3.43	3.30	3.15	3.00	2.84	2.76	2.67	2.58	2.49
20	8.10	5.85	4.94	4.43	4.10	3.87	3.70	3.56	3.46	3.37	3.23	3.09	2.94	2.78	2.69	2.61	2.52	2.42
25	7.77	5.57	4.68	4.18	3.85	3.63	3.46	3.32	3.22	3.13	2.99	2.85	2.70	2.54	2.45	2.36	2.27	2.17
30	7.56	5.39	4.51	4.02	3.70	3.47	3.30	3.17	3.07	2.98	2.84	2.70	2.55	2.39	2.30	2.21	2.11	2.01
40	7.31	5.18	4.31	3.83	3.51	3.29	3.12	2.99	2.89	2.80	2.66	2.52	2.37	2.20	2.11	2.02	1.92	1.80
60	7.08	4.98	4.13	3.65	3.34	3.12	2.95	2.82	2.72	2.63	2.50	2.35	2.20	2.03	1.94	1.84	1.73	1.60
120	6.85	4.79	3.95	3.48	3.17	2.96	2.79	2.66	2.56	2.47	2.34	2.19	2.03	1.86	1.76	1.66	1.53	1.38

Adapted from Pearson, E. S. and Hartley, H. O. *Biometrika Tables for Statisticians, Vol. 1, 3rd edition*, Cambridge, Cambridge University Press, 1996.

specification, this evidence is not enough to justify rejection of the batch. This conclusion, however, is based on only three results, that is, on only two degrees of freedom. To make a more precise decision we need more information.

So we perform five more titrations, and obtain the next five values in Table 2.1. Now we have eight results, with $\bar{x} = 3.818\%$ and $s = 0.1286\%$. Substituting these values into Eq. (2.21), this time with $n = 8$ and $t_7 = 2.365$, we arrive at the inequality

$$3.71\% < \mu < 3.93\%.$$

The interval became narrower, and no longer includes the reference value. Now we can say that there is evidence, at the 95% confidence level, that the mean concentration of acetic acid in the examined batch is indeed below 4% (that is, we can reject the null hypothesis).

Actually, since 4% is the minimum concentration required by law, not the average concentration, we must adapt our calculations to this fact. A simple way of doing this is to take 4% as the *lower* limit of an interval about the mean. Assuming that $\sigma \cong 0.15\%$, which is the standard deviation of all the values in Table 2.1, and adding three times this quantity to 4%, we obtain 4.45% as a new mean to be tested. If $\mu = 4.45\%$, then 99.87% of the observations are expected to occur above 4% (because this limit is now three standard deviations below the mean). Now, we only need to repeat the test taking 4.45% as the new mean. If we cannot reject this hypothesis, 99.87% of the samples are expected to have a concentration of acetic acid of 4% or more, and therefore be in compliance with the law.

Exercise 2.20. Use the new reference mean for acetic acid, 4.45%, and repeat the test with the three last values of Table 2.1.

Exercise 2.21. A chemist is testing a new method for assaying iron. Performing four analyses on a standard whose true concentration is 14.3%, she obtains 13.7, 14.0, 13.9 and 14.1% iron. How would you evaluate the accuracy of the new method, at the 95% confidence level? Do the four determinations come from a distribution with a mean of 14.3%?

2.7.2. Determining sample size

We have just seen a practical example of how we can reduce the width of a confidence interval from a t-test by increasing the number of values contained in the sample. A similar argument helps us determine how large a sample should be to detect a variation of a certain magnitude in a mean, or to estimate the value of a parameter with a certain precision.

Continuing with the titration example, suppose that we wish to obtain an estimate of the concentration with a precision of 0.1%. How many replicate titrations should we perform?

Since our t-test intervals are given by

$$\mu \pm t_{\nu} \frac{s}{\sqrt{n}},$$

and we want to estimate the concentration mean within $\pm$ 0.1 %, we need to perform a number of titrations n such that

$$\left| t_{\nu} \frac{s}{\sqrt{n}} \right| \leq 0.1\%,$$

or

$$n \geq \left(\frac{t_{\nu} s}{0.1\%} \right)^2 .$$

Here we have a problem. The s value must be calculated from the sample, yet we do not know how many titrations should be performed. In practice, fortunately, this problem is not as grave as it seems, because past measurements can often be used to furnish a "historical" s-value. This is what usually happens in routine procedures, such as quality control. In our example, we can use the standard deviation of all 20 titrations in Table 2.1, that is $s = 0.1509\%$, and write

$$n \geq \left(\frac{t_{19} \times 0.1509\%}{0.1\%} \right)^2 .$$

Since the standard deviation was calculated from 20 observations, the t value is the one corresponding to 19 degrees of freedom, independent of the value of n, which is to be used to calculate the average. This contributes to reducing the width of the confidence interval. Substituting $t_{19} = 2.093$, we have

$$n \geq 9.98.$$

So we must perform at least 10 titrations to obtain the desired precision.

When we have an estimate of the standard deviation obtained from a historical record of reasonable size, the difference between the t and normal distributions loses its importance. This is a very common situation in analysis laboratories, where the same procedures are repeated many times, day after day. In these cases, the sample size can be determined using the simpler expression

$$n \geq \left(\frac{z\sigma}{L} \right)^2 , \tag{2.24}$$

where L is the desired precision, σ the "population" standard deviation and z the point of the standard normal distribution for the selected confidence level.

Exercise 2.22. An analysis laboratory performs determinations with a historical standard deviation of 0.5%. A client sends in a sample, whose concentration she would like to know within $\pm 0.2\%$. Use Eq. (2.24) to estimate how many replicate determinations the analyst needs to perform to obtain a result with the desired precision at the 95% confidence level.

Exercise 2.23. Suppose we wish to determine a 95% confidence interval for the mass of a bean so that the difference between the extreme values of the interval is one sample standard deviation. How many beans must we weigh?

2.7.3. Statistically controlling processes

Imagine a chemical industry of some complexity, such as polymer manufacturing. The engineers hired to plan and build the plant have to make sure that it will be capable of producing polymers with the characteristics required by its customers. To accomplish this they need to consider carefully all of the relevant variables — usually more than just a few — and project the plant so that these variables are kept under control.

After many calculations and tests in laboratories and pilot plants, the full-scale factory is constructed. Some time is then spent fine-tuning the large-scale process until regular operation finally begins. From this time on, to certify that everything is running according to plan, that is, that the process remains under control, the plant operators continue to systematically monitor the characteristics of the polymer being produced.

One of the most important variables used to control polymer production is viscosity. From time to time, a polymer sample is collected from the production line and sent to the laboratory, where its viscosity is measured. The values obtained in this way — or, more commonly, their averages — are systematically plotted as a function of time. If the process is completely under control, in the absence of crass or systematic errors, what kind of distribution should the points have?

You are right: a normal distribution, for the individual observations, or Student's t-distribution, for the averages.[19] When the process is under control, its variability is only due to random errors, and for this reason its responses should follow a normal distribution or a distribution closely related to it. This is the basic principle of quality control — again, another consequence of the central limit theorem.

Fig. 2.12 shows 80 viscosity values, in the order they were obtained, at regular intervals during the process. The units are arbitrary. One can observe that the values are well behaved, distributing themselves randomly about an average value, 45, with a standard deviation of 1.67.

[19]This is strictly true if the variable is continuous, like viscosity. For other kinds of variables other distributions are more adequate. You can find out more about this in quality control textbooks.

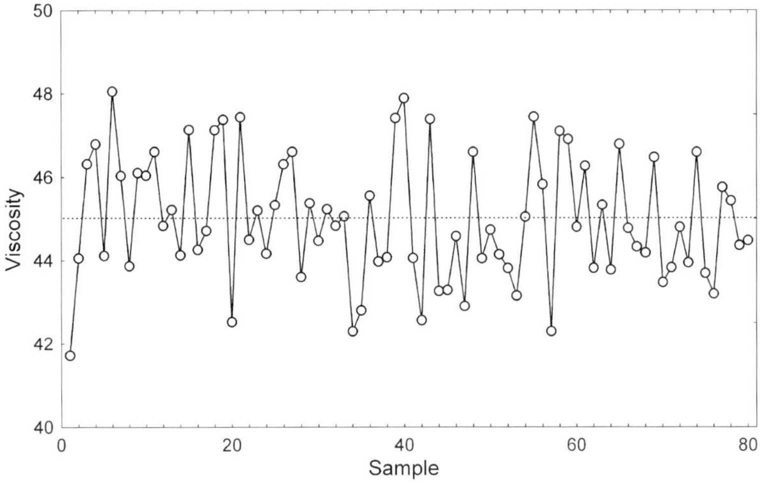

Fig. 2.12. Plot of viscosity against time, for a process under control.

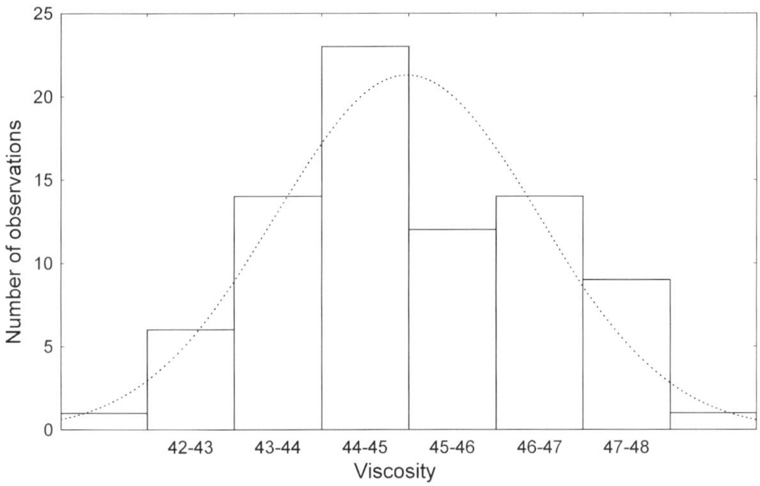

Fig. 2.13. Histogram of the data in Fig. 2.12.

In Fig. 2.13, which shows a histogram of these values, we see that their distribution is quite accurately represented by a normal distribution. This ideal situation is the dream of every production engineer.

In practice, graphs such as Fig. 2.12 – called control charts – are traced point by point, in real time, by the production line personnel, and serve as an important tool to detect problems that could be perturbing the process. As each point is added, the graph is analyzed. Any anomalous pattern, suggesting deviations from normality, is a warning to those responsible for the process to take the proper measures for bringing it back under control.

The commonest control chart is identical to the graph in Fig. 2.12, with three parallel horizontal lines that define the characteristics of the process when it is running without problems. The central line corresponds to the process mean, $\mu = 45$ in our example. The other two lines are drawn three standard deviations above and below the average. The line corresponding to $\mu+3\sigma = 48$ is the upper control limit. The line corresponding to $\mu-3\sigma = 42$ naturally is the lower control limit. Between these two limits, as we already know, 99.73% of all the individual values are expected to fall, if everything is performing perfectly well.

Fig. 2.14 shows a control chart with upper and lower limits, but with some groups of points illustrating two of the most commonly occurring anomalous situations. As soon as one of these appears, the technicians should intervene and take the necessary steps to bring the process back under control. To do this they will have to rely on their technical knowledge about how the plant works, as well as on the evidence obtained from the control chart. A level change in the value being monitored, for example, could be associated with a change in work shift, and be due to different ways of operating the process. A systematic trend could indicate equipment deterioration or progressive contamination of a reagent or catalyst.

There are several rules of thumb to help the operators detect anomalous situations. For example, the following occurrences are considered signs

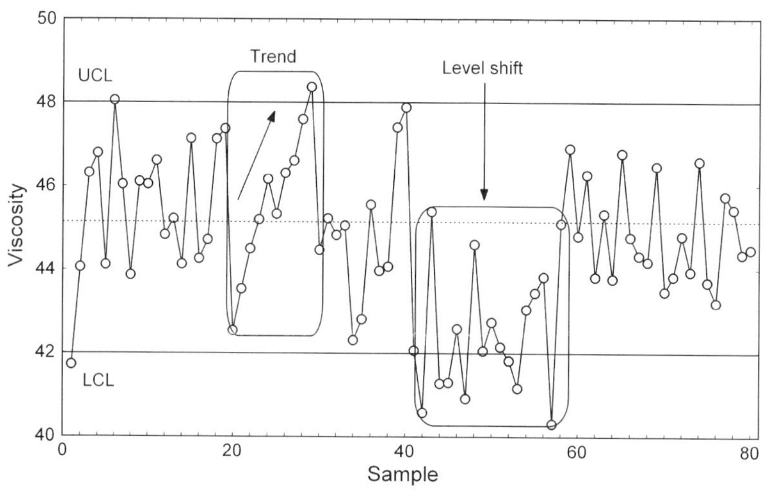

Fig. 2.14. Control chart with patterns of anomalous behavior. UCL and LCL indicate the control limits. The central line corresponds to the process mean.

that a process is starting to escape control:

- One or more points located outside the control limits
- Four of five successive points situated more than one standard deviation from the mean, all on the same side of the central line
- Six consecutive ascending or descending points
- Nine successive points on the same side of the central line

These are not hard-and-fast rules but vary from process to process. Also, different authors recommend variations of these rules. You can find many others in the references recommended for reading at the end of the book.

Quality is always defined by customer satisfaction. The customers decide what characteristics they want the product to have. A tire manufacturer, for example, could tell the rubber manufacturer that he will only buy his product if its viscosity is 45. This is the nominal specification value. However, since it is not realistic to expect that all the batches have exactly this value, the tire manufacturer, in practice, accepts a variation in the viscosity value of, say, three units. These values define tolerance limits for the viscosity specification: no less than 42, and no more than 48.

Once the tolerance limits have been agreed, the tire manufacturer regularly inspects samples from the batches of rubber he receives. Batches with values outside the 42–48 tolerance range are rejected and returned to the supplier. Clearly, the supplier has the greatest interest in developing an efficient and stable process, capable of satisfying the demands of his clients.

Now, the fact that a process is under control only means that it is operating in a consistent way, and not necessarily that it will satisfy the specifications. It is also necessary to compare the behavior of the process, that is, its control parameters, with the specification limits. Only thus will we have a measure of process capability.

One of the most used capability indexes is the quality index C_{pk}, defined as the smaller of the fractions

$$\frac{\text{USL} - \mu}{3\sigma} \quad \text{and} \quad \frac{\mu - \text{LSL}}{3\sigma},$$

where USL and LSL are the upper and lower specification limits, and μ and σ are reliable estimates of the mean and standard deviation of the process. For example, if the customer wants rubber batches with viscosities between 42 and 48, and the process is operating with $\mu = 46$ and $\sigma = 1.6$, then

$$\frac{\text{USL} - \mu}{3\sigma} = \frac{48 - 46}{3 \times 1.6} = 0.42,$$

$$\frac{\mu - \text{LSL}}{3\sigma} = \frac{46 - 42}{3 \times 1.6} = 0.83.$$

This means that $C_{pk} = 0.42$, which is a very poor value. So poor, in fact, that no client would accept it. Some of the most advanced companies in the world, like Motorola and General Electric, have already established a C_{pk} value of 2 as the minimum quality standard for their own processes and also for those of their suppliers (Bhote, 1996). To satisfy this demand, a process must be perfectly centered on its nominal specification value, while operating with a standard deviation of not more than 1/12 of the tolerance range. In this situation, only two values out of a billion will fall outside the tolerance limits. It is this goal that is the basis of the quality management movement known as Six Sigma[20] (Breyfogle, 1998).

What we have presented here is only a small portion, and very simplified at that, of the extensive array of concepts and techniques that constitute statistical process control. It is not our aim to exhaust this subject, but only to discuss it a little as an application of the normal distribution. Deeper treatments can be found in any of many books entirely dedicated to quality or statistical process control. To learn more about these important tools you can consult, for example, Oakland and Followell (1990) or Montgomery (1997).

2.7.4. Comparing two treatments

Comparing two averages. Two chemists, S. Arrhenius and J. Berzelius, were given the task of analyzing vinegar batches sent in by five different manufacturers. Each chemist analyzed a sample from each batch, obtaining the results given in Table 2.4, where we see that the average and standard deviation of the determinations made by Berzelius are smaller than those obtained by Arrhenius. To evaluate the performance of his two chemists, the head of the laboratory, A. Lavoisier, decided to do a statistical comparison of their results. Do the variations indicate a systematic difference between the techniques of the two analysts?

The data of Table 2.4 are an example of the simplest kind of experiment. There is only one factor, the analytical technique, and one response, the concentration obtained. The factor is being set at only two levels, represented by Arrhenius and Berzelius. Our objective is to discover if the response is affected by this change of level.

We can solve this problem by adapting Eq. (2.17) to the difference between two averages. We start by substituting $\bar{x}_A - \bar{x}_B$ for $\bar{x}$ and $\mu_A - \mu_B$ for μ in the numerator. In the denominator, we need the standard deviation of the difference between the two sample averages. Generalizing

[20]Can you guess why this quality trend is so named?

Table 2.4
Comparison of results obtained by two chemists titrating vinegar samples from five different producers. Concentrations of acetic acid in %

Sample	Arrhenius	Berzelius	$d = x_A - x_B$
1	3.77	3.62	0.15
2	3.85	3.69	0.16
3	4.07	4.10	– 0.03
4	4.83	4.70	0.13
5	5.05	4.89	0.16
Average	4.314	4.200	0.114
Standard deviation	0.5871	0.5772	0.0814
Degrees of freedom	4	4	4

the result of Exercise 2.13 for the case in which $\bar{x}_A$ and $\bar{x}_B$ are obtained respectively from n_A and n_B independent observations, we can write

$$\widehat{V}(x_A - \bar{x}_B) = \widehat{V}(\bar{x}_A) + \widehat{V}(\bar{x}_B) = \frac{s_A^2}{n_A} + \frac{s_B^2}{n_B}.$$

If we assume that s_A^2 and s_B^2 are estimates of the same population variance, we can combine them into a single estimate s^2 (with more degrees of freedom) and write

$$\widehat{V}(\bar{x}_A - \bar{x}_B) = s^2\left(\frac{1}{n_A} + \frac{1}{n_B}\right)$$

Therefore the expression for the t-test becomes

$$\frac{(\bar{x}_A - \bar{x}_B) - (\mu_A - \mu_B)}{s\sqrt{\frac{1}{n_A} + \frac{1}{n_B}}} \approx t_\nu, \tag{2.25}$$

and from this we arrive at the confidence interval for the difference between the two population means:

$$(\mu_A - \mu_B) = (\bar{x}_A - \bar{x}_B) \pm t_\nu s\sqrt{\frac{1}{n_A} + \frac{1}{n_B}}. \tag{2.26}$$

To obtain a pooled estimate of the standard deviation of an observation, s, we take an average of the variances of the two samples, weighted by their respective number of degrees of freedom:

$$s^2 = \frac{(n_A - 1)s_A^2 + (n_B - 1)s_B^2}{(n_A - 1) + (n_B - 1)}. \tag{2.27}$$

The number of degrees of freedom for the t-test, as we know, is the one used to calculate s. In this example, $\nu = n_A + n_B - 2 = 8$. With the appropriate numerical values, we obtain $s = 0.5822\%$ from Eq. (2.27). The estimate of the standard deviation of the difference between the averages then becomes

$$s\sqrt{\frac{1}{n_A}+\frac{1}{n_B}} = 0.5822\% \times \sqrt{\frac{1}{5}+\frac{1}{5}} = 0.3682\%.$$

At the 95% confidence level, $t_8 = 2.306$. Combining everything, we can write

$$\begin{aligned}\mu_A - \mu_B &= (4.314\% - 4.200\%) \pm (2.306 \times 0.3682\%).\\ &= 0.114\% \pm 0.849\%\\ &= [-0.735\%,\ 0.963\%]\end{aligned}$$

Conclusion: because the interval includes the zero value, we cannot say that the means of the results obtained by Arrhenius and Berzelius are really different, at this confidence level.

This test is probably the most used of all statistical tests. It is very valuable when systematic differences between samples are caused by a single factor, and is entirely appropriate for comparing two *independent* averages. Arrhenius and Berzelius, however, made determinations on samples from five different vinegar producers. It is natural to suspect that these samples have different concentrations of acetic acid, and that samples from the same manufacturer are likely to be more similar to each other than to those coming from different producers. For this reason alone the results of the analyses are already expected to vary, blurring a possible difference in analytical technique. Since we are interested in a possible difference *between the analysts*, we need an improved method, which eliminates the variation caused by the different producers on the final result.

Exercise 2.24. The quantity of α-PbO_2 in an automobile battery plate was determined by X-ray spectroscopy. Several spectra were repeatedly registered, applying or not baseline corrections. The results are shown below. Is there a systematic difference between these two ways of analyzing the plate?

	% α-PbO_2	
Spectrum	With correction	Without correction
1	16.2	19.0
2	16.7	19.8
3	17.3	18.5

Making paired comparisons. Even though we have just done an incorrect statistical analysis (because the samples were not independent), the experiments to compare the performances of Arrhenius and Berzelius were, in fact, properly executed. A procedure called blocking was used, which allows us to eliminate the influence of factors that do not interest us.

Blocking is one of the fundamental principles of good experimental technique. The 10 samples of Table 2.4 can be seen as five blocks of two samples, each pair coming from the same producer. In the present example, we are solely interested in the influence of experimental technique on the results of the analyses. The origin of the vinegar also affects the results, but it only complicates the comparison we wish to make. If we just randomly distribute five samples to one of the analysts and five to the other, the differences in the results between the two will be affected not only by the difference in their analytical procedures but also by variations in the concentrations of acetic acid owing to the different sample origins. The effects of these two factors are confounded, and we would have no way of separating the effect of one factor from the effect of the other. With blocking, we vary the factor of interest only within each block, and thus exclude the effect of the other factor (manufacturer) from the calculations.

In obtaining the data in Table 2.4, two samples were taken from each batch, one being assigned to Arrhenius and the other to Berzelius. Each line of the table therefore refers to two samples from the same manufacturer. If the results of the analyses differ in a systematic way *within the lines* or *blocks*, it can only be due to the way in which the chemical analyses were performed.

If there is really no difference in the analytical techniques of Arrhenius and Berzelius, then the average of the five differences in the last column of the table will be a point on a t distribution with mean $\Delta = 0$. To test this hypothesis, we need the standard deviation of the average of the variations d_i within each line:

$$\frac{s_d}{\sqrt{n}} = \frac{0.0814\%}{\sqrt{5}} = 0.0364\%.$$

This value is about 10 times smaller than the standard deviation of the difference between the overall averages of Arrhenius and Berzelius, since now the variation due to the different producers was excluded from the test.

The confidence interval for Δ is given by

$$\Delta = \bar{d} \pm t_{\upsilon} \frac{s_d}{\sqrt{n}}. \tag{2.28}$$

Using the values in the table (and $t_4 = 2.776$), we obtain $\mu = [0.014\%, 0.215\%]$, and conclude that there is indeed a systematic difference between the results of the two analysts. The analyses performed by Berzelius tend to have lower results than those of Arrhenius, and the difference has a 95% probability of being between 0.014 and 0.215%. Note that this result does not tell us which analyst has the better technique, since we do not know the true concentration values. We can only state that there is evidence of a systematic difference between their results.

To block or not to block? Unfortunately this is not always clear before we perform our experiments. Yet the randomization details of the experimental procedure depend on this decision. If we are sure there is only one major source of variation in the data, no gain is achieved by blocking. In fact, blocking would then result in lowering the number of degrees of freedom, as we have just seen in the above example, and this in turn would widen the confidence interval being calculated. If, on the other hand, the experimenter feels there might be sources of variation extraneous to those that interest him, it is wise to block, because confounding cannot be undone by increasing sample size. Increasing sample size only helps to narrow (correctly calculated) confidence intervals, improving the precision of our estimates, as we have discussed before.

An alternative way to do this test is to use the observed values of the samples to estimate a point on the t distribution and compare it with the tabulated value (under the hypothesis that $\Delta = 0\%$). Doing it this way, we would have

$$\widehat{t} = \frac{|\bar{d} - \Delta|}{s_d/\sqrt{n}} = \frac{0.114\% - 0\%}{0.0814/\sqrt{5}} = 3.13. \tag{2.29}$$

Since this value is larger than $t_4 = 2.776$, we conclude that the chances it occurred randomly, without there being a real difference in analytical technique, are less than 2.5%, and we reject the null hypothesis, that is, the hypothesis that $\Delta = 0\ \%$. This procedure — hypothesis testing — is preferred by a majority of statisticians, and can be extended to other distributions as well. In the next section we show how to test the hypothesis that two variances are equal. This is a test we will use later, to evaluate the quality of fit of an empirical model.

The philosophy underlying hypothesis testing is easy to understand. The term that appears in the denominator of Eq. (2.29) is an example of a standard error (in this case, of the average of the differences $x_A - x_B$). The estimate $\hat{t}$ is the deviation of the sample value relative to the population value corresponding to the null hypothesis, measured in standard error units. The larger this deviation, the less are the chances that the null hypothesis is true. Confidence intervals can always be transformed into hypothesis tests, for which the numerator is an estimate of the parameter of interest and the denominator is the corresponding standard error. For the difference between two averages, for example, the standard error is, from Eq. (2.26),

$$s\sqrt{\frac{1}{n_A}+\frac{1}{n_B}}.$$

Exercise 2.25. What would be the hypothesis test for comparing an average with a reference value?

Exercise 2.26. Redo Exercise 2.24 testing the null hypothesis, instead of calculating the confidence interval. How much would we need to change the confidence level, to reverse our conclusion?

Comparing two variances. To compare the averages of two independent samples, we combine the two sample variances to form a single pooled estimate. Since the pooled value has more degrees of freedom, the confidence interval becomes narrower and the test more sensitive, that is, it becomes capable of detecting smaller systematic differences. Obviously it only makes sense to combine sample variances if they are estimates of the same population variance. To justify calculating a pooled estimate, we need to test the null hypothesis that s_A^2 and s_B^2 are estimates of identical population variances, $\sigma_A^2 = \sigma_B^2$. This can be done by performing an F-test, which is based on Eq. (2.23). If the two population variances are equal, Eq. (2.23) becomes

$$\frac{s_A^2}{s_B^2} \approx F_{\nu_A,\nu_B}.$$

Now we only need to compare the ratio of the two sample variances with the tabulated value for the F distribution with the appropriate numbers of degrees of freedom. Using Table A.4 we see that $F_{4,4} = 6.39$, at the 95% confidence level. To reject the null hypothesis, the variance ratio will have to be larger than this value. This would also imply that the pooled estimate should not be calculated. Since in our example

$$(s_A^2/s_B^2) = (0.5871/0.5722)^2 = 1.035,$$

nothing is wrong with the calculations we have done.

Note that the smallest value in Table A.4 is 1.000. This means that in the F-test the numerator is always the larger of the two variances. According to a much used rule of thumb that dispenses with Table A.4, we can always pool variances to obtain a single variance estimate as long as the ratio of the larger and smaller variances does not exceed four.

2A. Applications

2A.1. From home to work

One of the authors never wanted to learn to drive.[21] Since he lives about 12 km from his university, he usually takes the bus to work. The bus spends a little over an hour to reach the campus, passing underneath a passageway that crosses over the busy BR-101 highway. From there to the bus terminal on the other side of the campus there are 16 bus stops. Our researcher normally uses one of three different paths to arrive at the department where he works:

A. Gets off the bus at the first stop after the passageway, uses it to cross the highway, and follows along one side of the campus until the entrance giving access to the department. This is the most deserted path, and also the most exposed to weather.
B. Gets off at the third stop after the passageway, crosses BR-101 on foot, and walks on a diagonal path across the campus. In spite of the risk of being run over by a creative Brazilian driver, this path is used by many people and has several pleasant shady spots.
C. Gets off at the final point on the other side of the campus and takes a diagonal path in the direction opposite to that in item B. This is the most pleasant and secure path and is used by many more people than the other two.

Table 2A.1 contains the results of 32 experiments performed by the researcher, in which the time taken to travel these paths was measured. The stopwatch was started just as the bus passed beneath the passageway and stopped exactly when our hiker reached the entrance of the building housing his department. The experiments were performed in the same order as given in the table. The order was not random, for reasons to be discussed later. The objective of the experiments was to quantify the differences in the time spent taking each of the three paths. Fig. 2A.1 shows all the times measured, in the same order as in the table.

One of the times in the table (experiment 9) was excluded from further consideration right from the start. On the day experiment 9 was performed a strong tropical rain fell, practically bringing the city to a standstill. The unfortunate researcher spent more than four hours to arrive at his destination. The value given in the table, 56.3 min (timed from the passageway close to the campus), evidently is not typical of the times for path C. It is called an anomalous or outlying point. No deprecation is intended here. "Anomaly" only means that this value cannot be viewed as coming from the same distribution that produced the

[21]BBN, of course. As an American citizen, REB would be very hard pressed if he could not drive a car. ISS, a young professional and a mother, needs a car to conciliate housekeeping with university duties.

Table 2A.1
Experimental data

Experiment	Departure time (h)	Weekday	Path	Time (min)
1	10:55	Monday	C	18.3
2	11:20	Wednesday	C	18.9
3	10:40	Friday	B	10.9
4	11:25	Monday	C	20.7
5	12:50	Friday	B	11.4
6	11:30	Wednesday	C	22.9
7	11:25	Wednesday	B	12.1
8	7:35	Tuesday	A	12.8
9	8:10	Monday	C	56.3
10	7:00	Tuesday	A	13.3
11	8:10	Thursday	B	10.9
12	17:00	Friday	A	13.1
13	15:00	Wednesday	A	12.7
14	12:30	Monday	C	20.6
15	7:30	Tuesday	C	18.9
16	12:30	Wednesday	B	11.0
17	8:15	Friday	B	10.3
18	7:05	Thursday	A	13.0
19	12:50	Monday	C	18.6
20	7:35	Tuesday	A	13.0
21	8:00	Thursday	B	10.6
22	9:20	Wednesday	B	10.4
23	7:15	Thursday	C	21.5
24	8:15	Friday	B	10.9
25	8:40	Monday	B	10.9
26	8:40	Wednesday	B	11.0
27	9:00	Friday	C	19.1
28	10:00	Wednesday	C	16.1
29	9:10	Friday	B	12.1
30	9:15	Wednesday	C	18.1
31	11:15	Monday	B	12.2
32	14:30	Friday	C	19.2

other points, and for this reason it does not make sense to compute any population estimate including it. Many tests to detect anomalies are available. Later we will discuss two of them.

A more orthodox statistical analysis would probably start with a procedure called analysis of variance. First we would decide *if there is* any difference among the times for the three paths. Only later would we try to discover where the differences come from and how large they are. The

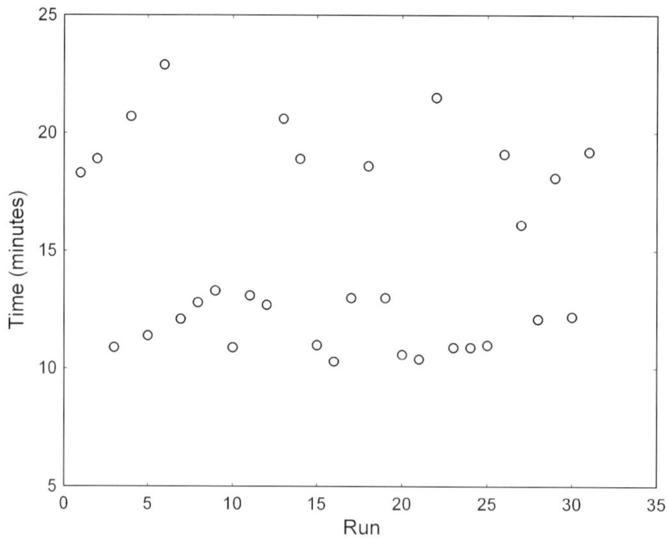

Fig. 2A.1. Experimental times, in the chronological order they were recorded. The time of run 9 was excluded (see text).

authors of this book, however, are inclined to follow what could be called the *Yogi Berra School of Data Analysis.* We are very fond of another of Yogi's sayings that goes likes this: "you can observe a lot by watching". As for the analysis of variance, we will return to it in Chapter 5, in another context.

Fig. 2A.2 shows the same data as Fig. 2A.1, with a difference. Now the times are stratified, that is, grouped according to the path used. (This procedure is one of the seven basic quality management tools, as mentioned before.) We do not need any statistics to see immediately that path C takes the most time and that path B is the quickest, closely followed by path A. It also becomes evident that the dispersion of the values is much larger for path C than for the other two.[22]

Someone less enthusiastic about graphical representations than we are might argue that the same conclusions could be reached by an inspection of the numerical values themselves, especially if they were ordered according to path. This might be true, but it would require more work, and the difference in the dispersions for the different paths would not be so clear. Besides, with the current level of instrument automation it is so easy to produce so much data in so little time that graphical procedures to filter or concentrate information now are rather a necessity than a convenience.

[22]You can discover why, rereading the descriptions of the three paths.

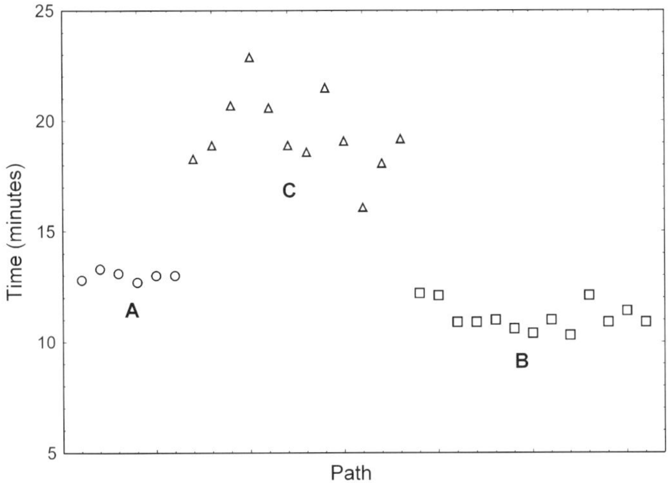

Fig. 2A.2. Experimental data, stratified according to the path.

So there is no doubt that paths A and B are fastest. Is the difference between the two statistically significant? The answer lies in a t-test, similar to the one discussed in Section 2.7.4. To perform it, we need some of the values in Table 2A.2. Using Eq. (2.26), we write

$$\begin{aligned}(\mu_A - \mu_B) &= (\bar{x}_A - \bar{x}_B) \pm t_\nu s\sqrt{\frac{1}{n_A} + \frac{1}{n_B}} \\ &= (12.7 - 10.3) \pm t_{17} s\sqrt{\frac{1}{6} + \frac{1}{13}} \\ &= 2.4 \pm (2.11 \times 0.5464 \times 0.4936) = 2.4 \pm 0.57 = (1.83,\ \ 2.97).\end{aligned}$$

The standard deviation value, 0.5464, is a pooled estimate, with 17 degrees of freedom, calculated using Eq. (2.27). Since this interval does not include zero (at the 95% confidence level), we can conclude that path B indeed takes less time — between 1.83 and 2.97 min — than path A.

This is a good place to stress a fundamental point: *statistical significance is one thing, practical importance is quite another.* The result of a statistical test only indicates if a certain numerical value is a consequence of some systematic characteristic of the phenomenon we are studying, or a random fluctuation. If this finding is or is not important will have to be decided by the researcher, probably based on evidence of another nature. In our example, the data indicates that by using path B the professor arrives earlier at his destination. This path, however, is the riskiest, because it includes crossing a busy Brazilian highway. The second choice, path A, has very few shelters from the weather. Moreover, the average difference in time between this path and path C is only about

Table 2A.2
Descriptive statistics of the times of Table 2A.1, excluding experiment 9

Path	Runs	Minimum	Maximum	Average	Standard deviation
A	6	12.7	13.3	12.98	0.2137
B	13	10.3	12.2	11.13	0.6356
C	12	16.1	22.9	19.41	1.7799

7 min. As a result, unless he is in a big hurry, BBN prefers to continue using path C, which is by far the most comfortable.

Several other facts can be discovered upon a closer examination of the experimental data. One that springs to mind is the extraordinary flexibility in the times the professor leaves for work. Consider, however, Fig. 2A.3, where the times of departure are plotted against the days of the week. For Tuesday and Thursdays the times are never later than 8:00 in the morning. Probably the professor has early classes on these days.

Fig. 2A.4 shows how path choice is related to departure time. We see that path A was chosen only when our researcher left early for work or when he left in the late afternoon. The explanation is simple: this path is the most exposed to the elements and the day is less warm at these times. It was for this reason, by the way, that the experiments were not done in random order. The researcher's comfort, which we have failed to mention so far, actually was one of the determining factors for the whole enterprise.

We now return to the apparently anomalous value. Many tests to detect outlying data have been proposed. One of the most frequently used in chemistry is Dixon's Q test, which assumes that the values being tested are normally distributed. Actually there are several tests identified as Dixon's, all based on comparing differences between the suspect value and the other values of the sample. You can obtain more information about these tests in Skoog et al. (1996), and in Rorabacher (1991). Here we will limit our discussion to the following question: can we consider the 56.3 min. time of experiment 9 as an element of the same distribution that produced the other times recorded for path C?

To perform Dixon's test, we start by calculating the smallest and largest differences between the suspect element and the rest of the values of the data set. Then we take the ratio of the smallest difference to the largest, and compare the result with a tabulated critical Q-value that depends on the desired confidence level and the total number of elements in the sample. If the calculated ratio is larger than the tabulated value, we can consider the suspect value as an outlier. In our example, there exists only one suspect value (the 56.3 min. time obtained in the ninth experiment), so we will use the Q-values for a one-sided tail test. Since the extreme

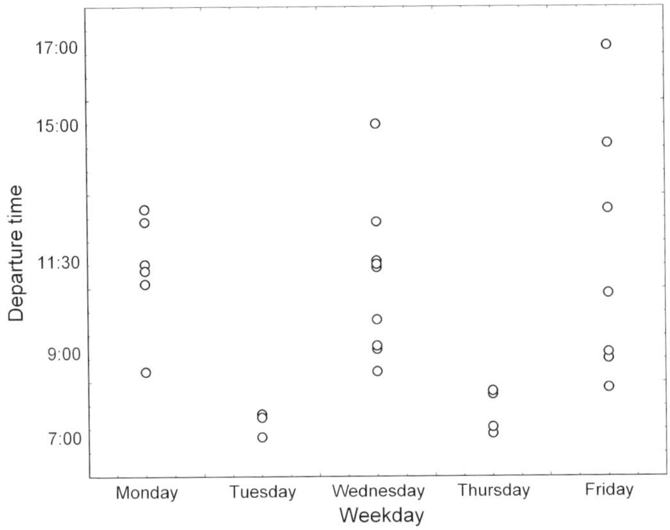

Fig. 2A.3. Departure time as a function of weekday.

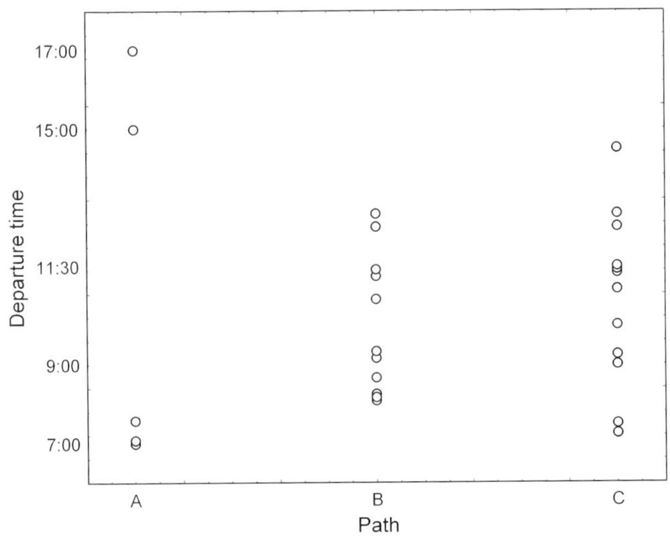

Fig. 2A.4. Departure time as a function of the path taken.

values of the other twelve times corresponding to path C are 16.1 and 22.9 min (Table 2A.2), we may write

Largest difference = 56.3–16.1 = 40.2

Smallest difference = 56.3–22.9 = 33.4

$$\text{Ratio} = \frac{33.4}{40.2} = 0.831.$$

Critical Q-value[23] for $n = 13$:

0.479 (95% confidence level);
0.579 (99% confidence level).

The calculated ratio is much larger than the two tabulated values for Dixon's test. This result shows, as we already suspected, that the time of experiment 9 is really different from the others. In fact, we already *knew* that the conclusion would be this one, since the weather on that experiment's day was completely atypical.

Another very popular test for outliers is Grubbs's test, which is the one recommended by the International Organization for Standardization (the famous ISO, the acronym that companies love to place on their advertisements). Grubbs's test also assumes a normal distribution and compares the distance from the average of the data set to the suspect value, measured in standard deviations. (The suspect value is included in the calculation of the average and the standard deviation.) If that distance is larger than a certain tabulated critical limit, the suspect value is considered an outlier. Using all 13 results corresponding to path C, we have

$$G = \frac{|x_a - \bar{x}|}{s} = \frac{56.3 - 22.25}{10.37} = 3.28,$$

which is larger than the tabulated G-value at the 95% confidence level, and therefore the presence of an outlier is again confirmed. Note that including the anomalous value in the calculations makes the average and the standard deviation much larger than the values given for path C in Table 2A.2.

The study of outliers is a vast and complex subject. Barnett and Lewis (1984), for example, discuss 47 different equations proposed as tests for outliers. The ideal procedure would be to repeat the suspect observation, in order to check if it is really valid. Since this is not always feasible, researchers are increasingly becoming interested in using non-parametric (also called robust) tests, which are less sensitive to possible outliers.

2A.2. Bioequivalence of brand-name and generic medicines

The World Health Organization is currently stimulating the adoption of a generic medicine policy, in order to guarantee universal access to health care and rationalize the use of medicines. A generic medicine is a formulation that, after patent expiry or renouncing, can be manufactured using the same active ingredients, the same dosage levels, the same pharmaceutical form, is

[23]Taken from a reference table.

administered in the same way and has the same therapeutic indications as the reference medicine. Finding out if there is bioequivalence between a generic medicine and its analog manufactured under patent protection is a very important public health issue. One of the measures commonly used to test for bioequivalence is the area under the curve describing how the blood concentration of the active ingredient varies with time.

A bioequivalence study involving 21 volunteers was carried out in two stages. In one stage each person took the generic medicine (A) and in the other the reference brand-name medicine (B), its presumed equivalent (Cavalcante, 1999). The order of administering the medicines was chosen at random, and neither the volunteers nor the research workers in charge of the test knew which of the two formulations was being supplied at a given time. For obvious reasons, this is called a double-blind test The values of the areas under the curves determined in the experiments are reproduced in Table 2A.3.

To test if the two formulations are equivalent we should use the same procedure employed in Section 2.7.4, where we made paired comparisons of the analytical capabilities of Arrhenius and Berzelius. Applying Eq. (2.28) to the values in the last column of Table 2A.3, we have

$$\Delta = \bar{d} \pm t_v \frac{s_d}{\sqrt{n}} = 54.62 \pm 2.086 \frac{1906.3}{\sqrt{21}} = 54.62 \pm 867.8.$$

Since this interval includes the zero value, we can consider the two medicines as bioequivalent, at least regarding the area below the curve. One point deserves comment, though. When we treated the data of Arrhenius and Berzelius, we saw that the paired comparison interval was much narrower, because the variation owing to sample origin had been eliminated. In the present example, the data behave in another way. The standard deviation of the differences is similar to the standard deviations of the averages of the two treatments. In Fig. 2A.5 we can see both the change in the average values and the similarity of the dispersions. We can also notice that the points seem to deviate a little from a normal distribution, but this should not be a cause for concern. As we have already said, the t-test is quite robust to such deviations.

2A.3. Still more beans?

In the best spirit of Rothamsted Experimental Station,[24] where R.A. Fisher did a great deal of his most important work, the authors decided to pursue their own agricultural studies and measured the masses of samples of two different kinds of beans. Each sample contained six beans randomly chosen from their respective populations, two 1-kg packages

[24]Also for lack of a better idea.

Table 2A.3
Bioequivalence test results

Volunteer	Area under curve[a]		
	A	B	Difference
1	12,761	10,983	1778
2	10,241	8211	2030
3	8569	9105	−536
4	13,321	12,508	813
5	11,481	12,114	−633
6	14,061	11,520	2541
7	12,287	11,983	304
8	14,696	14,454	242
9	12,526	11,246	1280
10	9060	10,740	−1680
11	12,329	10,879	1450
12	13,244	13,818	−574
13	7864	7156	708
14	9684	12,297	−2613
15	11,811	12,279	−468
16	10,109	9751	358
17	10,966	9895	1071
18	10,485	15,579	−5094
19	11,899	9296	2603
20	13,200	16,163	−2963
21	12,368	11,838	530
Average	11,569.6	11,515	54.62
Standard deviation	1827.2	2265.7	1906.3

[a] Of the blood concentration of the active ingredient against time, in the 0–8 h period.

bought in a supermarket. The masses obtained are given in Table 2A.4. The *carioca* bean seems to be heavier than the *roxinho* bean. Are these samples of only six elements large enough to give us much confidence in this conclusion?

The answer, once more, resides in a t-test, identical to the one used for comparing the times recorded in the walking professor experiment. However, we are going to do things a little differently this time and test the hypothesis that the two mean weights are identical, to see if we can reject it. We start by adapting Eq. (2.25), which we have seen in Section 2.7.4, to the null hypothesis of equality of the means. We therefore write

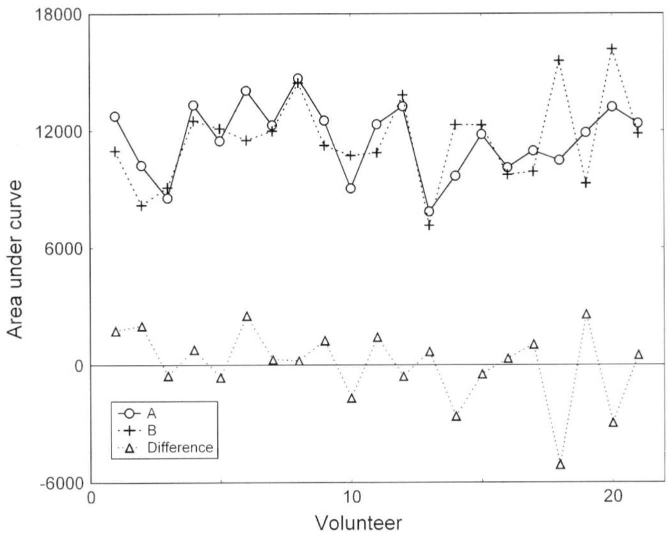

Fig. 2A.5. Areas under the curve, from Table 2A.3.

Table 2A.4
Masses of two kinds of beans

Variety	Mass (g)						$\bar{x}$	S
Carioca	0.2580	0.2607	0.2854	0.2895	0.2712	0.2766	0.27357	0.012779
Roxinho	0.2175	0.2205	0.2260	0.2261	0.2119	0.2146	0.21943	0.005875

$$\hat{t} = \frac{(\bar{x}_C - \bar{x}_R) - 0}{s\sqrt{\dfrac{1}{n_C} + \dfrac{1}{n_R}}}.$$

Substituting the appropriate values and using as the standard deviation the value obtained by pooling the variances of the two samples, we have

$$\hat{t} = \frac{(0.27357 - 0.21943) - 0}{0.0099456\sqrt{\frac{1}{6} + \frac{1}{6}}} = \frac{0.05414}{0.0057421} = 9.429.$$

The t-value with 10 degrees of freedom is 2.228, at the 95% confidence level. Even at the highest confidence level in Table A.2, 99.95%, the t value is only 4.587, which is still much smaller than the estimate obtained from the masses of the beans. We can say then, practically with certainty, that the mean mass of a *carioca* bean is larger than the mean mass of a *roxinho* bean.

The average mass of the six *roxinho* beans, 0.21943 g, is similar to the average mass of the 140-bean sample that we extensively discussed in this chapter (0.2024 g). It is clear that we cannot treat a package of black beans and a package of *roxinho* beans[25] as belonging to the same population, but could we not describe the two populations using the same parameter values?

We start by testing the variances, recalling that the standard deviation of the sample of black beans was 0.0363 g. The ratio between the sample variances is given by

$$\frac{s_1^2}{s_2^2} = \frac{(0.0363)^2}{(0.005875)^2} = 38.18.$$

The value of the corresponding tabulated point of the F distribution, at the 95% confidence level, is $F_{139,\,5} \cong 4.38$ (making a minor approximation, since the table does not contain values for 139 degrees of freedom). Since the value calculated for the variance ratio is much larger, we cannot assume that the two populations have the same variance. This implies that we do not have the right to combine the two variances in order to obtain a pooled estimate and then proceed to compare the means, as we have done so far. There is not an exact t-test for this situation, but we can use an approximate procedure that is a modification of the test we have used several times (see e.g. Montgomery and Runger, 1999, Chapter 9). The necessary estimate for the test can be approximated by

$$\hat{t}_v^* = \frac{\bar{x}_A - \bar{x}_B}{\sqrt{\dfrac{s_A^2}{n_A} + \dfrac{s_B^2}{n_B}}}.$$

Substituting the appropriate values, we have

$$\hat{t}_{144}^* = \frac{0.21943 - 0.2024}{\sqrt{\dfrac{(0.005875)^2}{6} + \dfrac{(0.0363)^2}{140}}} = \frac{0.01703}{0.003894} = 4.373$$

This result must be compared with the point of the t distribution with $n_A + n_B - 2$ degrees of freedom, at the desired confidence level. It is even larger than the value in Table A.2 at the 99.95% confidence level, $t_{120} = 3.373$ (we are being conservative regarding the number of degrees of freedom). Conclusion: we are faced with very strong evidence that the

[25]For one thing, *roxinho* beans are purple.

mean masses of these two kinds of beans — black and *roxinho* — are also different.

2A.4. Marine algae productivity

Agar-agar, a gel prepared from cell walls of several red algae, is used as a culture medium in laboratories and also as a thickener, stabilizer or adhesive in the food, cosmetic and pharmaceutical industries. Geyer et al. (1990) studied how the amount of agar-agar extracted from the alga *Pterocladia capillacea* (Rhodophyceae) varied with the location where the samples were collected, offshore Arraial do Cabo beach in the state of Rio de Janeiro. Table 2A.5 contains results obtained at two locations, one with little urban activity but with industrial activity (A), and the other with much urban activity, but without factories (B). Does this difference in environment affect the amount of agar-agar extracted?

First let us see if we can pool the variances in the traditional way. Using the standard deviations from the table, we have

$$\frac{s_B^2}{s_A^2} = \left(\frac{5.4948}{4.9077}\right)^2 = 1.254.$$

Since $F_{8,\,9} = 3.23$ (95% confidence), we conclude that we can pool the variances, which results in a standard deviation estimate of 5.1923%, with 17 degrees of freedom. Now we use Eq. (2.26) to obtain a confidence interval for the difference between the two mean amounts of agar-agar:

$$\begin{aligned}(\mu_A - \mu_B) &= (\bar{x}_A - \bar{x}_B) \pm t_{17} s \sqrt{\frac{1}{10} + \frac{1}{9}} \\ &= (33.866 - 37.784) \pm 2.110 \times 5.1923 \times 0.4595 \\ &= -3.918 \pm 5.034\%\end{aligned}$$

It is evident that the interval contains the zero value, and therefore we do not have evidence, at this confidence level, that the change in the kind of activity – from urban to industrial or vice versa – alters the agar-agar amount of the algae.

Table 2A.5
Amounts of agar-agar extracted from marine algae collected at two locations

Location	Samples	Agar amount extracted (%)					$\bar{x}$	s
A	10	39.75	36.40	33.88	27.85	31.42	33.866	4.9077
		34.40	36.62	36.50	38.04	23.80		
B	9	42.37	45.23	34.14	37.00	29.96	37.784	5.4948
		31.82	34.58	42.58	42.38			

CHAPTER 3

Changing everything at the same time

One of the most common problems faced by the experimental researcher is to assess the influence of one or more variables on another variable of interest. For example, studying a certain chemical reaction, the chemist who did the vinegar titrations might want to know how the yield would be affected by varying the temperature or using a different catalyst. In statistical parlance, we say that he is interested in discovering how the **response** (the reaction yield) depends on two **factors** – temperature and catalyst. We can approach this problem as a particular case of the situation shown in Fig. 3.1. A certain number of factors, $F_1, F_2, \ldots, F_k$, acting on the process under study, produce the responses $R_1, R_2, \ldots, R_j$. The system acts as a function – unknown in principle, or there would be no need to perform the experiments – that operates on the input variables (the factors) and yields the observed responses as output. The objective of the person performing the experiments is to discover this function, or at least a satisfactory approximation to it. With this knowledge, he will be able to understand more about the nature of the

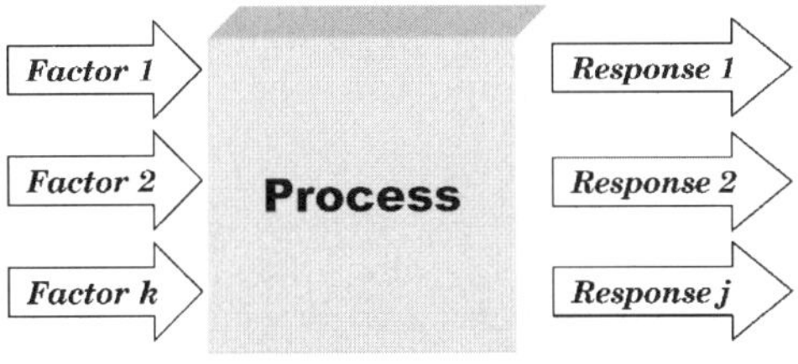

Fig. 3.1. A system can be represented by a function (in principle unknown) relating the factors (the input variables) to the responses (the output variables).

reaction under study, and thus select better conditions for operating the process.

The first thing we must do in planning any experiment is decide which variables are the factors and which are the responses. In general, the factors are those variables that the experimenter can control.[26] They can be qualitative, like the type of catalyst, or quantitative, like temperature. Sometimes, in a certain experiment, we know that some other factors can affect the responses but we have no way of — or interest in — controlling them. One example is the origin of the vinegar, as encountered in the comparison between Arrhenius and Berzelius. We must be careful with factors of this kind, to avoid that their effects be confounded with those of factors that do interest us. One of the techniques that can be used to deal with this situation is blocking, as we have already seen. Another important one is randomization, to be discussed later.

The responses are the output variables of the process in which we are interested, and they will be affected — or not — by modifications in the factors (the manipulations). They also can be qualitative or quantitative. Depending on the problem, several responses might have to be considered simultaneously.

After identifying all the factors and responses, the next step is to define, as clearly as possible, the objective we would like to attain with the experiments, so that we can choose the most appropriate experimental design. For example, our chemist might only want to know whether substituting the catalyst by a less expensive one will not decrease the reaction yield. Or perhaps he would like to discover what temperature should be used to give him a maximum yield. Or still, what are the limits for changing certain factor levels without compromising the yield or the quality of the final product. The experimental design, that is, the detailed specification of all the experimental operations to be performed, will depend on the particular objective the experimentalist has in mind. Different objectives will require different designs.

In this chapter we will study two-level factorial designs. They are very useful in preliminary investigations, when we want to know if certain factors affect a response or not, and we are not yet interested in describing these possible influences in detail (Box et al., 1978, Chapter 10). These designs are very simple to conduct, and can easily be augmented to form a more sophisticated design, if we wish to know in more detail the relation between the response and the important factors. On the other hand, if we

[26]Many engineers prefer to call the alterations in the factors “manipulations”, instead of “control”. Control, for them, is what you want to do with the *dependent* variable (i.e., the response) when the factors are manipulated. It makes some sense, but we will stick to our usage, which is traditional in the literature of experimental design.

are considering a relatively large number of factors, it is quite likely that some of them will not have significant influences on the response. In this case, running a complete factorial design would be a waste of time, money and effort. The best strategy is to do a screening first, to find out which factors deserve closer scrutiny. This can be accomplished with an incomplete factorial design, such as a fractional factorial design, a subject we will discuss in the next chapter.

Exercise 3.1. Think of an experiment, preferably in your area of research interest, with a qualitative response. Which factors would you like to examine, in order to determine their possible influences on the response? Which factors could be confounding? Which factors could contribute to noise — that is, random fluctuations — in the response?

3.1. A 2^2 factorial design

To run a factorial design, we start by specifying the levels at which each factor will be studied, that is, the factor values (or types, for qualitative factors) that we intend to use in the experiments. For example, we might wish to study the effect of temperature at four levels, 50, 60, 70 and 80 °C, and the effect of a catalyst at three levels, say, types A, B and C. To run a full factorial design, we must perform experiments at all possible combinations of the factor levels. Each of these experiments, carried out at a specified set of levels (for example: 60 °C temperature and type A catalyst), is an experimental run. With four levels for one factor and three for the other, as in this example, $4 \times 3 = 12$ different runs are needed, and the design is called a 4×3 factorial. In general, if there are n_1 levels of factor **1**, n_2 of factor **2**, ..., and n_k of factor **k**, the design is an $n_1 \times n_2 \times \cdots \times n_k$ factorial. This does not necessarily mean that only $n_1 \times \cdots \times n_k$ experiments will be performed. This is the minimum number needed for a complete factorial design. We may wish to estimate experimental error from replicate runs, and in this case we will have to perform more than this minimum number of experiments.

To study the effect of any factor on a certain response, we have to change its level (manipulate it, right?) and observe the consequences of this variation on the response. To do this, the factor must be set at least at two different levels. The simplest designs, then, are those in which every factor is studied at only two levels.

For k factors, that is, k variables controlled by the experimentalist, a complete two-level design requires $2 \times 2 \times \cdots \times 2 = 2^k$ different runs and is called a 2^k factorial design. In this section, we will examine the effects of increasing the temperature and changing the catalyst on the yield of a reaction. First, we show how to construct a 2^2 factorial design, and then how to analyze the data obtained from the runs of this design. This rather

Table 3.1
Results of a 2^2 factorial design for studying the effects of temperature and catalyst on the yield of a reaction

Run	Temperature (°C)	Catalyst	Yield (%)		Average
1	40	A	57	61	59
2	60	A	92	88	90
3	40	B	55	53	54
4	60	B	66	70	68

simple example allows us to introduce a number of fundamental concepts that later will be applied to designs involving any number of factors.

We start by selecting the factor levels: let us say that they are 40 and 60 °C for the temperature, and A and B for the catalyst types. Since this is a fictitious process devised for learning purposes, these level choices are arbitrary. In real life, we would base our choices on the available knowledge about the system under study, and think hard before deciding which levels should be investigated.

To perform a 2^2 design, we must carry out experimental runs and record the responses (reaction yields, in this case) observed for all four possible combinations of the chosen levels: (40 °C, A), (40 °C, B), (60 °C, A) and (60 °C, B). The list of these combinations, called the design matrix, is presented in Table 3.1, together with the yields obtained in the experiments. Note that all the experiments were done in duplicate, producing a total of eight responses. Thanks to this, we can estimate the experimental error of the individual response values. The size of this error is important for deciding whether significant effects result from changing the factor levels.

Exercise 3.2. Besides the temperature and catalyst, at the levels described above, our chemist would like to study the effect of pressure at three levels: 1, 5 and 10 atm at the same time using a full factorial design. What is the minimum number of runs that must be carried out?

3.1.1. Calculating the effects

According to Table 3.1, when we use catalyst A and raise the temperature from 40 to 60 °C (runs 1 and 2), the average yield changes from 59% to 90%. That is, an increase of 90–59 = 31% occurs. When the catalyst is type B (runs 3 and 4), the yield only increases by 68–54 = 14%. This shows that the effect of temperature — what happens to the reaction yield when we increase the temperature from 40 to 60 °C — depends on the catalyst level. The effect of the catalyst, on the other hand, also depends on the temperature level. At 40 °C (runs 1 and 3) the change in catalyst decreases the average yield by 5%. At 60 °C (runs 2 and 4), this reduction becomes 22%. When the effect of one variable depends on the level of another, as in this case, we say the two variables interact, and we can calculate an interaction effect for them, as we will presently see.

The main effect of the temperature is, by definition, the average of the temperature effects at the two levels of the catalyst. Using a bold face letter **T** to represent this effect, and with $\bar{y}_i$ being the average response observed for the ith run, we can write

$$\begin{aligned}\mathbf{T} &= \frac{(\bar{y}_2 - \bar{y}_1) + (\bar{y}_4 - \bar{y}_3)}{2} \\ &= \frac{(90 - 59) + (68 - 54)}{2} \\ &= \frac{(31) + (14)}{2} \\ &= 22.5\%. \end{aligned} \tag{3.1}$$

This value indicates that the yield of the reaction increases 22.5%, on average, when the temperature is changed from its lower level (40 °C) to the higher one (60 °C). However, this statement is incomplete. As we have just seen, the temperature and the catalyst interact, and we cannot talk about the temperature effect without also saying something about the type of catalyst used. We must actually interpret the effects of the two factors together, so as to leave no doubt about their interaction. We will return to this point in Section 3.1.4

For two-level designs it is customary to identify the higher and lower levels with (+) and (–) signs, respectively. Using this notation, we see that runs 2 and 4 in Table 3.1 correspond to the (+) level of the temperature, while runs 1 and 3 correspond to the (–) level. These algebraic signs can also be assigned to the levels of qualitative factors. In our example, we shall associate the (+) level with catalyst B. The choice is arbitrary and does not affect our conclusions.

Equation (3.1) can be rewritten as a difference between two averages:

$$\mathbf{T} = \left(\frac{\bar{y}_2 + \bar{y}_4}{2}\right) - \left(\frac{\bar{y}_1 + \bar{y}_3}{2}\right). \tag{3.1a}$$

Since $\bar{y}_2$ and $\bar{y}_4$ belong to the (+) level and $\bar{y}_1$ and $\bar{y}_3$ belong to the (–) level of the temperature, we see that the main effect **T** is the difference between the average response at the higher level and the average response at the lower level of this factor:

$$\mathbf{T} = \bar{y}_+ - \bar{y}_-. \tag{3.2}$$

This expression is valid for any main effect of any complete two-level factorial design and can be considered as an alternative definition.

For the catalyst, with our sign choice, the higher level corresponds to runs 3 and 4 and the lower one to runs 1 and 2. Therefore, according to Eq. (3.2) the main effect of the catalyst will be given by

$$\mathbf{C} = \bar{y}_{+} - \bar{y}_{-} = \left(\frac{\bar{y}_3 + \bar{y}_4}{2}\right) - \left(\frac{\bar{y}_1 + \bar{y}_2}{2}\right)$$
$$= -13.5\%. \tag{3.3}$$

Note that this effect is negative. When we change from catalysts A to B the yield *drops* 13.5%, on average. If we had inverted our sign choice and chosen catalyst A as the higher level, instead of B, the calculated effect would be $\mathbf{C} = +13.5\%$. In practice, the conclusion would remain the same: there is a difference between the yields observed with the two catalysts and the results obtained with catalyst B are, on average, 13.5% lower.

If there were no interaction, the temperature effect would be the same with either catalyst. However, we already know that this is not the case. The temperature effect is +31% with catalyst A, but falls to +14% when B is used. Since, in the absence of interaction, these two values would be identical (except for experimental error), we can take the difference between them as a measure of the interaction between the **T** and **C** factors. Actually, to be consistent with the definition of the main effects, *half* of the difference is taken to be the definition of the interaction effect between the two factors. Using $\mathbf{T} \times \mathbf{C}$, or simply **TC**, to represent this effect, we can write

$$\mathbf{T} \times \mathbf{C} = \mathbf{TC} = \frac{14 - 31}{2} = -8.5\%.$$

Note that the difference is obtained by subtracting the value of the **T** effect corresponding to the lower level of the catalyst (which is A, by our sign convention) from the value corresponding to the higher level (B), in an order like that of Eq. (3.2). Identifying the response values by the runs in which they were obtained, we can write

$$\mathbf{TC} = \left(\frac{\bar{y}_4 - \bar{y}_3}{2}\right) - \left(\frac{\bar{y}_2 - \bar{y}_1}{2}\right) = \left(\frac{\bar{y}_1 + \bar{y}_4}{2}\right) - \left(\frac{\bar{y}_2 + \bar{y}_3}{2}\right). \tag{3.4}$$

Equations (3.1), (3.3) and (3.4) show that all the observed response values enter into the calculation of any effect. Each effect is the difference between two averages. Half of the observations contribute to one of the averages, and the other half to the other. This is an important characteristic of two-level factorial designs. *All* response values are used to calculate *all* effects.

Exercise 3.3. We have calculated a measure of the interaction between the **T** and **C** factors using the difference in the temperature effects. It might be asked why we did this instead of calculating the difference between the catalyst effects at the two different temperature levels. Use algebraic arguments to show that these two measures are identical. Remember that, with the sign convention we adopted, the calculation that you should make is ((Catalyst effect at 60 °C)–(Catalyst effect at 40 °C)), and not the contrary.

3.1.2. Geometrical interpretation of the effects

We can give a geometrical interpretation to the effects we have just calculated. To do this, we represent the factorial design in a Cartesian coordinate system, with one axis for each factor. Since we have only two factors, the space defined by them is a plane. Appropriately choosing the axis scales, we can place the four runs at the vertices of a square (Fig. 3.2). Assigning algebraic signs consistent with those of Eqs. (3.1a), (3.3) and (3.4), we see that the main effects are contrasts — that is, differences between averages[27] — of values situated on opposite sides and perpendicular to the corresponding factor axis, as shown in Figs. 3.2a and b. The interaction effect, on the other hand, is a contrast between the two diagonals, in which the diagonal joining the (− −) and (++) runs contains the positive values (Fig. 3.2c). It was for this reason, we divided by 2, when we calculated the **TC** effect. In this way it can also be interpreted geometrically as a difference between averages.

3.1.3. Estimating the error of an effect

The runs in Table 3.1 were performed in duplicate, so that we would have a way of estimating experimental error, and from its value assess the statistical significance of the effects. To do this properly, the repeated runs must be genuine replicate experiments, that is, they must include again all the steps in the run, from the very beginning, such as cleaning the glassware, until separation and analysis of the final product. This point is very important. If we perform the replicates improperly, taking shortcuts here and there, some of the variability in the process will not be included in the results and the errors will appear smaller than they really are. This could lead us to conclude that some effects are statistically significant when in fact they do not exist at all.

To guard against statistical distortion in our results, that is, to avoid that atypical deviations become associated with certain level combinations, we should perform the runs in random order. Suppose, for example, that the chemical reaction of Table 3.1 could be influenced by light, and that we had to do part of the experiments during the day and the rest at night. If we use only catalyst A in daytime and catalyst B at night, we would be confounding the effect due to catalyst change with a possible

[27]A contrast is more formally defined as a linear combination of the responses whose coefficients add up to zero.

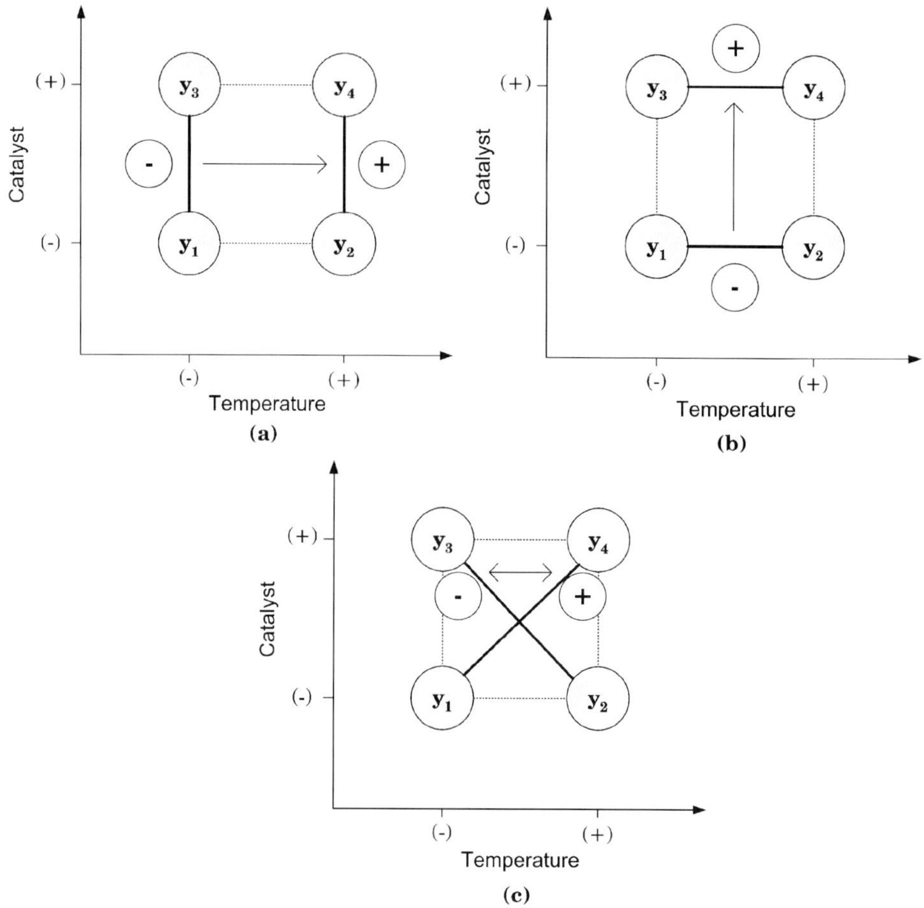

Fig. 3.2. Geometrical interpretation of the effects of a 2^2 factorial design. The main effects are contrasts between opposing sides (a) and (b). The interaction effect is a contrast between the two diagonals (c).

effect owing to the change in lighting conditions. To minimize this problem, we should randomly choose the order of performing the runs, that is, do the complete randomization we mentioned earlier.

Randomization is another very important experimental principle. It helps us prevent undesirable factors, of which we are unaware, contaminate the effects we wish to investigate. If we randomly choose the order of the runs, the probability that any of these uncontrolled factors affects a certain response will be the same for all responses, and in this way their influences on the calculated effects are diminished. Numbering the runs as in Table 3.1 is only a convenient way of identifying the different level combinations and has nothing to do with the order in which the experiments are actually performed. This, as we have just seen, should preferably be random.

It is important to distinguish randomization from blocking which is discussed at the end of this chapter. In randomization, we are taking preventive measures against factors that could influence results, but of whose presence we are unaware. In blocking we *know* right form the start that certain factors can affect the response, but we are not interested in measuring their effects. This is taken into account by specifying the design in such a way that confounding is avoided or minimized. Thus, for example, if we already know that lighting conditions might affect the reaction yield, we should include it as one of the factors, together with the others used to build the design. If, for some reason, this effect does not interest us, we can try to discount its influence by blocking.

Another important point is that the replicate runs should reflect the variability of the process in the whole range studied, not at only certain level combinations. If it is not feasible to repeat all the runs in the design, we must choose level combinations covering a large portion of the experimental range studied for the replicate experiments. This precaution might appear unwarranted when we have only four runs, but it becomes increasingly important as more factors are added to the design matrix.

From the replicate observations made for a certain level combination we can obtain an estimate of experimental error. For example, the yields observed for the two replicates of run no. 1 were 57% and 61%. Since they are genuine replicates and were performed in random order, we can take the variance of this pair of values, which is 8, as an estimate of the typical variance of our experimental procedure. Strictly, it is an estimate relative to the combination of levels from which the two results were obtained – at 40 °C temperature and catalyst A. However, if we assume that the variance is the same for the entire region investigated, we can combine the information from all of the runs and obtain a variance estimate with more degrees of freedom. In practice, this assumption customarily works quite well. If any doubt arises, we can always perform an F test to verify its validity.

Each one of the runs was performed twice, and therefore furnishes a variance estimate with only 1 degree of freedom. To obtain a pooled estimate, with 4 degrees of freedom, we generalize Eq. (2.27) and calculate an average of all the estimates, weighted by their respective numbers of degrees of freedom. Including the variances calculated for the other three runs (8, 2 and 8, respectively), we have

$$s^2 = \frac{(1 \times 8) + (1 \times 8) + (1 \times 2) + (1 \times 8)}{1 + 1 + 1 + 1} = \frac{8 + 8 + 2 + 8}{4} = 6.5.$$

Taking the square root of this value, we have an estimate, with 4 degrees of freedom, of the standard deviation associated with an observation in this experimental region, that is, the characteristic experimental error — often called the standard error — of our response values:

$$s = \sqrt{6.5} = 2.55\%.$$

When the number of replicates is the same for all runs, the pooled estimate of the experimental variance is simply the arithmetic average of the observed variances of the individual runs, as in this example. In general, if each run is repeated n_i times and there are m different runs, the pooled estimate of the experimental variance is given by

$$s^2 = \frac{\nu_1 s_1^2 + \nu_2 s_2^2 + \cdots + \nu_m s_m^2}{\nu_1 + \nu_2 + \cdots + \nu_m}, \qquad (3.5)$$

where $\nu_i = n_i - 1$ is the number of degrees of freedom of s_i^2, the variance estimate of the ith run.

Each of the calculated effects in Eqs. (3.1)–(3.4) is a linear combination of four $\bar{y}_i$ values, with a_i coefficients equal to +1/2 or –1/2. Because the replicates are genuine and the order of the runs is random, these values should be statistically independent. Assuming further that they have the same population variance $\sigma_{\bar{y}}^2$, we can apply Eq. (2.15), with $a_i^2 = 1/4$, to calculate the variance of an effect:

$$\hat{V}(\text{effect}) = \left(\frac{1}{4} + \frac{1}{4} + \frac{1}{4} + \frac{1}{4}\right)\sigma_{\bar{y}}^2 = \sigma_{\bar{y}}^2.$$

Recalling that each $\bar{y}_i$ value in this example is really an average of two independent observations, we can again apply Eq. (2.15) and write $\sigma_{\bar{y}}^2 = \sigma^2/2$, where σ^2 is the variance of an individual observation. Using our estimate $s^2 = 6.5$ instead of σ^2, we obtain an estimate, with 4 degrees of freedom, of the standard error of an effect in our experiment:

$$s(\text{effect}) = \sqrt{\frac{\sigma^2}{2}} = 1.80\%.$$

Another way of obtaining the standard error of an effect is by using Eq. (3.2). Since an effect is a contrast between two averages, that is,

$$\text{effect} = \bar{y}_+ - \bar{y}_-,$$

we can write

$$\hat{V}(\text{effect}) = \hat{V}(\bar{y}_+ - \bar{y}_-) = \hat{V}(\bar{y}_+) + \hat{V}(\bar{y}_-) = \frac{s^2}{4} + \frac{s^2}{4} = \frac{s^2}{2},$$

since in this case we have four response values with positive signs and another four with negative ones. Making the appropriate changes in the denominators, a similar expression holds for any two-level factorial design, since an effect is always a contrast between two averages, with half of the responses in each average.

We can construct confidence intervals for the effect values with the standard error, using Student's distribution:

$$\hat{\eta} - t_\nu \times s(\text{effect}) < \eta < \hat{\eta} + t_\nu \times s(\text{effect}). \qquad (3.6)$$

To avoid confusion with the notation already used for averages, we use the Greek letter η, in this equation, to represent the true value of the effect, that is, its population value. The circumflex mark (^) on η indicates that it is an estimate of this value obtained from the runs comprising the design, i.e., a sample value. In practice, this equation implies that only those effects having estimates with absolute values larger than the product of the standard error by the appropriate point of the t distribution should be considered significant. In that case, their confidence intervals will not include the zero value.

Exercise 3.4. Demonstrate that for any pair of numerical values $s^2 = d^2/2$, where d is the difference between the two values. Use this result and show that for a set of n duplicate runs (i.e., each run repeated only once, as in Table 3.1) the pooled estimate of the variance is

$$s^2 = \sum_i d_i^2 / 2n.$$

3.1.4. Interpreting the results

Table 3.2 contains the results of our analysis of the data in Table 3.1, and includes the grand average of all the yields in the design, which, like the effects, is also a linear combination of all observations. In the first place, we must find out which of the calculated effects are significantly different from 0 and therefore deserve interpretation. According to Eq. (3.6), only effects whose absolute values are larger than $t_4 \times s(\text{effect}) = 2.776 \times 1.8\% = 5.0\%$ should be considered statistically significant, at the

Table 3.2
Calculated effects for the 2^2 factorial design of Table 3.1

Grand average	67.75 ± 0.9
Main effects	
T	22.5 ± 1.8
C	-13.5 ± 1.8
Interaction effect	
TC	-8.5 ± 1.8

Note: The standard error of the global average is half the standard error of the effects.

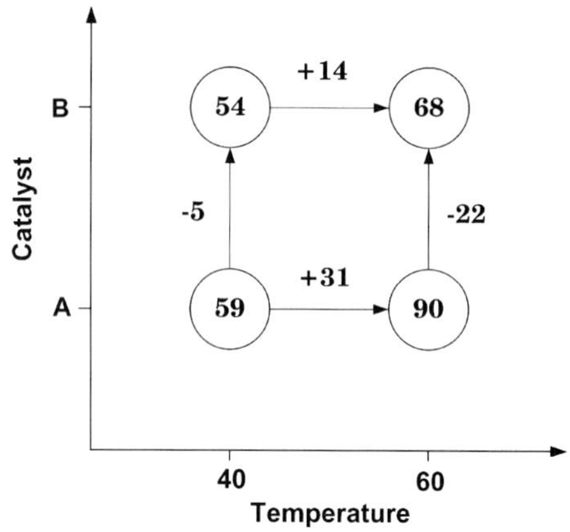

Fig. 3.3. Diagram for interpreting the results of the 2^2 factorial design. The values at the vertices of the square are average response values (percentage reaction yields).

95% confidence level. Applying this criterion to the values in Table 3.2, we see that all of them are significant, that is, all these effects really exist. We are entitled, therefore, to try to understand what they mean in practice.

Since the interaction effect is significant, the temperature and catalyst effects must be interpreted together. The best way to do this is to draw a diagram containing the average response values for all the level combinations of the variables, as in Fig. 3.3. Examining this diagram, we can conclude that

1. Raising the temperature increases the yield of the reaction, but this effect is much more pronounced with the type A catalyst than with the type B one (+31% compared with +14%).
2. Changing the catalyst from types A to B decreases the reaction yield, and this effect is much more significant at 60 °C than at 40 °C (–22% compared with –5%).
3. The largest yields (90%, on average) are obtained with type A catalyst at 60 °C.

Exercise 3.5. According to Table 3.2, the standard error of the grand average is half the standard error of the effects. Use Eq. (2.15) to show that this has to be true.

Exercise 3.6. The observations listed below were recorded in genuine replicates of the different runs. Calculate a pooled estimate of the experimental error associated with these observations. How many degrees of freedom does this estimate have?

Run	Observation							Average	Variance
1		20		25		23		22.7	6.33
2			40		37			38.5	4.50
3	10		8		12		7	9.3	4.92
4				31				31	–
5		53		49		45		49.0	16.00

3.1.5. An algorithm for calculating the effects

In this very simple example, we can calculate the effects directly from their definitions. This procedure, however, becomes very laborious and subject to (human) error as the number of factors increases. In this section, we present an alternative procedure that can be used to calculate any effect with little difficulty, no matter what the design size is.

We start by rewriting the design matrix of Table 3.1, substituting the elements by algebraic signs that identify the associated level as high or low. Thus, 40 and A are substituted by minus signs and 60 and B by plus signs:

$$\begin{array}{cc} \mathbf{T} & \mathbf{C} \end{array} \qquad \begin{array}{cc} \mathbf{T} & \mathbf{C} \end{array}$$
$$\begin{bmatrix} 60 & \text{A} \\ 60 & \text{A} \\ 40 & \text{B} \\ 60 & \text{B} \end{bmatrix} = \begin{bmatrix} - & - \\ + & - \\ - & + \\ + & + \end{bmatrix}.$$

Now, we augment this matrix with a column of positive signs — the first one — plus another whose signs are the products, element by element, of the signs in the **T** and **C** columns. This gives us a 4 × 4 matrix, which we call the table of contrast coefficients:

$$\begin{array}{cccc} \mathbf{M} & \mathbf{T} & \mathbf{C} & \mathbf{TC} \end{array}$$
$$\begin{bmatrix} + & - & - & + \\ + & + & - & - \\ + & - & + & - \\ + & + & + & + \end{bmatrix}.$$

To calculate the effects, we choose the appropriate column, apply its signs to the corresponding response values, do the algebraic sum and finally divide the result by 2. The first column, which only contains positive signs, is used to calculate the average of all the runs, and obviously has a divisor of 4.

For the catalyst effect, for example, we use the third column:

$$\mathbf{C}$$
$$\begin{bmatrix} - \\ - \\ + \\ + \end{bmatrix}.$$

Applying these signs to the column of responses,

$$\begin{bmatrix} \bar{y}_1 \\ \bar{y}_2 \\ \bar{y}_3 \\ \bar{y}_4 \end{bmatrix} = \begin{bmatrix} 59 \\ 90 \\ 54 \\ 68 \end{bmatrix},$$

and performing the rest of the operations, we can write:

$$\begin{aligned} \mathbf{C} &= \frac{-\bar{y}_1 - \bar{y}_2 + \bar{y}_3 + \bar{y}_4}{2} \\ &= \frac{-59 - 90 + 54 + 68}{2} \\ &= -13.50\%. \end{aligned}$$

We leave the task of confirming that the other columns also produce the correct results to the reader.

Including the unit symbol in the table of contrast coefficients, that is, letting

$$\begin{bmatrix} + & - & - & + \\ + & + & - & - \\ + & - & + & - \\ + & + & + & + \end{bmatrix} = \begin{bmatrix} 1 & -1 & -1 & +1 \\ 1 & +1 & -1 & -1 \\ 1 & -1 & +1 & -1 \\ 1 & +1 & +1 & +1 \end{bmatrix}, \tag{3.7}$$

we can calculate all the effects with a single matrix equation. Each effect is given, except for a divisor, by the scalar product of the appropriate vector in the matrix of contrast coefficients by the vector of response values. Thus, for example, the catalyst effect that we have just calculated is given by

$$\mathbf{C} = \frac{1}{2}\begin{bmatrix} -1 & -1 & +1 & +1 \end{bmatrix} \begin{bmatrix} 59 \\ 90 \\ 54 \\ 68 \end{bmatrix} = -13.50\%.$$

Traditionally, vectors and matrices are represented in boldface. Letting

$$\mathbf{x}_C = \begin{bmatrix} -1 \\ -1 \\ +1 \\ +1 \end{bmatrix} \text{ and } \mathbf{y} = \begin{bmatrix} 59 \\ 90 \\ 54 \\ 68 \end{bmatrix},$$

we can determine the catalyst effect with the equation

$$\mathbf{C} = \frac{1}{2}\mathbf{x}_C^t\mathbf{y}, \tag{3.8}$$

where x_C^t is the row vector that we obtain transposing the column vector $\mathbf{x}_C$. With this concise formula (which can be extended to any two-level factorial design) and making use of a linear algebra software, calculating the effects becomes very easy.

In the general case of a two-level factorial design with k factors, for which we must perform a total of 2^k runs, the divisor will be 2^{k-1} for the effects and, of course, 2^k for the average. If $\mathbf{X}$ is the matrix of all contrast coefficients, the effects will be, except for the divisors, the elements of the $\mathbf{X}^t\mathbf{y}$ product, which is a column vector. For our 2^2 factorial

$$\mathbf{X}^t\mathbf{y} = \begin{bmatrix} +1 & +1 & +1 & +1 \\ -1 & +1 & -1 & +1 \\ -1 & -1 & +1 & +1 \\ +1 & -1 & -1 & +1 \end{bmatrix} \begin{bmatrix} 59 \\ 90 \\ 54 \\ 68 \end{bmatrix} = \begin{bmatrix} 271 \\ 45 \\ -27 \\ -17 \end{bmatrix}.$$

Dividing the first element by 4 and the rest by 2, we finally obtain

$$\begin{bmatrix} \mathbf{M} \\ \mathbf{T} \\ \mathbf{C} \\ \mathbf{TC} \end{bmatrix} = \begin{bmatrix} 67.75 \\ 22.5 \\ -13.5 \\ -8.5 \end{bmatrix}.$$

Note that the rows and columns of the $\mathbf{X}$ matrix are orthogonal, that is, the scalar product of any two lines or any two columns is 0. Consequently, when we perform the $\mathbf{X}^t\mathbf{y}$ product and apply the appropriate divisors, we obtain, from four statistically independent values (the elements of $\mathbf{y}$), four new values that are also independent — the grand average, the two main effects and the interaction effect.

We are using a square 4×4 $\mathbf{X}$ matrix because we are basing our calculations on the four average response values, not on the eight individual values. If we prefer, we can perform the same calculation by directly using the eight responses. In this case, the $\mathbf{X}$ matrix would be of order 8×4, and the divisor would be 4, instead of 2. The $\mathbf{C}$ effect, for example, would be calculated as follows:

$$\mathbf{C} = \frac{1}{4}[-1 - 1 + 1 + 1 - 1 - 1 + 1 + 1]\begin{bmatrix} 57 \\ 92 \\ 55 \\ 66 \\ 61 \\ 88 \\ 53 \\ 70 \end{bmatrix} = -13.50\%.$$

Exercise 3.7. The data below were obtained in a study of the influence of two factors on the initial curing time of plaster of Paris, that is, the time when the plaster of Paris starts to harden after the powder is mixed with water (M.F. Pimentel and B.B. Neto, Anais do XXXI Congresso Brasileiro de Química, Recife, 1991). The runs were carried out in duplicate and in random order. Calculate all effects and their standard errors. Interpret the results.

Factor **1**: Granulation: 100–150 *mesh* (–), 150–200 *mesh* (+)
Factor **2**: Residual water: 6.6% (–), 7.5% (+)
Response: Initial curing time (min)

i	Factor **1**	Factor **2**	Response		$\bar{x}_i$	s_i^2
1	–	–	12.33	13.00	12.67	0.224
2	+	–	10.52	10.57	10.55	0.0013
3	–	+	10.33	9.75	10.04	0.168
4	+	+	9.00	8.92	8.96	0.0032

3.1.6. The statistical model

In the algorithm used to calculate the effects, +1 or –1 substitute the real values of the factor levels. This corresponds to a coding of the original variables, similar to what was used in the last chapter to standardize a random variable. To transform the 40 and 60 °C values, for example, into –1 and +1, we subtract from each value the average, 50 °C, and divide the result by half the range of variation, that is, the difference between the higher and lower values:

$$\frac{40-50}{(60-40)/2}=\frac{-10}{10}=-1,$$

$$\frac{60-50}{(60-40)/2}=\frac{10}{10}=+1.$$

Obviously, this means that the origin of the temperature axis is now placed at the average value, 50 °C, and its scale is such that one unit represents 10 °C. In the same way, the coding process centers the origin of the other axis halfway between types A and B catalysts, in a hypothetical "catalyst zero state" that has no physical meaning but can be algebraically treated just like the origin of the temperature axis. This transformation is illustrated in Fig. 3.4, where the temperature and catalyst variables are, respectively, labeled x_1 and x_2 after coding. The (hypothetical) response value corresponding to the origin of the new coordinate system, at the center point of the design, is, by symmetry, the average of all the observed responses, 67.75%.

After coding, each effect always corresponds to a variation of two units in the corresponding factor, since the factor level varies from –1 to +1. The effects per unit measure of x_1 and x_2, consequently, are one half the values calculated with Eqs. (3.1)–(3.4). To say that the temperature effect is 22.50% when T changes from 40 to 60 °C is the same as saying that this effect is 11.25% for a unit change in x_1. Dividing the three effects calculated using Eqs. (3.1)–(3.4) by 2, we obtain the values per unit of x_1 and x_2: 11.25% (x_1, temperature) –6.75% (x_2, catalyst) and –4.25% (the $x_1 \times x_2$ interaction).

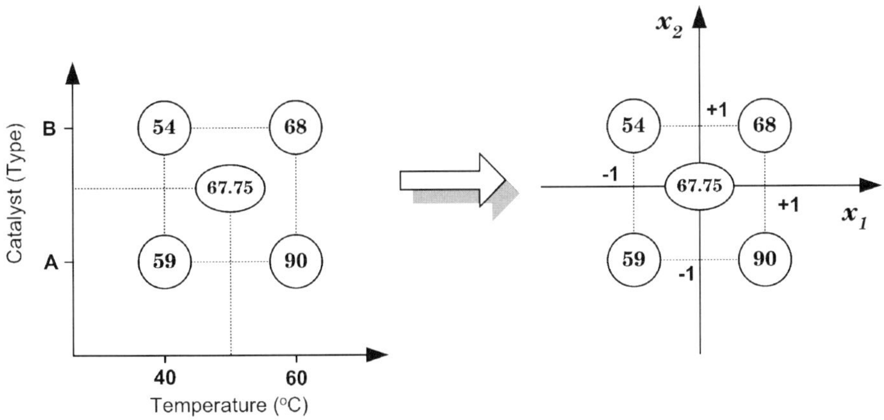

Fig. 3.4. Variable coding for the 2^2 factorial design. The origin of the new system is located at the average of all the response values. The unit measure for each axis is half the range of variation of the corresponding factor.

The statistical model used to describe the responses of the factorial design is expressed in terms of unit variations in x_1 and x_2. For the 2^2 design, the observed response at the (x_1, x_2) level is considered a random variable $y(x_1,x_2)$. This variable is distributed about a certain population mean $\eta(x_1,x_2)$, with a certain population variance $\sigma^2(x_1,x_2)$. Therefore, we can write

$$y(x_1,x_2) = \eta(x_1,x_2) + \varepsilon(x_1,x_2), \tag{3.9}$$

where $\varepsilon(x_1,x_2)$ represents the random error with which the response values fluctuate about the population mean defined by the values of x_1 and x_2.

When we perform run no. 1 in duplicate, we obtain 57% and 61% as response values. We can imagine this pair of values as a sample — of two elements — of the population of all possible yields when $(x_1,x_2) = (-1,-1)$. The population mean of yields for this combination of factor levels is a certain value $\eta_1 = \eta(-1,-1)$ that we cannot determine with perfect accuracy but for which the two observations furnish the estimate $\bar{y}_1 = (57 + 61)/2 = 59\%$. The variance with which the yields are distributed about $\eta(-1,-1)$ is not known either, but can be estimated from the duplicate observations. With the 57 and 61 replicate values observed for run no. 1 we obtain the variance estimate $s_1^2 = 8$.

The statistical model also requires that the random errors $\varepsilon(x_1,x_2)$ be independently distributed, with zero mean and equal variance, at all level combinations of the factorial design. This is exactly what we assumed when we combined all observations to obtain a pooled variance estimate. If we wish to perform a t- or an F-test on our results, we must also assume that the errors are normally distributed (Section 2.6).

In a 2^2 factorial design, our model postulates that the population mean $\eta(x_1,x_2)$ is adequately represented by the expression

$$\beta_0 + \beta_1 x_1 + \beta_2 x_2 + \beta_{12} x_1 x_2,$$

where β_0 is the population value of the average of all the response values of the design, and β_1, β_2 and β_{12} are population values of the two main effects and the interaction effect per units of x_1 and x_2. Substituting this expression into Eq. (3.9), we see that an individual observation made at any run can be represented by

$$y(x_1,x_2) = \beta_0 + \beta_1 x_1 + \beta_2 x_2 + \beta_{12} x_1 x_2 + \varepsilon(x_1,x_2), \tag{3.10}$$

where $\varepsilon(x_1,x_2)$ is the random error associated with the $y(x_1, x_2)$ response. This description is valid for the vertices of the design. From a strictly mathematical point of view, we should not interpret it as an equation that can be extrapolated or even interpolated, since the coded values of x_1 and x_2 are restricted, by their very definitions, to +1 or −1. In many applications, however, we will set mathematical rigor aside lest our models will not be of much practical use.

Since the β's are population values, to determine them exactly we would have to perform an infinite number of experiments. Obviously, the values that we calculate from the results of our eight experiments are only estimates of these values. Using the Latin alphabet for the symbols of the estimates, we may write

$$\hat{y}(x_1,x_2) = \hat{\eta}(x_1,x_2) = b_0 + b_1x_1 + b_2x_2 + b_{12}x_1x_2, \tag{3.11}$$

with the circumflex mark on y reminding us that we are not dealing with a population value but merely with an estimate of it. The b_0, b_1, b_2 and b_{12} coefficients are called estimators of the population parameters β_0, β_1, β_2 and β_{12}.

Eq. (3.11) also can be written as a scalar product:

$$\hat{y}(x_1,x_2) = \begin{bmatrix} 1 & x_1 & x_2 & x_1x_2 \end{bmatrix} \begin{bmatrix} b_0 \\ b_1 \\ b_2 \\ b_{12} \end{bmatrix}.$$

For run no. 1, for example, we have

$$\hat{y}(-1,-1) = \begin{bmatrix} 1 & -1 & -1 & 1 \end{bmatrix} \begin{bmatrix} 67.75 \\ 11.25 \\ -6.75 \\ -4.25 \end{bmatrix} = 59\%,$$

which is the estimate given by our model for the population yield of run no. 1. It is, as we have seen, the average of the two values observed for this run: $(57 + 61)/2 = 59\%$. The elements in the row vector are exactly those corresponding to run no. 1 in the matrix of contrast coefficients (Eq. (3.7)). You can easily verify that similar calculations, using other rows of the matrix, can be made for the other runs.

Recalling that we used $\mathbf{X}$ to stand for the complete matrix, and choosing $\mathbf{b}$ to represent the row vector of the four coefficients of Eq. (3.11), we can write the following matrix equation to calculate the estimates of all the runs:

$$\hat{\mathbf{y}} = \mathbf{Xb}, \tag{3.12}$$

where $\hat{\mathbf{y}}$ is a column vector containing the $\hat{y}_1, \ldots, \hat{y}_4$ estimates given by the model for the yields of the four runs. Since the estimates are the averages observed for each run, the $\hat{\mathbf{y}}$ vector in Eq. (3.12) is identical to the $\mathbf{y}$ vector of Eq. (3.8).

Using the appropriate $\mathbf{X}$ and $\mathbf{b}$ matrices, we can apply Eq. (3.12) to any two-level factorial design. Its usefulness, however, is much more far-reaching. We will see in Chapter 5 that we can apply it to *any* model whose parameters (i.e., the β's) figure only as linear coefficients. This includes an enormous number of models.

Our estimate of the yield of run no. 1 is $\hat{y}_1 = 59\%$, while the values actually observed in the individual experiments were $y_1 = 57\%$ and $y'_1 = 61\%$. The prediction made by the model for run no. 1 therefore leaves the two residuals $y_1 - \hat{y}_1 = 57 - 59 = -2\%$ and $y'_1 - \hat{y} = 61 - 59 = 2\%$.

Differences between observed and predicted values always appear when we use models containing fewer parameters than the total number of observations employed in their determination. In our case, as shown by Eq. (3.11), the estimate of an observation (to be precise, the estimate of the average value of the run) is a function of two independent variables, x_1 and x_2, with four coefficients to be determined (b_0, b_1, b_2 and b_{12}) that in turn estimate the four parameters of the model. The residuals appear because this model is fitted to eight observations. If there were only four, the fit would be perfect and all residuals would be zero.

It is important to stress that these residuals are not all independent. The equations used to calculate the parameter estimates eliminate 4 degrees of freedom from the original observations, leaving only 4 degrees of freedom for the set of eight residuals.

An analysis of the residuals is fundamental for evaluating the fit of any model to the observations. An abnormally high individual residual, for example, can indicate the presence of an anomalous observation, perhaps caused by a gross error, and thus could lead us to repeat the corresponding run. In a well-fitted model, the behavior of the residuals should not be incompatible with what we expect for the random errors $\varepsilon(x_1,x_2)$ we included in Eq. (3.10). Analyzing the residuals, however, only makes sense when the number of degrees of freedom of the set of residuals is relatively high. We will return to this subject in Chapter 5, where we discuss model fitting by the least squares method.

Exercise 3.8. If we include the divisors, the matrix for calculating the effects becomes

$$\mathbf{A} = \begin{bmatrix} +1/4 & +1/4 & +1/4 & +1/4 \\ -1/2 & +1/2 & -1/2 & +1/2 \\ -1/2 & -1/2 & +1/2 & +1/2 \\ +1/2 & -1/2 & -1/2 & +1/2 \end{bmatrix},$$

such that

$$\mathbf{Ay} = \begin{bmatrix} +1/4 & +1/4 & +1/4 & +1/4 \\ -1/2 & +1/2 & -1/2 & +1/2 \\ -1/2 & -1/2 & +1/2 & +1/2 \\ +1/2 & -1/2 & -1/2 & +1/2 \end{bmatrix} \times \begin{bmatrix} 59 \\ 90 \\ 54 \\ 68 \end{bmatrix} = \begin{bmatrix} 67.75 \\ 22.5 \\ -13.5 \\ -8.5 \end{bmatrix} = \mathbf{e},$$

where $\mathbf{e}$ is a column vector containing the grand average and the original effects, before they are divided by 2. Multiplying this vector on the left by the inverse of $\mathbf{A}$, we get back our original observations, that is, the **y** vector

$$\mathbf{A}^{-1}\mathbf{e} = \mathbf{A}^{-1}\mathbf{A}\mathbf{y} = \mathbf{I}_4\mathbf{y} = \mathbf{y},$$

where $\mathbf{I}_4$ is the identity matrix of order 4. Determine $\mathbf{A}^{-1}$ (recall that the rows of the matrix of contrast coefficients are orthogonal and use your good sense; the calculation is in fact very simple) and show that the $\mathbf{A}^{-1}\mathbf{e}$ product is identical to the $\mathbf{Xb}$ product of Eq. (3.12). To understand why this is so, compare the $\mathbf{A}^{-1}$ and $\mathbf{X}$ matrices and the **b** and **e** vectors.

3.2. A 2^3 factorial design

Let us now add a third factor to the temperature and the catalyst. Imagine we wish to also study the concentration of a reagent at the 1.0 M (–) and 1.5 M (+) levels. The complete factorial design now requires $2^3 = 8$ unique runs. The yields observed for these runs, performed in random order and in duplicate, are presented in Table 3.3.

The runs are listed in this table in what is called the standard order. All columns begin with the (–) level and then the signs alternate: one by one in the first column, $- + - + \cdots$, then two by two, $- - + + \cdots$, and finally four negative signs followed by four positive ones in the third column. If there were a fourth factor, its column would start with eight negative signs, followed by eight positive signs. (The total number of runs

Table 3.3
Results of a 2^3 factorial design

Factors	(–)	(+)
1: Temperature (°C)	40	60
2: Catalyst (type)	A	B
3: Concentration (M)	1.0	1.5

Run	**1**	**2**	**3**	Yield (%)		Average
1	–	–	–	56 (7)	52 (12)	54.0
2	+	–	–	85 (9)	88 (10)	86.5
3	–	+	–	49 (11)	47 (15)	48.0
4	+	+	–	64 (2)	62 (1)	63.0
5	–	–	+	65 (13)	61 (5)	63.0
6	+	–	+	92 (6)	95 (16)	93.5
7	–	+	+	57 (14)	60 (3)	58.5
8	+	+	+	70 (8)	74 (4)	72.0

Note: The number within parentheses after the yield indicates the order in which the experiments were performed.

would, of course, be 16.) For a full two-level factorial with k factors, the last column would begin with a sequence of 2^{k-1} negative signs, followed by another consisting of 2^{k-1} positive signs. Keeping this in mind, we can easily write the design matrix for any two-level factorial design.

Starting with the design matrix we can build a table of contrast coefficients in the same way we did for the 2^2 design, multiplying the signs in the appropriate columns to obtain the columns corresponding to the interactions. Now we have, besides the three main effects **1** (temperature), **2** (catalyst) and **3** (concentration), three two-factor interactions, **12**, **13** and **23**. There is, moreover, a new type of effect. Since the design contains three factors, the interaction of any two of them can depend on the level of the third factor. The **12** interaction, for example, will have a certain value when factor **3** is at the (+) level, and possibly another value when it is at the (–) level. Reasoning in the same way we did for the 2^2 factorial, we will take half of the difference between these two values as a measure of the interaction between factor **3** and the **12** interaction. Since the argument is symmetric, this value also measures the interaction between **13** and **2** and between **23** and **1** (Exercise 3.9). This quantity, as you have probably suspected, is the three-factor interaction effect, which we denote by **123**. The signs needed to calculate its value are obtained by multiplying the signs of the three columns, **1**, **2** and **3**. Finally, adding the column of positive signs used to compute the grand average to the matrix, we will have a total of $2^3 = 8$ columns of contrast coefficients (Table 3.4).

3.2.1. Calculating the effects

Table 3.4 contains all the signs we need for calculating the effects. The divisor is 8 for the average and 4 for the effects. Using the appropriate

Table 3.4
Contrast coefficients for the 2^3 factorial design

Average	**1**	**2**	**3**	**12**	**13**	**23**	**123**	$\bar{y}$
+	–	–	–	+	+	+	–	54.0
+	+	–	–	–	–	+	+	86.5
+	–	+	–	–	+	–	+	48.0
+	+	+	–	+	–	–	–	63.0
+	–	–	+	+	–	–	+	63.0
+	+	–	+	–	+	–	–	93.5
+	–	+	+	–	–	+	–	58.5
+	+	+	+	+	+	+	+	72.0

Note: The last column contains the average values of the yields obtained in the duplicate runs.

signs to perform the algebraic sum of the average response values (given in the last column of the table) and applying the divisors, we calculate the seven effects and the grand average. All response values are used in each effect calculation, as happened for the 2^2 design.

Every effect column in Table 3.4 has four positive and four negative signs. This means that we can interpret any effect as the difference between two averages, each containing half of the observations, exactly as in the case of the 2^2 factorial design.

Transforming the table of contrast coefficients into an **X** matrix with +1 or −1 elements, we can calculate all effects (except for the divisors) by means of the $\mathbf{X}^t\mathbf{y}$ product, where **y** is a column vector containing the average yields of the runs. Thus, we obtain

$$\mathbf{X}^t\mathbf{y} = \begin{bmatrix} 538.50 \\ 91.50 \\ -55.50 \\ 35.50 \\ -34.50 \\ -3.50 \\ 3.50 \\ 0.50 \end{bmatrix}. \quad (3.13)$$

Dividing the first element by 8 and the rest by 4 we obtain the vector of effect values (rounded off to the second decimal place):

$$\begin{bmatrix} \bar{\mathbf{y}} \\ \mathbf{1} \\ \mathbf{2} \\ \mathbf{3} \\ \mathbf{12} \\ \mathbf{13} \\ \mathbf{23} \\ \mathbf{123} \end{bmatrix} = \begin{bmatrix} 67.31 \\ 22.88 \\ -13.88 \\ 8.88 \\ -8.63 \\ -0.88 \\ 0.88 \\ 0.13 \end{bmatrix}$$

Note that the effects appear in the same order as the columns of the **X** matrix.

Exercise 3.9. For each of the two levels of variable **3** there is a complete 2^2 factorial in variables **1** and **2**. Calculate, from the values in Table 3.3, the **12** interaction at both levels of variable **3**. Take the difference between these two values, divide by 2, and call the result the interaction of factor **3** with the **12** interaction. Repeat the whole process, starting from the values of the **23** interactions at the two levels of factor **1**. You will then have the interaction of factor **1** with the **23** interaction.

Compare the results obtained in these two calculations with the value of the **123** interaction given in the text.

3.2.2. Estimating the error of an effect

Since all the individual experiments were performed in duplicate, we can use the result of Exercise 3.4 to calculate a pooled estimate of the variance of an individual response as

$$\hat{V}(y) = s^2 = \sum d_i^2/2n,$$

where d_i is the difference between the two observations corresponding to the ith run. This expression can also be written as a scalar product. Denoting by $\mathbf{d}$ the column vector of the differences between responses obtained in duplicate experiments, we can write

$$s^2 = \frac{\mathbf{d}^t\mathbf{d}}{2n}.$$

Substituting the numerical values and making $n = 8$, we obtain $s^2 \cong 5.2$.

In the 2^3 factorial design, each effect is a linear combination of eight values, each with coefficients equal to $\pm 1/4$. Assuming that these values are statistically independent,[28] we can apply Eq. (2.15) to obtain an estimate of the variance of the effects. In this case we let $a_i^2 = 1/16$, for $i = 1, 2, \ldots, 8$. Each value of the linear combination is in turn an average of two others, since the experiments were made in duplicate. If the variance of an individual observation is estimated as 5.2, the variance of the average of two observations will be 5.2/2. Combining all of this, we arrive at the estimate

$$\begin{aligned}\hat{V}(\text{effect}) &= \left(\frac{1}{16} + \cdots + \frac{1}{16}\right) \times \left(\frac{5.2}{2}\right) \\ &= \left(\frac{8}{16}\right) \times \left(\frac{5.2}{2}\right) = 1.30.\end{aligned}$$

The standard error of an effect is the square root of this value, approximately 1.14%. The standard error of the grand average is half this value, 0.57%, because the coefficients of the linear combination in this case are all equal to 1/8, instead of $\pm 1/4$. Table 3.5 contains the values calculated for all the effects and their standard errors.

Exercise 3.10. Use Eq. (2.15) to calculate the variance of the effects of a 2^3 factorial design without replicates, starting from Eq. (3.2).

[28]This is why it is so important to randomize the order of the experimental runs and to perform genuine replicates.

Table 3.5
Calculated effects for the 2^3 factorial design of Table 3.3 and their standard errors (in %)

Average	67.3 ± 0.6
Main effects	
1 (Temperature)	22.9 ± 1.1
2 (Catalyst)	–13.9 ± 1.1
3 (Concentration)	8.9 ± 1.1
Two-factor interactions	
12	–8.6 ± 1.1
13	–0.9 ± 1.1
23	0.9 ± 1.1
Three-factor interaction	
123	0.1 ± 1.1

3.2.3. Interpreting the results

Analyzing the values in Table 3.5, we can see that including the concentration in the factorial design results only in one more significant main effect. There is no evidence of an interaction of the concentration with either of the other two factors. The main effects of the temperature and the catalyst, as well as their interaction, are practically the same as those found in the 2^2 design, and can be interpreted in the same way as before, using the average values shown in Fig. 3.5. The main effect of the concentration can be interpreted all by itself, because there is no interaction of this factor with any of the others. The interpretation is simple:

- As the concentration is increased from 1.0 to 1.5 M, an average increase of about 9% in the yield occurs. There is no evidence that this increase depends on the levels of the other two factors, at least for the factor ranges studied in this design.

Exercise 3.11. As an exercise in a chemometrics course, M.R. Vallim and V.F. Juliano analyzed data obtained by a research worker in a series of experiments involving the synthesis of polypyrrole in an EPDM matrix. Three factors were studied: reaction time (**t**), oxidant concentration (**C**) and particle size (**P**). The response observed was the reaction yield. Using the data presented below, calculate the values of the effects and their standard errors. Before doing this, however, carefully examine the set of response values, taking into account the signs of the design matrix. Is it possible to anticipate which variable has the largest influence on the yield?

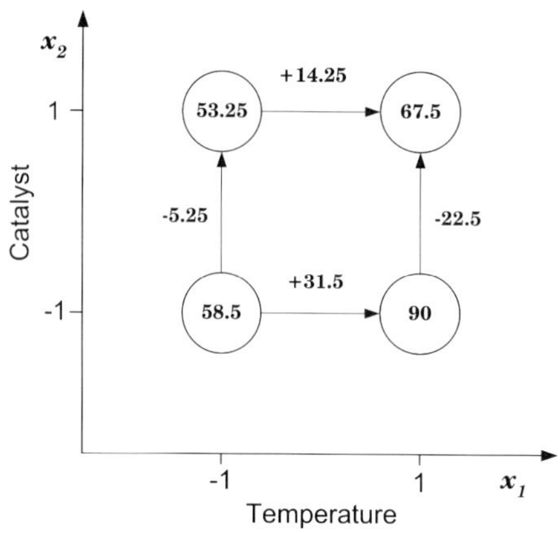

Fig. 3.5. Diagram for interpreting the temperature and catalyst effects, for the 2^3 factorial design. Compare this figure with Fig. 3.3.

i	**t**	**C**	**P**	Yield (%)		$\bar{x}_i$	s_i^2
1	−	−	−	4.39	4.73	4.56	0.058
2	+	−	−	6.21	5.75	5.98	0.106
3	−	+	−	14.51	13.45	13.98	0.562
4	+	+	−	19.57	21.11	20.34	1.186
5	−	−	+	2.09	1.93	2.01	0.013
6	+	−	+	3.15	3.39	3.27	0.029
7	−	+	+	11.77	12.69	12.23	0.423
8	+	+	+	19.40	17.98	18.69	1.008

The effects calculated for the 2^3 factorial design can also be interpreted as geometric contrasts. With three factors instead of two, the basic figure will be a cube instead of a square. The eight runs in the design matrix correspond to the vertices of the cube (Fig. 3.6). The main and the two-factor interaction effects are contrasts between two planes, which we can identify by examining the contrast coefficients in Table 3.4. For example, to calculate the main effect of factor **1** (the temperature), runs 1, 3, 5 and 7 enter the algebraic sum with negative signs, whereas the rest have positive ones. We can see in Fig. 3.6 that the runs with negative signs are all on the face of the cube that is perpendicular to the axis of factor **1** and is located at the lower level of this factor. The other runs are on the opposite face, which corresponds to the higher level. The main effect of factor **1** is therefore the contrast between these two faces of the cube, as shown in Fig. 3.6a. The other two main effects are also contrasts between opposite faces of the cube, situated perpendicularly to the axis of the

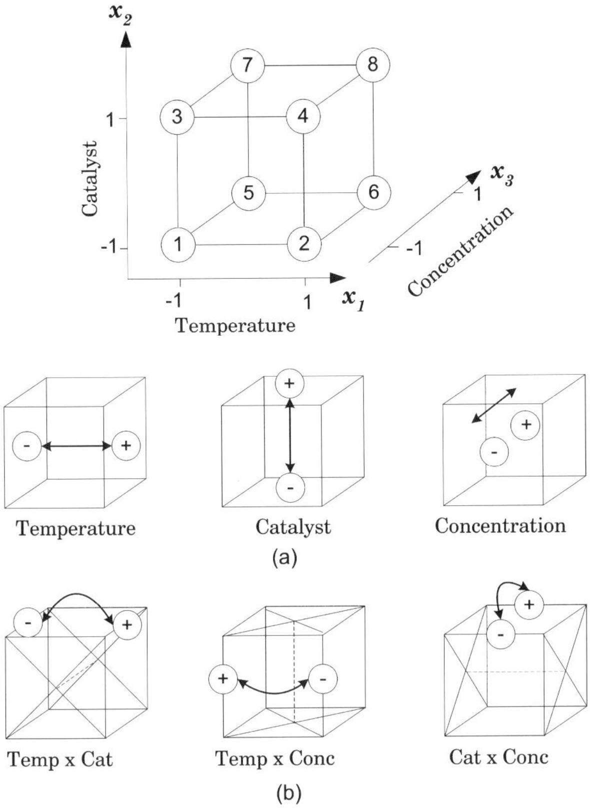

Fig. 3.6. Geometrical interpretation of the effects of the 2^3 factorial design. The main effects are contrasts between opposite faces (a), and the interaction effects are contrasts between diagonal planes (b).

corresponding factor. The two-factor interaction effects are contrasts between the two diagonal planes that are perpendicular to the plane defined by the axes of the two factors involved in the interaction, as shown in Fig. 3.6b.

Run no. 8, corresponding to the (+++) level, enters into the calculation of every effect with a positive sign, as we can see in the last row of Table 3.4. As a result, this run is always included in the positive plane of the geometric interpretation of the effect values. You can confirm this fact in Fig. 3.6. Remembering this will help you make the correct geometric interpretation of any effect.

Exercise 3.12. What conclusions can you draw from Fig. 3.5?

Exercise 3.13. What is the geometric interpretation of the **123** interaction in the 2^3 factorial design?

3.2.4. The statistical model

The statistical model for a 2^3 design can be constructed by analogy with Eq. (3.10), except that now we have three coded variables, x_1, x_2, and x_3. Using the customary notation, we write

$$\begin{aligned} y(x_1, x_2, x_3) = {} & \beta_0 + \beta_1 x_1 + \beta_2 x_2 + \beta_3 x_3 \\ & + \beta_{12} x_1 x_2 + \beta_{13} x_1 x_3 + \beta_{23} x_2 x_3 \\ & + \beta_{123} x_1 x_2 x_3 + \varepsilon(x_1, x_2, x_3). \end{aligned} \tag{3.14}$$

The coefficients in this equation (the β's) represent population values of the effects per unit changes of the coded variables. Substituting the data from Table 3.4 into Eq. (3.13) and dividing everything by 8, we obtain the following coefficient estimates:

$$\mathbf{b} = \frac{\mathbf{X}^t \mathbf{y}}{8} = \begin{bmatrix} b_0 \\ b_1 \\ b_2 \\ b_3 \\ b_{12} \\ b_{13} \\ b_{23} \\ b_{123} \end{bmatrix} \cong \begin{bmatrix} 67.3 \\ 11.4 \\ -6.9 \\ 4.4 \\ -4.3 \\ -0.4 \\ 0.4 \\ 0.1 \end{bmatrix}.$$

Using these values we can obtain an estimate of a response at the (x_1, x_2, x_3) combination of levels:

$$\begin{aligned} \hat{y}(x_1, x_2, x_3) = {} & b_0 + b_1 x_1 + b_2 x_2 + b_3 x_3 \\ & + b_{12} x_1 x_2 + b_{13} x_1 x_3 + b_{23} x_2 x_3 \\ & + b_{123} x_1 x_2 x_3. \end{aligned} \tag{3.15}$$

Exercise 3.14. Our analysis of the results of the 2^3 factorial design indicates that the interactions **13**, **23** and **123** can be neglected. Exclude the terms corresponding to these interactions from Eq. (3.15) and then estimate the yields for the eight runs. Calculate the differences between the estimated and observed values, and compare these differences — which are the residuals of the simplified model — with the observed average values.

3.3. A 2^4 factorial design

Now, let us add a fourth variable to our design, the pH of the reaction medium, at the neutral (7) and slightly acidic (6) levels. With this addition, the total number of runs increases to 16. The results obtained in the experiments, this time performed without replication, are given in Table 3.6. With them we can calculate, besides the grand average, 15 effects: four main effects, six two-factor interactions, four three-factor interactions and an interaction involving all four factors.

3.3.1. Calculating the effects

To obtain the necessary algebraic signs, we proceed in the usual manner, multiplying the columns of the design matrix, element by element. First two columns are multiplied for each binary effect, then three columns for the tertiary ones and finally four columns are multiplied to give the signs needed to calculate the four-factor effect.

Table 3.6
Results of a 2^4 factorial design for studying the influence of four factors on the yield of a reaction

Factors	(–)	(+)
1: Temperature (°C)	40	60
2: Catalyst (type)	A	B
3: Concentration (M)	1.0	1.5
4: pH	7.0	6.0
Response	Yield (%)	

Run	**1**	**2**	**3**	**4**	Response
1	–	–	–	–	54
2	+	–	–	–	85
3	–	+	–	–	49
4	+	+	–	–	62
5	–	–	+	–	64
6	+	–	+	–	94
7	–	+	+	–	56
8	+	+	+	–	70
9	–	–	–	+	52
10	+	–	–	+	87
11	–	+	–	+	49
12	+	+	–	+	64
13	–	–	+	+	64
14	+	–	+	+	94
15	–	+	+	+	58
16	+	+	+	+	73

Table 3.7
Contrast coefficients for the 2^4 factorial design

M	1	2	3	4	12	13	14	23	24	34	123	124	134	234	1234
+	−	−	−	−	+	+	+	+	+	+	−	−	−	−	+
+	+	−	−	−	−	−	−	+	+	+	+	+	+	−	−
+	−	+	−	−	−	+	+	−	−	+	+	+	−	+	−
+	+	+	−	−	+	−	−	−	−	+	−	−	+	+	+
+	−	−	+	−	+	−	+	−	+	−	+	−	+	+	−
+	+	−	+	−	−	+	−	−	+	−	−	+	−	+	+
+	−	+	+	−	−	−	+	+	−	−	−	+	+	−	+
+	+	+	+	−	+	+	−	+	−	−	+	−	−	−	−
+	−	−	−	+	+	+	−	+	−	−	−	+	+	+	−
+	+	−	−	+	−	−	+	+	−	−	+	−	−	+	+
+	−	+	−	+	−	+	−	−	+	−	+	−	+	−	+
+	+	+	−	+	+	−	+	−	+	−	−	+	−	−	−
+	−	−	+	+	+	−	−	−	−	+	+	+	−	−	+
+	+	−	+	+	−	+	+	−	−	+	−	−	+	−	−
+	−	+	+	+	−	−	−	+	+	+	−	−	−	+	−
+	+	+	+	+	+	+	+	+	+	+	+	+	+	+	+

The complete set is shown in Table 3.7. Starting with the **X** matrix and the **y** vector, as done in the cases discussed so far, and applying the appropriate divisor (8, in the present case), we arrive at the effect estimates presented in Table 3.8. Finally, dividing these values (except the average) by 2, we obtain estimates of the coefficients of the statistical model for the yield, which now has 16 terms.

Exercise 3.15. Write the full equation of the statistical model corresponding to the 2^4 factorial design.

3.3.2. Estimating the error of an effect

This time the experiments were not replicated, and we cannot calculate error estimates in the same way as before. We have to find another way to do this.

You have undoubtedly noticed that, as the number of factors increases, the mathematical model is augmented to accommodate more terms. With two factors, we had only two main effects and an interaction. With three, the model was extended until it included a three-factor interaction term. Now, with four factors, the model has 16 terms, the last of these representing a four-factor interaction. With k factors, the full model will

Table 3.8
Calculated effects for the 2^4 factorial design

Average	67.188		
Main effects			
1 (Temperature)	22.875		
2 (Catalyst)	–14.125		
3 (Concentration)	8.875		
4 (pH)	0.875		
Two-factor interactions			
12	–8.625	**13**	–0.625
14	0.875	**23**	–0.625
24	0.875	**34**	0.375
Three-factor interactions			
123	0.875	**124**	–0.125
134	–0.625	**234**	0.375
Four-factor interaction			
1234	0.375		

grow until it includes an interaction involving all k factors, just like an expansion of a function as a power series.

Let us now imagine that the response surface, that is, the relation between the response and the factors in the experimental range we are investigating, is sufficiently smooth, so that small variations in the factors do not cause abrupt variations in the response. If this is so, we can expect the model coefficients to obey a certain hierarchy, where the lower-order terms are in principle more important than the higher-order ones. Returning to our example, we expect the main effects to be more important to the model than, say, the four-factor interaction. In general, we can expect that an interaction term will become less important as the number of factors involved in its definition increases. Otherwise, we would not be able to obtain a satisfactory model for our response surface with a finite number of terms.

Table 3.8 clearly shows that some effects are much more important than others. If we assume that the main and two-factor interaction effects are sufficient to adequately describe the response surface, we can use the remaining effects to estimate the error in the effect values arising from the experimental error in the response values. According to this assumption (which is equivalent to saying that the series expansion can be truncated after the second-order terms), the three- and higher-order factor interactions do not really exist. The values determined for **123**, **124**,

134, **234** and **1234** in Table 3.8, then, can only be attributed to the random fluctuations inherent in the measuring process, that is, the "noise" embedded in the response values. Taking the square of each one of these, we have an estimate of the variance of an effect, and the average of the five values gives us a pooled estimate, with 5 degrees of freedom (since the five values are independent). We have therefore

$$\hat{V}(\text{effect}) = \frac{(0.875)^2 + (-0.125)^2 + \cdots + (0.375)^2}{5} = 0.291.$$

The square root of this value, $s \cong 0.54$, is our estimate of the standard error of an effect.

Exercise 3.16. Interpret the values in Table 3.8, in the light of the error estimates we have just made.

3.4. Normal probability plots

The analysis by means of normal probability plots is an alternative technique for trying to find out, among the results calculated for a factorial design, those that correspond to real effects, as opposed to those caused only by noise. Its working principle is based on the idea of cumulative probability, which was introduced in Exercise 2.9. We shall start by discussing this concept in more detail.

A normally distributed random variable x obeys the following equation:

$$f(x)\mathrm{d}x = \frac{1}{\sigma\sqrt{2\pi}} \mathrm{e}^{-(x-\mu)^2/2\sigma^2} \mathrm{d}x.$$

As we saw in the last chapter, the graphical representation of this curve is bell-shaped, like that in Fig. 3.7a.

Now, let us consider a specific numerical value for x, say, x_1. The tail area to the left of x_1 (Fig. 3.7a) is what we call the cumulative probability of x_1. This area, which corresponds to the probability of observing a value of x less than or equal to x_1, increases steadily as x_1 is moved to the right. As x_1 tends towards $+\infty$, the cumulative probability approaches 1, the value for the total area under the curve in Fig. 3.7a. A plot of the cumulative probability for a normal distribution is shaped like a capital S[29] and increases monotonically and asymptotically from 0 on the far left to 1 on the far right, as shown in Fig. 3.7b.

Exercise 3.17. Suppose that x is a normalized standard variable. What are the cumulative probabilities corresponding to: (a) $x_1 = 0$; (b) $x_1 = 1$; (c) $x_1 = 1.96$?

[29]In other words, a sigmoid curve.

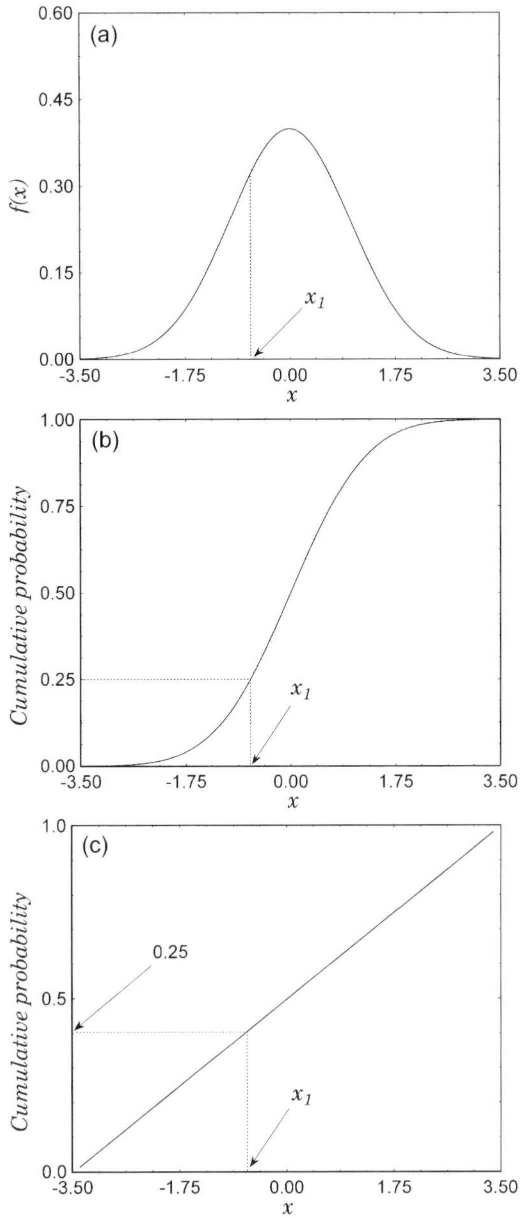

Fig. 3.7. (a) Plot of the probability density of a standard normal distribution. The cumulative probability corresponding to the x_1 value is the area under the normal curve to the left of x_1. (b) Cumulative probability on a standard Cartesian coordinate system. (c) The plot of (b), on a normal probability axis. Note that the scale of this axis is not linear. The point corresponding to the cumulative probability of x_1 (0.25) is not located halfway between 0.0 and 0.5, but closer to 0.5.

The curve in Fig. 3.7b can be turned into a straight line by a modification of its scale. This modification consists in symmetrically expanding the scale of the ordinate axis about the point representing 0.5 (i.e., 50% cumulative probability), so as to "unbend" the farthest parts of the S curve on both sides (Fig. 3.7c). The procedure is similar to that used to make logarithmic graph paper. The function $y = \log x$, which on standard graph paper is represented by a curve, becomes a straight line when plotted on logarithmic paper. In the same way, if x follows a normal distribution, the plot of its cumulative probabilities will be a sigmoid curve on the usual Cartesian scale, but is transformed into a straight line when we use the vertical scale of Fig. 3.7c. Note that the cumulative probabilities 0 and 1, which correspond, respectively, to $-\infty$ and $+\infty$ on the abscissa, cannot be represented on the graph, because its scale is finite.

Consider now a random sample of 10 elements extracted from a normal population. To represent this sample on a normal plot, the first thing we must do is arrange its elements in increasing order. Thus, the first element will be the smallest of all and the 10th one the largest. Using a subscript to indicate the order of each element, we will denote the smallest of them by x_1 and the largest by x_{10}. Since the sampling was random, we can imagine that each of these elements is representative of a 10% slice of the total area of the distribution. The first element, which is the smallest, would represent the first 10% of the left tail area; the second would represent the region between 10% and 20%, and so on. The 10th element, the largest of all, would fall in the tail area on the right, in the region stretching from 90% to 100% cumulative probability. This concept is illustrated graphically in Fig. 3.8, using the probability density curve.

The next step is to associate each point with the cumulative probability of the center of the interval it represents. Thus x_1, which belongs to the interval going from 0% to 10%, corresponds to a cumulative probability of 5%, x_2 corresponds to 15%, x_3 to 25% ,..., and finally x_{10} will be associated with a probability of 95%. If we plot these probabilities against the $x_1, \ldots, x_{10}$ values on a standard Cartesian system, the points are expected to fall approximately on a sigmoid curve, as in Fig. 3.9a. Changing the ordinate scale to a normal probability one, we expect the points to fit a straight line reasonably well (Fig. 3.9b). If these expectations are not confirmed, we should suspect that something is wrong with our assumptions. This argument will allow us to arrive at conclusions of great practical importance, as we will now demonstrate.

Let us return to the 2^4 design of last section. Imagine that none of the 15 effects we calculated actually exists, that is, their *true* values are really 0. If this assumption holds (one more example of the null hypothesis), the numerical values that we nevertheless obtained from our experiments reflect only the random errors of our process. Invoking the central limit

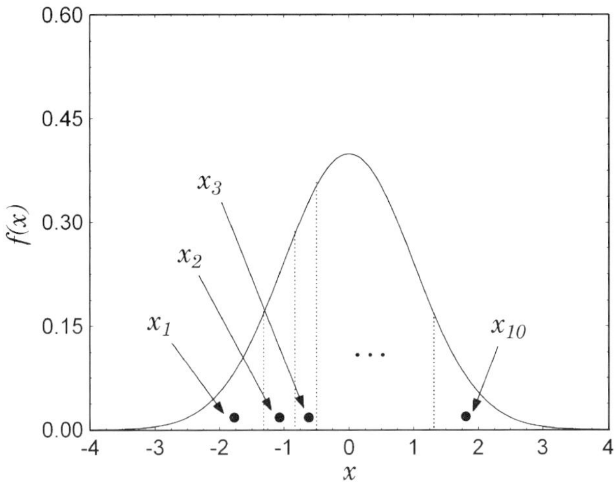

Fig. 3.8. Random sample of 10 elements taken from a standardized normal distribution. Each element represents a region whose area is equal to 1/10 of the total area under the curve.

theorem, we can consider them as a random sample drawn from an approximately normal distribution, with a population mean of zero.

Proceeding as in the example with 10 points, we can draw a normal plot of the 15 values and use it to test hypotheses about whether the effects exist or not. The data needed for this are listed in Table 3.9, where each effect of Table 3.8 is associated with a cumulative probability value. The easiest way to draw the normal plot of these values is to use one of the many statistical computer programs available. If you do not have access to one of these programs, you can use a standard linear scale, putting the values of the effects on the abscissa, but marking the values corresponding to the standardized normal distribution (z, in the last column of the table) for the center point of each interval on the ordinate axis.

The normal plot of the effects is shown in Fig. 3.10. We immediately notice that the central points are well fitted to a straight line that crosses the 50% cumulative probability level almost exactly at the zero point of the abscissa axis. It makes sense, therefore, to consider these points as coming from a normal population with zero mean. In other words, they represent "effects" devoid of any physical meaning.

The same cannot be said of the other points, labeled in the graph with the numbers of the effects they represent. It is very difficult to think of these points, so far removed from the straight line, as belonging to the same population that produced the central points. We must interpret them as really significant effects, and the farther removed from the central region, whether to the left or to the right, the more significant they

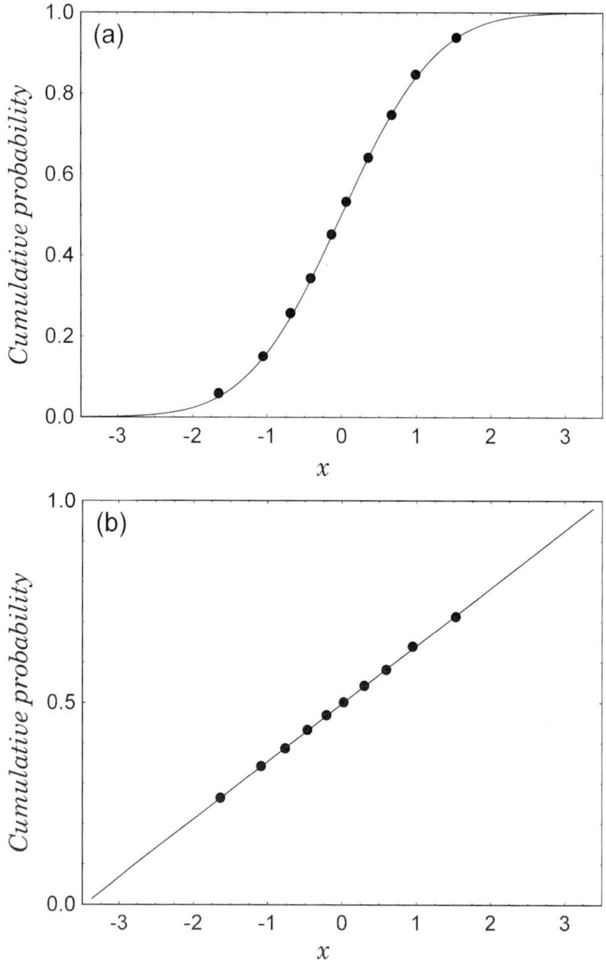

Fig. 3.9. (a) Cumulative probabilities for a sample of 10 elements randomly drawn from a standardized normal population, on a standard Cartesian scale. (b) The same points, on a normal probability plot.

are. This confirms what you already discovered, if you solved Exercise 3.16 and compared the calculated effects with the standard error estimate: only the **1**, **2** and **3** main effects and the **12** interaction effect are really significant.

Normal plots are also useful for evaluating the fit of any model, irrespective of the design used to obtain the raw data. If a model, whatever its nature, is well adjusted to the facts, it should be capable of adequately representing all of the systematic information contained in the data. The residuals left by such a model should represent only the random part, that is, the unavoidable noise contained in the measurements. The

Table 3.9
Correspondence between the calculated effects of the 2^4 factorial design and their cumulative probability values

Order	Effect	Region of cumulative probability (%)	Centre point	z
1	–14.125	0–6.67	3.33	–1.838
2	–8.625	6.67–13.33	10.00	–1.282
3	–0.625	13.33–20.00	16.67	–0.9673
4	–0.625	20.00–26.67	23.33	–0.7280
5	–0.625	26.67–33.33	30.00	–0.5244
6	–0.125	33.33–40.00	36.67	–0.3406
7	0.375	40.00–46.67	43.33	–0.1680
8	0.375	46.67–53.33	50.00	0.00
9	0.375	53.33–60.00	56.67	0.1680
10	0.875	60.00–66.67	63.33	0.3406
11	0.875	66.67–73.33	70.00	0.5244
12	0.875	73.33–80.00	76.67	0.7280
13	0.875	80.00–86.67	83.33	0.9673
14	8.875	86.67–93.33	90.00	1.282
15	22.875	93.33–100.00	96.67	1.838

appearance of a normal plot of these residuals should conform to what we expect from a random sample extracted from a normal distribution with zero mean. In other words, it should approximate a line passing through the $(x, z) = (0,0)$ point, as in the central region of Fig. 3.10.[30]

Exercise 3.18. Use the values of all the effects that fall on the line in Fig. 3.10 to calculate an estimate of the variance of an effect with 11 degrees of freedom. Use an F test to show that this estimate and the estimate obtained from the ternary and quaternary effects (with 5 degrees of freedom) can be considered as belonging to the same population.

3.5. Evolutionary operation with two-level designs

When an industrial process is still in the development stage, the technical staff usually proceeds to optimize it, making the necessary tests and adjustments in a pilot plant. When the process reaches the production line, however, the optimum conditions are not the same as those of the

[30]This depends, however, on the number of degrees of freedom of the residuals. The more they are, the better. We must not expect too much from a normal plot with only a few degrees of freedom (as, for example, in a 2^3 factorial design), because its values do not have many options to randomly distribute themselves.

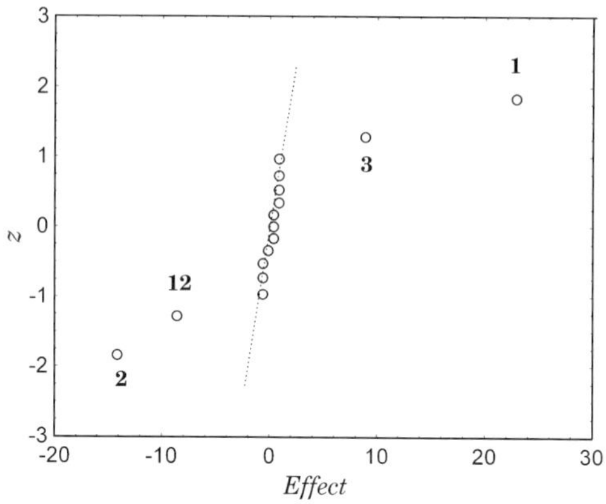

Fig. 3.10. Normal probability plot of the values in Table 3.9. Only the **1**, **2**, **3** and **12** effects appear to be significant.

pilot plant. In the first place, there is the scale-up problem. The effects of factors like agitation, mass and heat transfer, distiller efficiency and so on, vary with equipment size to an extent that is quite difficult to predict. Then there are the problems associated with routine production, like variations in raw material or storage conditions, substitution of spare parts or whole machines and, of course, equipment aging and deterioration. All of these factors conspire to make the large-scale optimum operating conditions quite different from those of the pilot plant or the laboratory. If we wish to discover what these optimum conditions are, we must use information obtained from the large-scale production process itself.

The truth is that considerations like this are rarely made in practice. In most cases, the industrial plant operates indefinitely under the same conditions established at its start-up, and process efficiency tends to decrease with time. This way of operating, which is without doubt the usual, might be called static operation, to contrast it with what is called evolutionary operation (Box, 1957), which we address in this section.

In evolutionary operation (better known under the acronym EVOP), the situation that we have just described is assumed to be true, and the large-scale operating conditions are deliberately modified, with the objective of extracting from the manufacturing process not only the desired product but also information about how this process can be improved. The adjective "evolutionary" was chosen on purpose, to emphasize that the fundamental idea is to subject the production settings to a process of variation and selection analogous to what governs the evolution of species in Darwinian theory. In this section, we limit our discussion to describing the basic features of EVOP. The interested reader can find much more information in the publications of G.E.P. Box, to whom we owe the conception and the first industrial application of EVOP, done in 1954 at the installations of the Imperial Chemical Industries, in England. The book *Evolutionary operation: a statistical method for process improvement* (Box and Draper, 1969) is particularly recommended.

The first step in performing an EVOP consists of introducing some variation in the operating environment, which is a prerequisite to evolution and adaptation of the process. Since EVOP is applied to the industrial process while it is in full operation, any perturbation must be made with the utmost care, to minimize the risk of manufacturing an unsatisfactory product. In this sense, EVOP is very different from experimental studies carried out in a research laboratory, where the factors can be varied as much as we please, with no regard for the acceptability of the final product. In EVOP, we are trying to fine tune a process that already works reasonably well, and every precaution should be taken. Besides being small enough, the perturbations we introduce should be made in a carefully planned way, to ensure that we can extract useful information about how the process can be improved from their effects.

The kind of variation recommended by Box is a two-level factorial design around the usual operating conditions, as illustrated schematically in Fig. 3.11. It is best to vary only two or three factors at a time, to facilitate the analysis of the results. The extent of the variations is a matter that should be pondered and decided by the production personnel themselves. They always have a good idea about the security margin of the process, and can make the necessary variations without running excessive risks.

Since the variations are small, their effects can be difficult to detect in the presence of noise in the response values (which, by the way, is usually larger in a full-scale process than in the laboratory). For this reason, we must run the process several times at a given set of conditions, and base our analysis on the final average result. Enough replicates should be made until it is possible to discern some pattern in the responses. The analysis of the results is made in the same way as described earlier in this chapter and should present no difficulty.

Consider the initial situation illustrated in Fig. 3.11a. Let us imagine, for example, that the analysis of the response values has shown that the operating conditions at point B produce more satisfactory results than those at the other points investigated, and that the worst of all results occurs at point D. In evolutionary terms, we say that point B appears more fit to survive, and point D less so. Operationally, we conclude that it seems worthwhile to change the production conditions from the initial situation, represented by point A, to those of point B. In this way, the process *evolves* in the direction A → B, obeying a kind of "natural selection" of the fittest, and thus becomes operated in a more efficient way.

After the displacement to the more favorable conditions is performed, the process is again repeatedly operated at the conditions defined by the

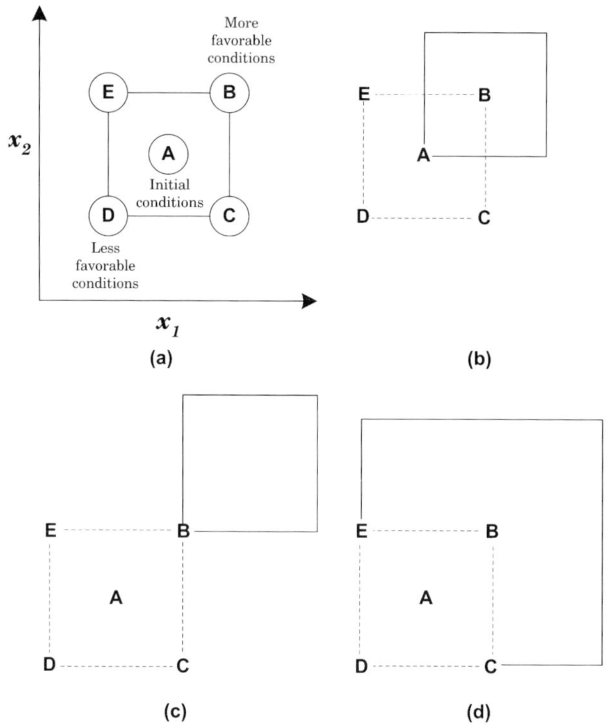

Fig. 3.11. (a) Factorial design with a center point, in the initial phase of an evolutionary operation. (b), (c), (d) some possible displacements relative to the initial phase.

new design, until the analysis of the new results permits a decision about a possible new displacement. The total procedure — design, analysis, displacement — is repeated until there is no evidence that it is still possible to improve process efficiency in this way. Even in this optimized phase, however, the operating settings should be continually subjected to controlled variations, because no one can guarantee that the optimum situation found today will still be the optimum situation some time from now. Treated in this way, the process can always evolve to a more favorable situation, if this proves feasible. EVOP should be considered as a permanent mode of operating the process. It is not an experiment done only for a limited time. In EVOP, the process must be prodded all the time to provide information about itself.

Owing to industrial security, detailed information about actual examples of EVOP are scarce in the literature. Even so, there are indications that these results have been highly significant, resulting in profits that reach millions of dollars/year, thanks to the large scale in which the consequences of small modifications in operating conditions are manifested (see, for example, Box et al., 1978, p. 365).

3.6. Blocking factorial designs

At the end of Chapter 2 we presented an experimental design for comparing analytical determinations performed by two chemists on vinegar samples of different origins. The objective of the study was to compare the performances of the analysts, not the variations due to the different manufacturers. The experiments were run in five blocks, each of which contained two samples from the same manufacturer. Thanks to this, we were able to separate the variance due to the different concentrations of acetic acid from that caused by the errors of each analyst, and consequently perform a more sensitive statistical analysis than would have been possible if the assignment of the sample to the analysts had been completely random.

Blocking can also be employed in factorial designs. Imagine, for example, that we decided to perform a 2^3 factorial design, but discovered that the raw materials in stock are not enough for all 8 runs. To complete the design, we will have to acquire a new batch of material, perhaps even from a different manufacturer. This clearly introduces a possible additional source of variation that we should take into account, so that the conclusions we reach in our investigation will not be distorted.

If we were interested in the effect of the origin of the material, we could include it as another factor in our experiments, transforming the design into a 2^4 factorial and doubling the number of runs to be executed. This is not the case here. We know the origin of the raw material might influence the results, but we are only interested in estimating the effects of the other three factors. What must we do, to minimize the effect of this fourth undesired factor without increasing the number of experiments?

Randomization of all the runs would not be advisable, since it could confound the effects of interest with the one owing to the change in raw material. With blocking, we can minimize this problem.

Suppose each batch of raw material is sufficient to execute four runs, that is, half of the design. In the 2^3 factorial, the less important effect, from a hierarchical point of view, is the **123** interaction. Since using two different batches is unavoidable, we can run the design so that batch change becomes confounded with the **123** effect. To accomplish this, we allot the samples of the two batches as in Table 3.10. The four samples of a block are assigned to the experimental runs according to the algebraic signs of the **123** interaction, and randomization is done only within each block, not for the entire design. Thus, when we calculate the interaction of the three factors, we know that a contribution owing to batch variation (which might or might not be important) will be included in the result. In compensation, the main and two-factor interaction effects, which in principle should be more important, are free from this contamination.

Table 3.10
2^3 factorial design performed in two blocks of four runs

Run	**1**	**2**	**3**	**123**	Block	Batch	Order
1	–	–	–	–	I	A	2
2	+	–	–	+	II	B	4′
3	–	+	–	+	II	B	1′
4	+	+	–	–	I	A	1
5	–	–	+	+	II	B	3′
6	+	–	+	–	I	A	4
7	–	+	+	–	I	A	3
8	+	+	+	+	II	B	2′

Note: The blocks are chosen according to the signs of the **123** interaction. The order of executing the runs is random within each block.

We can understand how this happens by examining the calculation of one of the main effects. For example, effect **1** is given by

$$\mathbf{1} = \frac{(y_2 - y_1) + (y_4 - y_3) + (y_6 - y_5) + (y_8 - y_7)}{4}.$$

Identifying the responses according to their batches, we can rewrite this equation as

$$\mathbf{1} = \frac{(\mathrm{B} - \mathrm{A}) + (\mathrm{A} - \mathrm{B}) + (\mathrm{A} - \mathrm{B}) + (\mathrm{B} - \mathrm{A})}{4}.$$

There are two differences associated with the A→B batch variation, and another two with the opposite change, B→A. If there is some systematic influence associated with batch change, it will be canceled in the calculation of effect **1**. The same thing happens to the other main effects and to the two-factor interaction effects, because all the columns of signs are orthogonal to each other and because we purposely confounded the **123** interaction with the batch effect. You can verify this easily in Fig. 3.12, which shows the geometric representation of the 2^3 design, with the runs identified according to their blocks. Note that the runs of blocks I and II occupy the vertices of two opposing tetrahedra (see Exercise 3.13).

Blocking can be extended to more complicated situations. One example is the division of a 2^3 factorial in four blocks, each containing two runs. The best design in this case would have two blocking variables, one corresponding to the **123** interaction and the other corresponding to one of the two-factor interactions. This second interaction clearly would be confounded with the block effect, and the interpretation of the results would become more involved. Box et al. (1978) (Chapter 10) and Montgomery (2004) (Chapter 8) are good references for those wishing to know more about blocking.

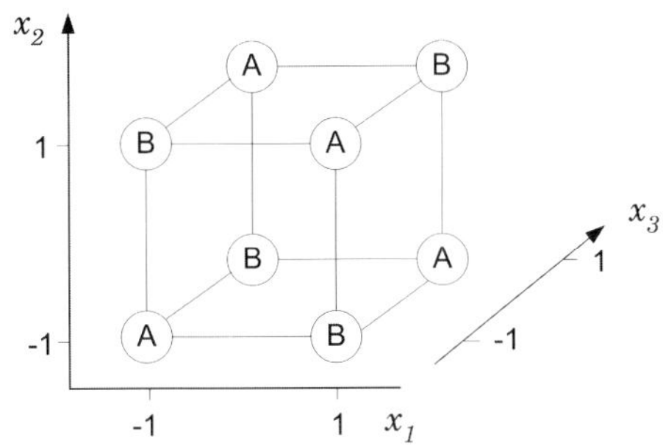

Fig. 3.12. Division of the 2^3 factorial in two blocks. Blocking is done so as to confound the block effect with the **123** interaction effect. In the calculation of every effect (except **123**) there are two differences B–A and two differences A–B that together cancel the block effect.

Exercise 3.19. Suppose that the **123** interaction effect does not really exist for the experiment we are discussing, even though its numerical value is found to be relatively high. How can we interpret this value?

Exercise 3.20. A 2^3 factorial was performed in two blocks. The runs of the second block were executed one month after those of the first one, and contain a contribution, h, caused by systematic errors that were absent from the response values of the first block. Show that the presence of this systematic difference in the second block does not affect the value calculated for the **23** interaction effect.

3A. Applications

3A.1. Resin hydrolysis

V.X. de Oliveira Jr. executed a 2^3 factorial design in duplicate to study the effect of three factors (hydrolysis time, temperature and type of catalyst) on the degree of substitution in a chloromethylated resin. His objective was to maximize the response. The data obtained in the experiments are presented in Table 3A.1, along with the calculated effects.

In the usual way, we calculate from the duplicate responses a pooled variance of 1.75×10^{-4}, which corresponds to a standard error of 1.323×10^{-2} in the response. For this design, the standard error of an effect is half the standard error of the response.[31] Multiplying this value by the t distribution point for 8 degrees of freedom, we find that the limits of the 95% confidence interval for the value of an effect are $\pm 1.525 \times 10^{-2}$. This means that only the main effects of factors **1** (time) and **3** (catalyst) and

[31]Why?

Table 3A.1
Experimental data and calculated effects

Factors	(–)	(+)
1: Hydrolysis time (h)	24	48
2: Temperature (°C)	130	160
3: Catalyst[a]	TFA	PA

Run	**1**	**2**	**3**	Degree of substitution		Average	Variance
1	–	–	–	0.52	0.54	0.530	0.00020
2	+	–	–	0.57	0.58	0.575	0.00005
3	–	+	–	0.55	0.54	0.545	0.00005
4	+	+	–	0.58	0.56	0.570	0.00020
5	–	–	+	0.47	0.45	0.460	0.00020
6	+	–	+	0.53	0.56	0.545	0.00045
7	–	+	+	0.52	0.53	0.525	0.00005
8	+	+	+	0.54	0.52	0.530	0.00020

Effects

1	**2**	**3**	**12**	**13**	**23**	**123**
0.040	0.015	**–0.040**	**–0.025**	0.005	0.010	–0.015

[a] TFA, trifluoroacetic acid, PA, propionic acid, both in 12 M HCl.

the **12** interaction (time × temperature) are significant, at this confidence level. According to these values, shown in boldface in Table 3A.1, to obtain the largest degree of substitution we should perform the hydrolysis for 48 h, using trifluoroacetic acid as a catalyst.

Fig. 3A.1 helps us visualize all the results of the experiments for this design. The responses obtained with trifluoroacetic acid (TFA) (the circles) are always larger. The **12** interaction can be inferred from the different slopes in the two plots: at 130 °C the effect of increasing the hydrolysis time is clearly larger than at 160 °C. Although this effect (shown on the abscissas) is diminished when the reaction is carried out at the higher temperature, the two largest responses were obtained with a reaction time of 48 h. This suggests that we should investigate longer hydrolysis times, perhaps at an intermediate temperature. As for propionic acid (PA), the results show that it can be discarded as a catalyst, in favor of TFA.

3A.2. Cyclic voltammetry of methylene blue

The electrochemical properties of methylene blue were investigated by cyclic voltammetry, with the objective of determining the experimental conditions providing the best reversibility of the redox process, i.e.,

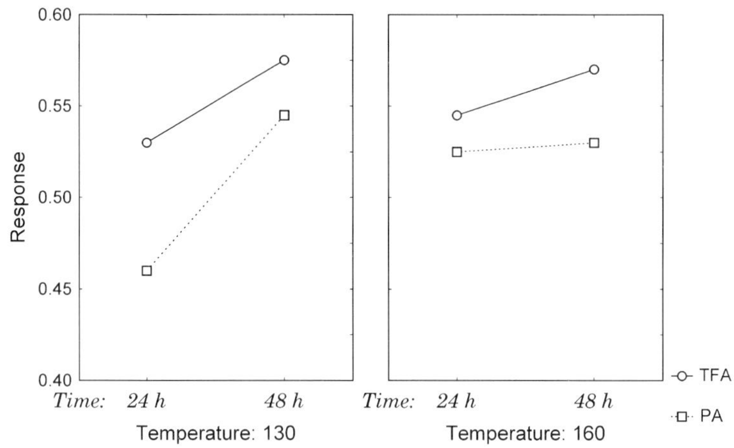

Fig. 3A.1. Response variation with the levels of the three factors.

minimization of the voltage difference (ΔE) of this process (Rocha et al., 1997). A 2^3 factorial design was executed, resulting in the data presented in Table 3A.2.

The calculated effect values are quite unusual. The largest effect is **23**, the pH × silica interaction, yet the main effect of the silica factor (**3**) is the smallest of all effect values. We can understand what happened by examining the cubic plot of the response values (Fig. 3A.2). The interaction between pH and silica is so strong that, when the pH is at its lower level (front face of the cube), the two individual effects of silica tend to increase ΔE. When the pH is at its higher level, the reverse occurs (back face). The main effect of silica thus winds up canceling itself, not because it really does not exist but because it is the average of two opposing effects. The behavior of concentration (factor **1**) is more orthodox. Raising the concentration almost always decreases the value of the voltage difference, ΔE. Since the objective of the experiments is to minimize the value of ΔE, it appears as though we should explore more thoroughly the region close to the settings of run no. 2 (+ − −), where the concentration is 0.5 mol L^{-1}, the modified silica is STM and pH is 4, and we observe the smallest response value for this design, $\Delta E = 98$ mV. We could arrive at the same conclusions by examining Fig. 3A.3. The sharply reversed directions of the line slopes in the two plots indicate the strong **23** interaction.

3A.3. Retention time in liquid chromatography

A 2^3 factorial design (Table 3A.3) was used to investigate how the retention time of a peak obtained with liquid chromatography is affected

Table 3A.2
Experimental data and calculated effects

Factors	(–)	(+)
1: Concentration (mol L^{-1})	0.1	0.5
2: Ph	4	7
3: Modified silica (type)	STM	STPM

Run	**1**	**2**	**3**	ΔE (mV)
1	–	–	–	106
2	+	–	–	98
3	–	+	–	139
4	+	+	–	141
5	–	–	+	137
6	+	–	+	123
7	–	+	+	119
8	+	+	+	103

Effects

1	**2**	**3**	**12**	**13**	**23**	**123**
–9.0	9.5	–0.5	2.0	–6.0	–28.5	–3.0

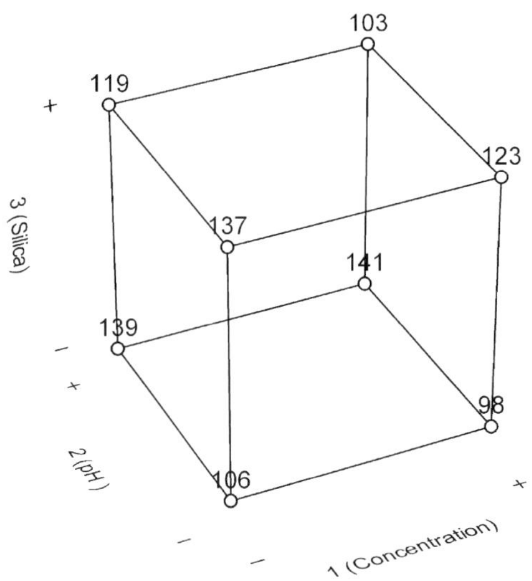

Fig. 3A.2. Cubic plot of the response values in Table 3A.2.

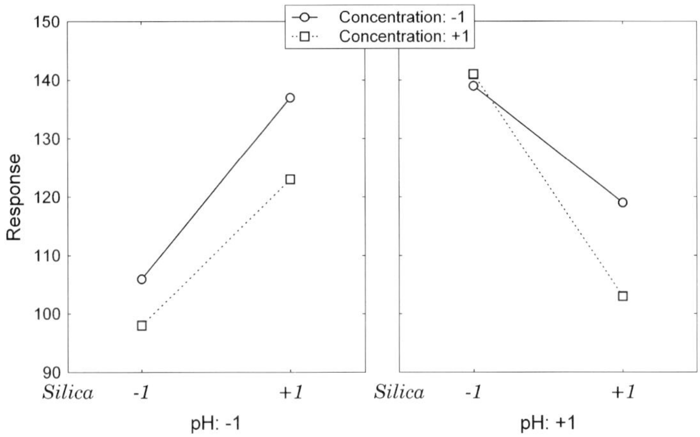

Fig. 3A.3. Response variations for the levels of the three factors.

by the percentage of ethanol, the column temperature and the gas flow rate (Ribeiro et al., 1999).

The main effects of the three factors are all negative. The temperature and flow rate main effects (**1** and **3**) are the most important, but the much smaller ethanol effect also seems significant. In all four comparisons of runs differing only in the ethanol level (1→3, 2→4, 5→7 and 6→8), the run performed at the higher level has a lower response value, as Fig. 3A.4 makes clear. If the main effect of alcohol (–4.54) is significant, then the **13** interaction (4.53) also is, and we should interpret the effects of temperature and flow rate together (Fig. 3A.5).

An increase in flow rate, whatever the temperature, reduces the average retention time. This reduction, however, is more pronounced (–24 min) at the lower temperature level than at the higher one (–15 min). We can also perceive the same result in Fig. 3A.4, by the difference in the slopes of the lines in the two graphs. If the objective of the experiment is to lower the retention time, we should work with experimental settings close to those of the run where the temperature and the flow rate — as well as the amount of ethanol — are at their higher levels (50 °C, 70% and 0.2 mL min^{-1}). In Fig. 3A.4, this run is represented by the rightmost square.

3A.4. Gas separation by adsorption

In a laboratory development of a process to enrich gases by adsorption, a 2^4 factorial design was used to investigate the influence of four factors on several responses, one of which was the productivity of an adsorbent (P). The data obtained are reproduced in Table 3A.4 (Neves, 2000).

Table 3A.3
Experimental data and calculated effects

Factors	(–)	(+)
1: Temperature (°C)	30	50
2: Percentage of ethanol (v/v)	60	70
3: Flow rate (mL min^{-1})	0.1	0.2

Run	**1**	**2**	**3**	Retention time (min)
1	–	–	–	49.26
2	+	–	–	31.27
3	–	+	–	42.20
4	+	+	–	26.61
5	–	–	+	23.81
6	+	–	+	15.07
7	–	+	+	19.57
8	+	+	+	12.86

Effects

1	**2**	**3**	**12**	**13**	**23**	**123**
–12.26	–4.54	–19.51	1.11	4.53	1.32	–0.09

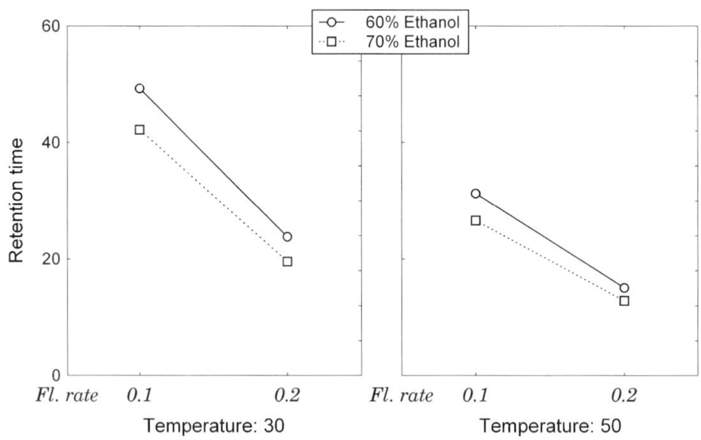

Fig. 3A.4. Response variations with the levels of the three factors.

Now the number of effects is large enough to justify drawing a normal plot (Fig. 3A.6). The adsorption time and feeding flow rate (**4** and **3**, respectively) are the most significant main effects, followed, at a certain distance, by their interaction effect and the adsorption pressure main

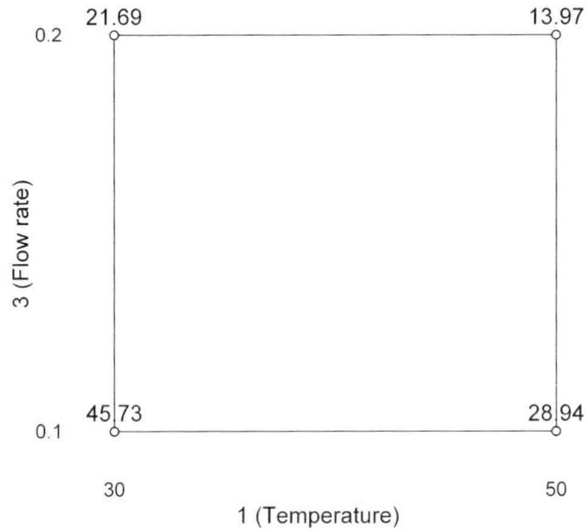

Fig. 3A.5. Plot of the average response values as a function of temperature and flow rate.

effect. Factor **2**, the desorption pressure, is of secondary interest, and it is natural to think about considering it inert, so that the design is converted into a factorial in duplicate in the other three factors, from which we can obtain an estimate of pure error. Note, however, that 10 out of the 11 calculated effects with points on the straight line close to the origin are positive, which is not in agreement with the hypothesis of a normal distribution with zero mean, as would be expected for a set of values representing pure error. If we examine the eight pairs of runs that only differ in the level of factor **2**, we see that for seven of them the higher level produces a slightly higher response. *Conclusion*: without doubt factor **2** is less important than the other three, but it also appears to act in the direction of increasing the productivity of the adsorbent. In fact, the lowest response corresponds to the first run, and the highest occurs for the last one, in exact contrast to what was observed in Application 3A.3, where all the main effects were negative.

3A.5. Improving wave functions

The development of high-quality wave functions is a crucial task in computational chemistry. There are several ways of trying to improve the quality of a wave function, and the response monitored in this process usually depends on the nature of the problem under study. Functions that have proven satisfactory for calculating certain properties are not necessarily the most indicated for other properties. In this application, the research workers were interested in evaluating how some parameters used to specify a wave function would affect the CH stretching frequency calculated from this function for the CH_3F molecule (Azevedo et al., 1996).

Table 3A.4
Experimental data and calculated effects

Factors	(–)	(+)
1: Adsorption pressure (bar)	1.40	2.40
2: Desorption pressure (bar)	0.05	0.20
3: Feeding flow rate ($m^3 h^{-1}$)	0.10	0.30
4: Adsorption time (s)	8	30

Run	**1**	**2**	**3**	**4**	P ($mol\,kg^{-1}\,cycle^{-1}$)
1	–	–	–	–	2.75
2	+	–	–	–	3.15
3	–	+	–	–	2.85
4	+	+	–	–	3.55
5	–	–	+	–	4.65
6	+	–	+	–	5.85
7	–	+	+	–	5.40
8	+	+	+	–	6.30
9	–	–	–	+	5.95
10	+	–	–	+	6.55
11	–	+	–	+	5.60
12	+	+	–	+	6.75
13	–	–	+	+	11.50
14	+	–	+	+	13.00
15	–	+	+	+	12.50
16	+	+	+	+	14.00

Effects

1	**2**	**3**	**4**	**12**	**13**	**14**	**23**
0.994	0.444	4.51	5.17	0.069	0.281	0.194	0.356
24	**34**	**123**	**124**	**134**	**234**	**1234**	
0.019	2.03	–0.144	0.069	0.031	0.181	0.0062	

The 2^4 factorial design shown in Table 3A.5 was chosen for this investigation.

The normal plot of Fig. 3A.7 clearly shows that only the **4** (correlation), **1** (valence basis set) and **24** interaction (polarization functions × electronic correlation) effects are significant. We again encounter a case where an interaction effect is important even though the main effect of one of its factors (**2**) is not significant. You can discover why this is so by examining

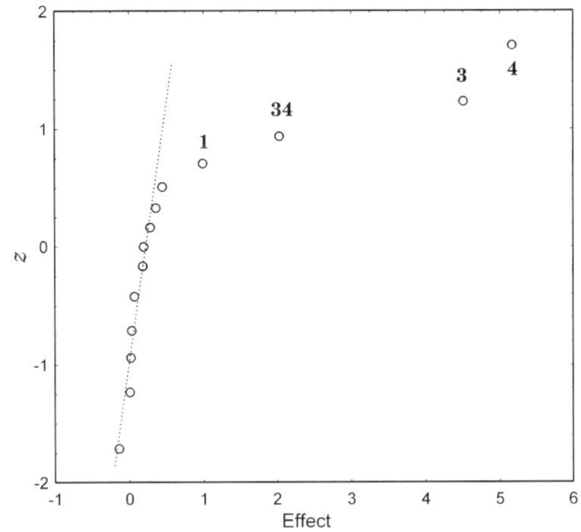

Fig. 3A.6. Normal plot of the effects.

the cubic plot of the response values as a function of the **1**, **2** and **4** factors (Fig. 3A.8).

3A.6. Performance of Ti/TiO_2 electrodes

The 2^5 factorial design whose results are presented below was used to investigate how five factors influenced the performance of Ti/TiO_2 electrodes in the electroreduction of nitrobenzene. The electrode performance was quantified in terms of cathodic charges measured (in $mC\,cm^{-2}$) during electrode processing (Ronconi and Pereira, 2001). The factors studied were the concentration of the titanium precursor in the precursor solutions used (**1**), the number of layers deposited on the electrode (**2**), the temperature at which the thermal treatment was carried out (**3**), the nitrobenzene concentration (**4**) and the scanning velocity of the applied potential (**5**). The factorial design and the responses obtained in the experimental runs are reproduced in Table 3A.6.

Judging by the normal plot (Fig. 3A.9), the main effects of factors **2–5** and the **45** interaction effect are the most significant ones. Since none of these involve factor **1**, it appears that we can consider this factor as inert and analyze the data as a 2^4 factorial in duplicate on the remaining factors, as shown in Table 3A.7.

The pooled variance calculated from the 16 "duplicate" runs is 0.9584. The variance of any effect will be one-eighth of this value,[32] which is 0.1198. The square root of this last value is the standard error of an effect. Multiplying it by t_{16}, we arrive at the limiting value for the statistical significance of the absolute value of an effect, 0.734 (95% confidence level).

[32]Why?

Table 3A.5
Data from the computational experiment and calculated effects

Factors	(–)	(+)
1: Basis set	6-31G	6-311G
2: Polarization functions	Absent	Present
3: Diffuse functions	Absent	Present
4: Electronic correlation	Hartree–Fock	MP2

Run	**1**	**2**	**3**	**4**	Frequency (cm^{-1})
1	–	–	–	–	3245.6
2	+	–	–	–	3212.4
3	–	+	–	–	3203.5
4	+	+	–	–	3190.3
5	–	–	+	–	3251.7
6	+	–	+	–	3209.4
7	–	+	+	–	3214.9
8	+	+	+	–	3193.5
9	–	–	–	+	3096.2
10	+	–	–	+	3049.3
11	–	+	–	+	3132.8
12	+	+	–	+	3087.6
13	–	–	+	+	3105.0
14	+	–	+	+	3050.4
15	–	+	+	+	3143.5
16	+	+	+	+	3093.5

Effects

1	**2**	**3**	**4**	**12**	**13**	**14**	**23**
–38.35	4.95	5.53	–120.38	5.90	–3.73	–10.83	2.28
24	**34**	**123**	**124**	**134**	**234**	**1234**	
34.18	1.10	0.48	–4.33	0.60	–0.60	0.25	

This result lends support to our conclusion about the five effects that do not fit the straight line in the normal plot. The values of two others, the **12** and **123** interaction effects, lie practically at the confidence limit. The conclusions do not change much in relation to the analysis of the full design, but it is important to recognize that to obtain an error estimate we are combining variances that differ by up to four orders of magnitude. This is a violation of the normal error hypothesis that is the basis for a

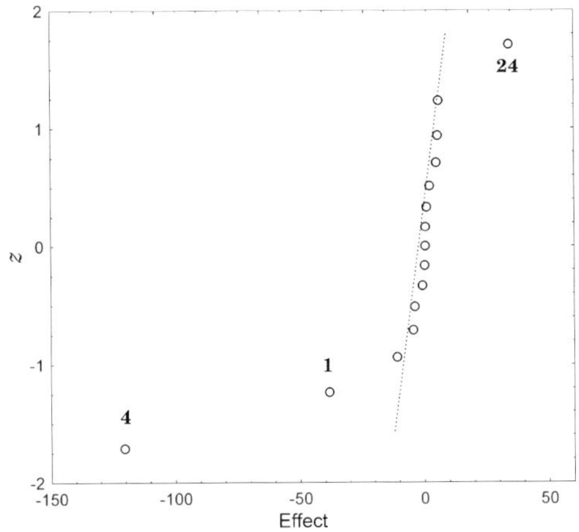

Fig. 3A.7. Normal plot of the effects.

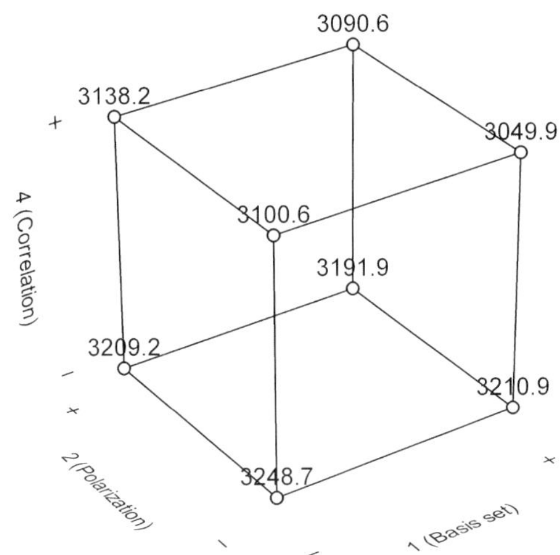

Fig. 3A.8. Cubic plot of the average response values as function of the **1**, **2** and **4** factors.

majority of statistical tests. Before attempting to determine the confidence limits, we should have used some transformation of the variance values (possibly logarithmic). For the details, which lie outside the scope of this book, a good reference is the book by Wu and Hamada (2000), which also presents alternative criteria for finding out the significant effects.

Table 3A.6
Experimental data and calculated effects

Factors	(−)	(+)
1: Precursor solution	1:4:16	1:2:8
2: Number of layers	2	10
3: Temperature (°C)	450	525
4: Nitrobenzene concentration (mM)	2	8
5: Scanning velocity ($mV.s^{-1}$)	50	200

Run	**1**	**2**	**3**	**4**	**5**	Performance
1	−	−	−	−	−	2.07
2	+	−	−	−	−	2.03
3	−	+	−	−	−	4.71
4	+	+	−	−	−	7.01
5	−	−	+	−	−	1.71
6	+	−	+	−	−	2.10
7	−	+	+	−	−	4.36
8	+	+	+	−	−	3.71
9	−	−	−	+	−	7.15
10	+	−	−	+	−	4.87
11	−	+	−	+	−	8.96
12	+	+	−	+	−	12.25
13	−	−	+	+	−	4.28
14	+	−	+	+	−	3.13
15	−	+	+	+	−	9.42
16	+	+	+	+	−	8.68
17	−	−	−	−	+	1.70
18	+	−	−	−	+	1.39
19	−	+	−	−	+	4.50
20	+	+	−	−	+	5.92
21	−	−	+	−	+	0.73
22	+	−	+	−	+	0.77
23	−	+	+	−	+	3.20
24	+	+	+	−	+	3.08
25	−	−	−	+	+	2.51
26	+	−	−	+	+	1.82
27	−	+	−	+	+	5.60
28	+	+	−	+	+	7.61
29	−	−	+	+	+	1.55
30	+	−	+	+	+	1.05
31	−	+	+	+	+	4.25
32	+	+	+	+	+	4.38

Effects

1	**2**	**3**	**4**	**5**	**12**	**13**	**14**
0.19	3.67	−1.48	2.41	−2.27	0.76	−0.52	−0.19
15	**23**	**24**	**25**	**34**	**35**	**45**	**123**
0.05	−0.45	0.68	−0.30	−0.27	−0.02	−1.47	−0.78
124	**125**	**134**	**135**	**145**	**234**	**235**	**245**
0.40	−0.15	−0.06	0.16	0.18	0.29	−0.22	−0.33
345	**1234**	**1235**	**1245**	**1345**	**2345**	**12345**	
0.20	−0.12	0.29	−0.18	−0.01	−0.32	0.10	

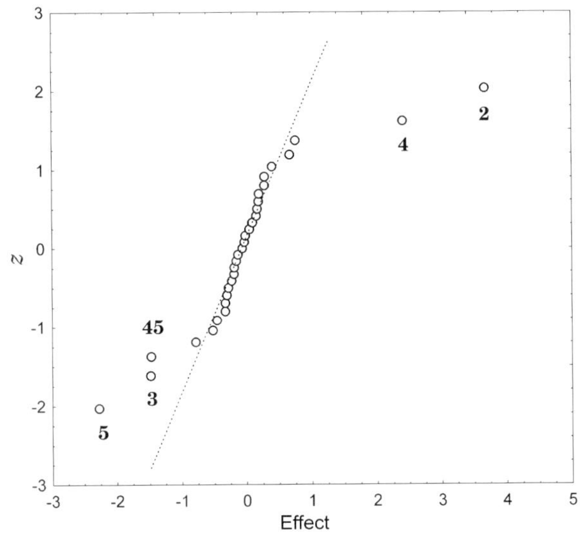

Fig. 3A.9. Normal plot of the effects.

Table 3A.7
2^4 factorial in duplicate obtained by eliminating factor **1** from the complete design of Table 3A.6

Run	**2**	**3**	**4**	**5**	Performance		Average	Variance
1	−	−	−	−	2.07	2.03	2.050	0.0008
2	+	−	−	−	4.71	7.01	5.860	2.6450
3	−	+	−	−	1.71	2.10	1.905	0.0761
4	+	+	−	−	4.36	3.71	4.035	0.2113
5	−	−	+	−	7.15	4.87	6.010	2.5992
6	+	−	+	−	8.96	12.25	10.605	5.4120
7	−	+	+	−	4.28	3.13	3.705	0.6613
8	+	+	+	−	9.42	8.68	9.050	0.2738
9	−	−	−	+	1.70	1.39	1.545	0.0481
10	+	−	−	+	4.50	5.92	5.210	1.0082
11	−	+	−	+	0.73	0.77	0.750	0.0008
12	+	+	−	+	3.20	3.08	3.140	0.0072
13	−	−	+	+	2.51	1.82	2.165	0.2380
14	+	−	+	+	5.60	7.61	6.605	2.0201
15	−	+	+	+	1.55	1.05	1.300	0.1250
16	+	+	+	+	4.25	4.38	4.315	0.0084

3A.7. Controlling detergent froth

The growing automation in cleaning processes for washing clothes, kitchenware and surfaces in general is increasing the demand for detergent products that not only remove dirt but also have low frothing powers, an important feature because of the high agitation rates at which washing machines work. One way of controlling the amount of froth produced by detergents is to introduce non-ionic surfactants, like the fatty EO/PO alcohols (that is, ethoxyl/propoxyl alcohols) into their formulations. Several other important properties, like the turbidity point, surface tension, dynamic froth height, detergency, flame point and thermal degradation temperature, are also usually monitored during the preparation of a detergent mixture. As an assignment in a chemometrics course, Ricardo Pedro decided to verify how the turbidity point varies with the ethylene oxide (EO) and propylene oxide (PO) submolecular units of the fatty alcohols, using a 2^2 factorial design augmented with a run in triplicate at the center of the design (Table 3A.8).

From the three experiments at the center point, we obtain an estimate of 0.40 for the standard error of the response, which in this case is the same as the standard error of an effect. Therefore, the limiting value for statistical significance of the absolute value of an effect (at the 95% confidence level) will be given by

$$t_2 \times s(\text{effect}) = 4.303 \times 0.40 = 1.72.$$

The effects of the number of EO units, 7.20, and the number of PO units, –5.30, are very significant. The interaction effect is not. These results can be visualized in Fig. 3A.10. The absence of significant interaction between the two factors is reflected in the almost perfect parallelism of the two lines. The turbidity point increases with the degree of ethoxylation and diminishes with the degree of propoxylation, being minimal for the A6O4 product (considering only the five alcohols evaluated, of course). Since surface-active agents with smaller turbidity points are also those with smaller frothing power, we can conclude that this alcohol is the one that should produce less froth.

If the response surface can be adequately represented by a plane, an elementary geometric argument, illustrated in Fig. 3A.11, tells us that the average value of the responses at the center point should coincide (within experimental error) with the average value of the responses at the vertices of the design. If the surface were quadratic, it can be shown that the difference between these two average response values is a measure of the surface's overall curvature, which is given by the sum of the coefficients of the quadratic terms x_1^2 and x_2^2 (Box and Draper, 1987). Using the data in the table, we have:

Table 3A.8
Experimental data and calculated effects

Run	Product[a]	**EO**	**PO**	Turbidity point (°C)
1	A4O4	–	–	32.10
2	A4O6	+	–	40.20
3	A6O4	–	+	27.70
4	A6O6	+	+	34.00
5	A5O5	0	0	35.00
6	A5O5	0	0	34.60
7	A5O5	0	0	35.40

Effects

EO	**PO**	**EO × PO**
7.2	–5.3	–0.9

[a] The numbers in the product name indicate the number of units of PO and EO, in this order.

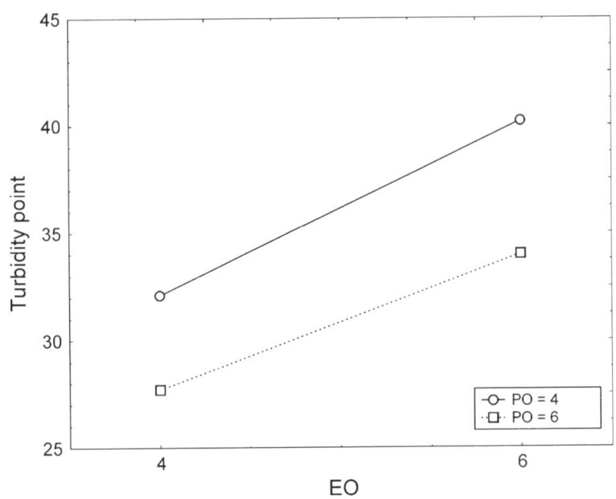

Fig. 3A.10. Response variation with the levels of the two factors.

$$\text{Center point}: \ \bar{y}_{\text{C}} = \frac{35.00 + 34.60 + 35.40}{3} = 35.00,$$

$$\text{Vertices}: \bar{y}_{\text{V}} = \frac{32.10 + 40.20 + 27.70 + 34.00}{4} = 33.50,$$

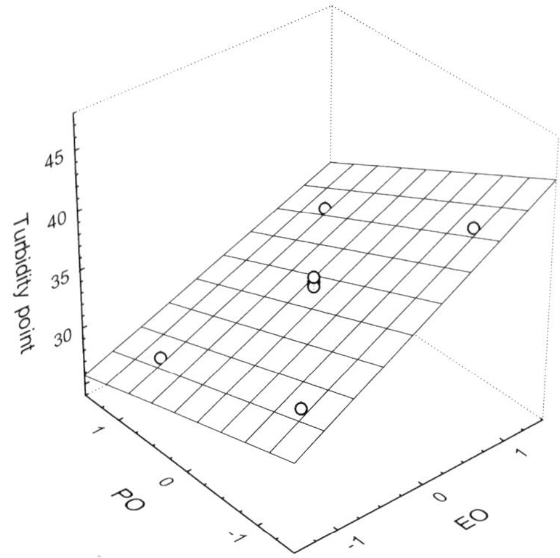

Fig. 3A.11. Response surface adjusted to the experimental data.

$$\text{Difference}: \Delta = \bar{y}_V - \bar{y}_C = 33.50 - 35.00 = -1.50.$$

To decide if this difference is significant, we need an estimate of its error. Using the procedure described in Section 2.5 to determine the variance of a linear combination, we can write

$$V(\bar{y}_V - \bar{y}_C) = V(\bar{y}_V) + V(\bar{y}_C) = \frac{s^2}{4} + \frac{s^2}{3} = \frac{7s^2}{12},$$

where s^2 is the variance of the responses at the center point. Substituting the value we obtained above and multiplying the result by t_2, we conclude that the limiting value for significance of this difference is 1.31, at the 95% confidence level. There is, then, evidence of a slight curvature in the response surface. Since the average response at the center point is larger than the average response at the vertices, the surface is slightly convex.

3A.8. Development of a detergent

The formulation of a detergent powder consists of a mixture of different ingredients, such as surfactants, polymers, whiteners and enzymes. One strategy to develop a more efficient product is to start with a standard formulation and vary in a controlled way the amounts of some additives. J.T. Bruns, a graduate chemical engineering student, decided to use a 2^3 factorial design to estimate how much the addition of three ingredients would affect the whitening power of the resulting detergent formulations. Each experiment was performed with the same amount of powder of the different formulations. The washings were performed in industrial simulators of laundry washers, using pieces of cloth that had been subjected to the same standard dirt solution, formulated to imitate typical domestic dirt. The results obtained are given in Table 3A.9, where the response is an optical density that measures the intensity of light reflected from the washed clothes. The objective of the experiment is to maximize this response. The ingredients and their levels are given in coded values, for industrial security reasons. The protocols of the industrial laboratory where the tests were carried out require that the experiments be performed in groups of 12 replicates. Some caution should be taken in interpreting these results, because the experiments were not run in completely random order.

One of the advantages of carrying out such a large number of experiments is that we can combine the variances (whose values are all very similar in this case) into a pooled estimate with many degrees of freedom (88, in this case), and thus obtain very precise estimates of the standard error of an effect. Combining the variance values according to the usual procedure, we obtain an estimate of 0.364 for the standard error of an effect. This means that we should only consider statistically significant, at the 95% confidence level, those effects whose absolute values are larger than $t_{88} \times 0.364 = 0.724$. With this criterion, only the main effects of additives **A** and **C** are significant, and therefore we can interpret the entire design as a two-level factorial in duplicate (Table 3A.10), with a model described by $\hat{y} = 78.94 + 0.65x_{\mathbf{A}} + 2.64x_{\mathbf{C}}$ (recall that the values of the model coefficients are always half of the effect values).

Table 3A.9
Effects of different formulations for detergent powder

Formulations[a]	**A**	**B**	**C**	Response[b]	Variance[c]
2	–	–	–	75.719	2.993
4	+	–	–	76.998	4.285
6	–	+	–	75.557	3.803
8	+	+	–	76.928	3.686
3	–	–	+	81.032	3.725
5	+	–	+	82.298	2.690
7	–	+	+	80.848	1.716
1	+	+	+	82.146	2.496

Effects

Average	**A**	**B**	**C**	**AB**	**AC**	**BC**	**ABC**
78.94	1.30	–0.14	5.28	0.03	–0.02	–0.03	–0.02

[a] Chronological order for the execution of each group of 12 replicate experiments.
[b] Average responses of the 12 replicate experiments.
[c] Variances of the 12 replicate experiments.

Table 3A.10
2^2 Factorial in duplicate embedded in the design of Table 3A.9

A	C	Average response
−	−	75.64
+	−	76.96
−	+	80.94
+	+	82.22

Since the objective of this experiment is to maximize the response, we can conclude that it is enough to add the **A** and **C** ingredients, at their higher levels, to the standard formulation. In terms of coded units, the **C** ingredient is four times more efficient than the **A** ingredient for increasing the reflected light intensity. As we will see in Chapter 6, we can use these results to determine a steepest ascent path along which we can run experiments that are expected to lead to even more satisfactory results. We must remember, however, that other important responses, like production cost, should also be taken into consideration.

Fig. 3A.12 shows the results of all 96 experiments as a function of the levels of the **A** and **C** additives. If such a large number of repetitions had not been made, it seems likely that the effect of the **A** additive would have gone unnoticed, in view of the relatively large variances observed for the replicate runs.

3A.9. A blocked design for producing earplugs

A blocked 2^4 factorial design was used to investigate the effects of four factors — the flow rates of solutions of polyol, pre-polymer, catalyst/extensor and reticulate — on the properties (apparent density, equilibrium force and retention time) of polyurethane earplugs manufactured on a small production line of a multinational company (Sato, 2002). In the manufacturing process, the four solutions are mixed simultaneously and their flow rates are expected to affect earplug properties. There was a suspicion that the quantity of water in one of the solutions also affected these properties and it was decided to add this factor to the study, without increasing the number of experiments. A blocked 2^4 full factorial was thus performed to study this process, instead of a 2^{5-1} fractional factorial design. High and low levels of the four flow rates were used to define the experiments as shown in Table 3A.11. Half of these were carried out using a standard polyol solution; the other half used the same solution mixed with 500 ppm water. The standard and the water-rich solutions were assigned to the experimental runs by confounding this factor with the **ABCD** four-factor interaction. As shown in the table, the standard

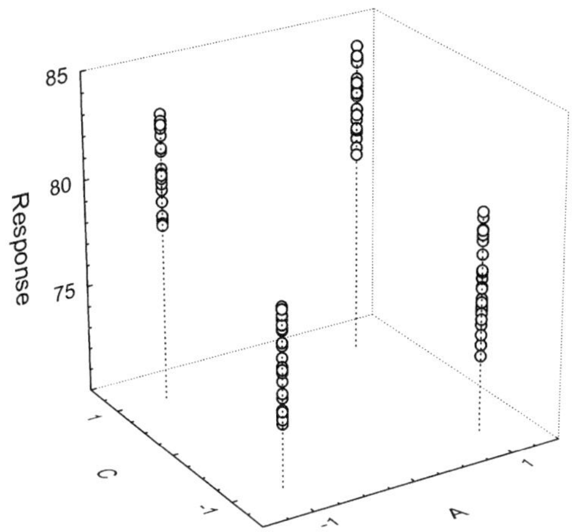

Fig. 3A.12. Response values as a function of the levels of the **A** and **C** additives.

solution was used in those experiments for which the signs of the factors **A–D** multiply to give a positive sign, i.e., block II. The other experiments were run with the water-rich solution.

Factor **D**, the flow rate of the reticulate solution, has a large negative effect on the apparent density of the earplugs. On average, changing this flow rate from 15 to 23 $g\,min^{-1}$ lowers the apparent density by $37.4 \times 10^{-3}\,g\,cm^{-3}$.

The second largest absolute effect value, $9.1 \times 10^{-3}\,g\,cm^{-3}$, occurs for the block effect. The researchers have the options of attributing this value to the water added to the polyol solution, to the **ABCD** interaction, or to a combination of both. Since four-factor effects are not commonly observed, the suspicion of the researchers seems confirmed: the quantity of water in the polyol solution has a distinct effect on the apparent density. This interpretation is consistent with the chemistry of the process, and the observation led to taking precautionary measures for buying, storing and manipulating polyol solutions, to improve the quality of the earplugs produced.

The other effect values are small. The normal probability plot shows that all of them fall on a straight line passing through the origin (Fig. 3A.13). These values can therefore be considered to come from a normal population of zero mean and can be attributed to experimental error, not to real effects.

As an exercise, the reader can analyze the effects of the four factors on the other responses, equilibrium force and retention time. The analysis for the equilibrium force is straightforward. The normal plot of the effect values for this response indicates that only the flow rate of the polyol solution is significant. For the retention time, in contrast, the normal plot of the effect values is more difficult to interpret. As will be seen in another

Table 3A.11
Apparent densities (AD), equilibrium forces (EF) and retention times (RT) for the blocked 2^4 factorial design for the production of protective earplugs

Factors		(-)	(+)
	A = Polyol solution flow rate ($g\,min^{-1}$)	446	454
	B = Pre-polymer solution flow rate ($g\,min^{-1}$)	318	324
	C = Catalyst/extensor solution flow rate ($g\,min^{-1}$)	32	36
	D = Reticulate solution flow rate ($g\,min^{-1}$)	15	23
Responses	AD = Apparent density ($g\,cm^{-3}$)		
	RT = Retention time (s)		
	EF = Equilibrium force (lb[a])		
Blocks	I = 500 ppm water added to standard solution		
	II = Standard solution		

Run	A	B	C	D	ABCD	Block	AD	EF	RT
1	–	–	–	–	+	II	0.170	0.212	2.35
2	+	–	–	–	–	I	0.156	0.136	5.05
3	–	+	–	–	–	I	0.163	0.187	2.23
4	+	+	–	–	+	II	0.174	0.249	1.71
5	–	–	+	–	–	I	0.161	0.167	3.98
6	+	–	+	–	+	II	0.172	0.187	2.62
7	–	+	+	–	+	II	0.175	0.237	1.73
8	+	+	+	–	–	I	0.162	0.170	3.17
9	–	–	–	+	–	I	0.122	0.042	41.26
10	+	–	–	+	+	II	0.135	0.045	43.06
11	–	+	–	+	+	II	0.130	0.063	20.38
12	+	+	–	+	–	I	0.126	0.046	36.16
13	–	–	+	+	+	II	0.130	0.045	40.72
14	+	–	+	+	–	I	0.130	0.034	104.29
15	–	+	+	+	–	I	0.127	0.043	44.17
16	+	+	+	+	+	II	0.134	0.054	27.21

Effect values for the AD response (units of $10^{-3}\,g\,cm^{-3}$)

A	**B**	**C**	**D**	**AB**	**AC**
1.4	1.9	1.9	–37.4	–1.1	–0.1
BC	**AD**	**BD**	**CD**	**ABC**	**ABD**
–0.6	2.6	–1.9	0.1	–3.1	–1.4
ACD	**BCD**	**ABCD+block effect**			
–0.4	1.1	9.1			

[a] 1 lb = 4.448 N.

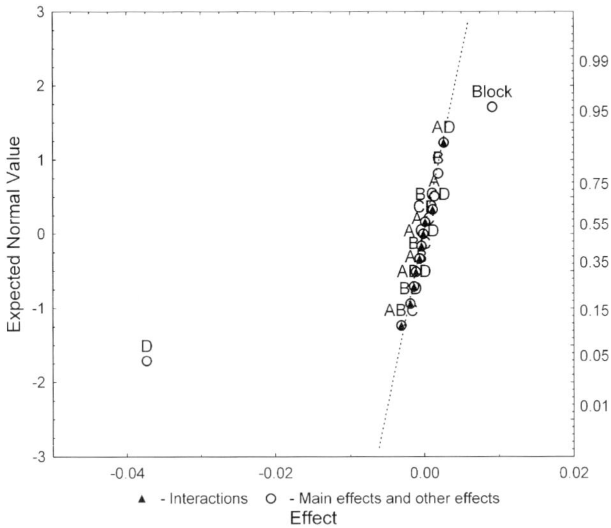

Fig. 3A.13. Normal probability plot of the blocked 2^4 factorial effect values.

case study at the end of Chapter 6, retention time has a complicated behavior, which is difficult to describe even when curvature effects are include in the model.

CHAPTER 4

When there are many variables

The number of experimental runs needed to perform a complete 2^k factorial design increases rapidly with k, the number of factors investigated. With seven factors, a full factorial requires no less than $2^7 = 128$ runs. Fortunately, in such a case, the desired information can be obtained from a much smaller number of runs, corresponding to a fraction of the number of runs of a complete factorial. This is possible for two reasons.

First, the number of higher-order interactions increases dramatically with the number of factors, as demonstrated in Table 4.1. In a majority of cases, these interactions have small values and consequently no practical importance. As in the series expansion of a function, the main effects (i.e., first-order ones) tend to be larger than the two-factor interactions (second order), which in turn are more important than the three-factor interactions, and so forth. This tendency becomes even more pronounced for effects of higher order. If these effects are not significant, it is not worthwhile to perform all the runs of a complete factorial design to estimate their values.

Second, as the number of factors increases, the chances also increase that one or more of them will not affect the response significantly, through main or interaction effects. Again, if the effects of these variables need not be determined, why perform all the experiments of a complete factorial?

On the other hand, in many experiments we cannot predict which variables will significantly affect the response. To avoid running the risk of excluding factors that could prove important, we must study, in the initial stage of an experiment, the largest number of possibly relevant factors. Using fractional factorial designs, instead of complete ones, we can do this without increasing the number of runs (Box et al., 1978, Chapter 12). In this chapter, we will see how to apply this strategy, taking

Table 4.1
The number of main and interaction effects, as a function of the number of factors, k. The order of an interaction is equal to the number of factors involved in its definition

	Order						
k	1st	2nd	3rd	4th	5th	6th	7th
3	3	3	1	–	–	–	–
4	4	6	4	1	–	–	–
5	5	10	10	5	1	–	–
6	6	15	20	15	6	1	–
7	7	21	35	35	21	7	1

as a first example an investigation from a research laboratory (Eiras, 1991; Andrade et al., 1991). After that, we will discuss a fractional factorial study performed outside the chemical laboratory to show how, with a minimum number of runs, we can screen all variables and distinguish the important ones from the rest.

4.1. Half-fractions of factorial designs

In the project we discuss in this section, the research workers were trying to optimize an analytical procedure for the determination of molybdenum traces in plants. The chosen method was based on the catalytic action of the Mo(VI) species on the oxidation of the I^- ion by H_2O_2, carried out in a monosegmented continuous flow system. Of all the factors considered important for the production of the analytical signal, four were selected for a preliminary study: the concentrations of H_2O_2, H_2SO_4 and KI, and the reaction time of these chemical species in the presence of Mo(VI). The influence of these factors on the intensity of the analytical signal was investigated with a 2^4 full factorial design. The design matrix for the 16 runs is given in Table 4.2, as well as the observed signal intensities. The effect values calculated from these results are presented in Table 4.3. From the normal plot of these values (Fig. 4.1), we can see that the main effects of [KI], [H_2O_2] and time and the interactions of [KI] with [H_2O_2] and [KI] with time appear significant. The sulfuric acid concentration does not seem to be involved in any important effect.

To execute the complete factorial, we needed to perform 16 runs. Let us imagine that, for economic reasons, the research workers had decided to perform only eight of them, and had chosen those indicated by a check mark in the design matrix of Table 4.2. In that case, they would have

Table 4.2
Results of the 2^4 complete factorial design performed to study the catalytic action of Mo(VI)

Factors	(–)	(+)
1 $[H_2SO_4]$ (mol L^{-1})	0.16	0.32
2 [KI] (mol L^{-1})	0.015	0.030
3 $[H_2O_2]$ (mol L^{-1})	0.0020	0.0040
4 Time (s)	90	130

	Run	1	2	3	4	Response[a]
√	1	–	–	–	–	52
	2	+	–	–	–	61
	3	–	+	–	–	124
√	4	+	+	–	–	113
	5	–	–	+	–	85
√	6	+	–	+	–	66
√	7	–	+	+	–	185
	8	+	+	+	–	192
	9	–	–	–	+	98
√	10	+	–	–	+	86
√	11	–	+	–	+	201
	12	+	+	–	+	194
√	13	–	–	+	+	122
	14	+	–	+	+	139
	15	–	+	+	+	289
√	16	+	+	+	+	286

[a] Analytical signal intensity × 1000.

obtained only the eight responses reproduced in the last column of Table 4.4.

Exercise 4.1. Use the data in Table 4.2 to confirm that the significant effect values in this design are really the same as those appearing in Table 4.3.

Multiplying the appropriate pairs of columns in the design matrix, we obtain the signs needed for calculating the six interaction effects involving two factors. Applying them to the response values, we arrive at the values also shown in Table 4.4, together with the four main effects and the overall average. All these values (except the average, of course) are contrasts between two halves of the set of eight responses. To calculate any one of them we use four response values with positive sign and four with negative sign. However, since the calculations no longer refer to a

Table 4.3
Analysis of the 2^4 factorial design to study the catalytic response of Mo(VI). The most significant values are italicized

Average = 143.31	**12** = −1.13
1 = −2.38	**13** = 2.88
2 = *109.38*	**14** = 1.13
3 = *54.38*	***23*** = *25.63*
4 = *67.13*	***24*** = *21.88*
	34 = 9.88
123 = 2.63	
124 = −2.63	**1234** = −8.88
134 = 5.38	
234 = 0.13	

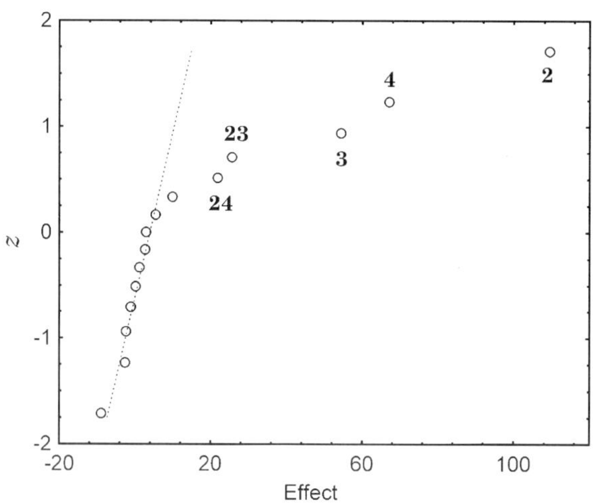

Fig. 4.1. Normal plot of the calculated effects for the 2^4 full factorial design used to study the catalytic action of Mo(VI).

complete design but to a fraction of it, we use the letter l to represent these contrasts and so distinguish them from the effects discussed in the last chapter.

Comparing the values of the effects obtained from the complete design (Table 4.3) with the contrasts calculated using only the half-fraction (Table 4.4), we see that the estimates of the average and the main effects are pretty much the same in both cases. This means that we can make accurate estimates of the average and the main effects using only eight runs, i.e., with half the effort needed to execute the full factorial.

Table 4.4
Analysis of a half-fraction of the 2^4 complete factorial design: a 2^{4-1} fractional factorial design. Example of the catalytic response of Mo(VI). Note that the order of the runs is not the same as in Table 4.2

Factors	(–)	(+)
1 $[H_2SO_4]$ (mol L^{-1})	0.16	0.32
2 [KI] (mol L^{-1})	0.015	0.030
3 $[H_2O_2]$ (mol L^{-1})	0.0020	0.0040
4 Time (s)	90	130

Run	I	1	2	3	4	12	13	14	23	24	34	Response
1	+	–	–	–	–	+	+	+	+	+	+	52
10	+	+	–	–	+	–	–	+	+	–	–	86
11	+	–	+	–	+	–	+	–	–	+	–	201
4	+	+	+	–	–	+	–	–	–	–	+	113
13	+	–	–	+	+	+	–	–	–	–	+	122
6	+	+	–	+	–	–	+	–	–	+	–	66
7	+	–	+	+	–	–	–	+	+	–	–	185
16	+	+	+	+	+	+	+	+	+	+	+	286

Average = 138.87	$l_{12} = 8.75$
$l_1 = -2.25$	$l_{13} = 24.75$
$l_2 = 114.75$	$l_{14} = 26.75$
$l_3 = 51.75$	$l_{23} = 26.75$
$l_4 = 69.75$	$l_{24} = 24.75$
	$l_{34} = 8.75$

The values of the significant interactions in the complete design (**23** and **24**) are also in good agreement with the values of the l_{23} and l_{24} contrasts calculated from the half-fraction. On the other hand, the interactions involving factor **1** are very overestimated by the l_{12}, l_{13} and l_{14} values. But what did we expect? It would be wonderful if we could reduce our work in half without having to pay a price. Unfortunately, things do not work this way.

The explanation is this. With the eight runs of the half-fraction we can estimate only eight independent quantities. After calculating the grand average and the four main effects, we are left with only 3 degrees of freedom. Obviously, this number is insufficient to furnish independent estimates of all six two-factor interactions. In Table 4.4, one can see that actually $l_{12} = l_{34}$, $l_{13} = l_{24}$ and $l_{14} = l_{23}$. This happens because the columns of signs for the **12**, **13** and **14** interactions are identical to the columns for the **34**, **24** and **23** interactions, respectively.

In spite of this, the situation can be salvaged. If we assume that the interactions involving factor **1** are not important (which seems reasonable, since its main effect is negligible), then we can conclude that $l_{12} \cong l_{13} \cong l_{14} \cong 0$. These new values, together with $l_{23} = 26.75$, $l_{24} = 24.75$ and $l_{34} = 8.75$, are very good estimates of the two-factor interactions calculated using all 16 runs of the 2^4 factorial.

The design with eight runs shown in Table 4.4 is a **half-fraction** of the full factorial of Table 4.2. It is customary to denote it by 2^{4-1}, that is, half of 2^4:

$$\frac{2^4}{2} = 2^{-1}2^4 = 2^{4-1} = 8.$$

This notation indicates that the design has four factors, each at two levels, but we perform only eight runs. The presence of the –1 superscript means the full factorial was divided by 2. If it had been divided into four parts, the exponent would be 4–2.

Exercise 4.2. Use the signs of Table 4.4 to calculate the contrasts corresponding to the **134** and **1234** interactions. Since we know that the degrees of freedom are not sufficient for this, these values will not be independent of the values already calculated. With what other contrasts are they confounded? Do you think it makes sense to interpret these values as really being estimates of the **134** and **1234** effects?

4.1.1. How to construct a half-fraction

We have presented the 2^{4-1} fractional factorial design of Table 4.4 as a mysteriously chosen subset of the complete design. Actually it was constructed in the following way:

1. First we constructed a complete 2^3 factorial in factors **1**, **2** and **3**.
2. Then we attributed to factor **4** the signs of the product of columns **1**, **2** and **3**.

The first consequence of this procedure is that the contrasts l_{123} and l_4 become identical, because they are calculated using the same signs. This is similar to the result obtained by blocking in Chapter 3. In addition, since other sign relations are used to define the two-factor interactions as well as the higher-order interactions (Exercise 4.2), we should suspect that other identities involving the contrasts exist. After all, the half-fraction only has 8 degrees of freedom, whereas the complete factorial, with all its effects, has 16. We could find the missing relations by taking all possible column products, but this is not necessary. We can discover them more easily by means of a very simple algebra.

We shall represent the columns of signs by boldface numbers (or products of numbers). Thus, for example, **123** indicates the column of signs obtained by multiplying the columns corresponding to the first three factors. This column, as we have seen, was defined as identical to that of factor **4**. We can thus write

$$\mathbf{4} = \mathbf{123}. \tag{4.1}$$

To find the relations between the various contrasts, we will employ two properties of the multiplication of columns of signs. The first is that multiplying a column of signs by itself, that is, elevating all the elements to their square powers, always produces a column containing only positive signs. This new column, when applied to another column, leaves it unchanged. It is, in other words, the identity element of our algebra, and for this reason we use the letter **I** to represent it. Thus, for example, we can write

$$\mathbf{11} = \mathbf{22} = \mathbf{33} = \mathbf{44} = \mathbf{I}.$$

The second property simply recognizes that the multiplication of columns of signs is commutative and associative. For example

$$\mathbf{123} = (\mathbf{1})(\mathbf{23}) = (\mathbf{23})(\mathbf{1}) = (\mathbf{12})(\mathbf{3}) = \mathbf{321} = (\mathbf{2})(\mathbf{31}) = \cdots$$

To obtain the relations between the various contrasts, we multiply the expression defining the fractional design, Eq. (4.1), by some product of columns and apply the properties we have just presented. When we want to know the equivalent of a certain contrast, we only need to make algebraic operations until it appears by itself on one of the sides of the rearranged equation.

For example, let us try to find the contrast that has the same signs as l_2. Examining Eq. (4.1), we see that it is possible to isolate factor **2** on the right-hand side by multiplying **123** by the **13** product, because this will transform **1** and **3** into identities. Of course, we also have to multiply the left-hand side by **13**, so that the equality remains valid:

$$(\mathbf{13})(\mathbf{4}) = (\mathbf{13})(\mathbf{123}) = (\mathbf{11})(\mathbf{33})(\mathbf{2}) = (\mathbf{I})(\mathbf{I})(\mathbf{2}) = \mathbf{2}.$$

On the left side of the equation we now have the **134** product, and from this we conclude that $l_{134} = l_2$. It is the same conclusion we can reach, though more laboriously, by doing Exercise 4.2. In statistical terminology, we say that the use of the half-fraction confounds the main effect, **2**, with the **134** interaction effect. The value of the calculated contrast, l_2 (or l_{134}), is actually an estimate of the sum of the two effects. You can confirm that this is true, by adding the values of the **2** and **134** effects in Table 4.3 and comparing the result with the value of l_2 in Table 4.4.

To show that the calculated contrast confounds the two effects and in fact estimates their sum, it is customary to use the notation

$$l_2 \to \mathbf{2} + \mathbf{134}.$$

Table 4.5
Relations between the contrasts of the 2^{4-1} half-fraction and the effects of the 2^4 full factorial. **M** is the average of all the response values

Relations between columns of signs	Contrasts of the 2^{4-1} half-fraction in terms of the effects of 2^4 factorial
$\mathbf{1} = \mathbf{234}$	$l_1 = l_{234} \rightarrow \mathbf{1} + \mathbf{234}$
$\mathbf{2} = \mathbf{134}$	$l_2 = l_{134} \rightarrow \mathbf{2} + \mathbf{134}$
$\mathbf{3} = \mathbf{124}$	$l_3 = l_{124} \rightarrow \mathbf{3} + \mathbf{124}$
$\mathbf{4} = \mathbf{123}$	$l_4 = l_{123} \rightarrow \mathbf{4} + \mathbf{123}$
$\mathbf{12} = \mathbf{34}$	$l_{12} = l_{34} \rightarrow \mathbf{12} + \mathbf{34}$
$\mathbf{13} = \mathbf{24}$	$l_{13} = l_{24} \rightarrow \mathbf{13} + \mathbf{24}$
$\mathbf{14} = \mathbf{23}$	$l_{14} = l_{23} \rightarrow \mathbf{14} + \mathbf{23}$
$\mathbf{I} = \mathbf{1234}$	$l_I \rightarrow \mathbf{M} + \frac{1}{2}(\mathbf{1234})$

The relations between the contrasts calculated from the 2^{4-1} half-fraction and the effects obtained from the 2^4 full factorial (the so-called **confounding patterns**) are shown in the second column of Table 4.5.

Exercise 4.3. How many experimental runs are there in a 2^{8-4} fractional factorial?

Exercise 4.4. Write the full algebraic expressions for calculating the effects **2** and **134** for the 2^4 factorial and show that the l_2 contrast calculated from the half-fraction really corresponds to the sum of these effects.

Exercise 4.5. All contrasts in Table 4.5 represent the sum of two effects except l_I, which estimates the overall average plus half of the **1234** interaction. Why half?

4.1.2. Generators of fractional factorial designs

Our half-fraction was obtained by setting $\mathbf{4} = \mathbf{123}$. The literature usually presents this relation in an equivalent form,

$$\mathbf{I} = \mathbf{1234}, \qquad (4.2)$$

obtained by multiplying both sides of Eq. (4.1) by **4**. In this form, with the identity element appearing alone on one side of the equation, the expression is known as the generating (or defining) relation of the fractional factorial. The **1234** interaction is called the **generator** of this fraction. The **generating relation** is sufficient to define the entire half-fraction, because it implicitly determines all the signs of the design matrix. Each possible fraction of a complete design is defined by a certain number of generating relations.

Consider now the rest of the runs in Table 4.2, those that are not checked and therefore were not used in the calculations of the fractional factorial. These eight runs also constitute a half-fraction of the complete design. You can easily confirm that the signs of factor **4** in these runs are the negatives of the signs of the **123** product. We can say then that this second fraction is originated by

$$\mathbf{4} = -\mathbf{123}, \tag{4.3}$$

or, equivalently, that its generating relation is

$$\mathbf{I} = -\mathbf{1234}. \tag{4.4}$$

By arguing along the same lines of Exercise 4.4, we can show that the contrasts are now estimates of the difference between two effects of the complete factorial. For example, now we have

$$l_2^* \rightarrow \mathbf{2} - \mathbf{134}.$$

The relations between the new contrasts (identified by asterisks) and the effects of the complete design are the same as those in the second column of Table 4.5, except that they have negative signs instead of positive ones.

We could also use the second half-fraction, which is called the alternate or complementary of the first half-fraction, to estimate the effects of the factorial. The final results would lead to the same conclusions. With the appropriate values and signs, we would calculate, for example,

$$l_2^* = 104.00,$$

which is also in good agreement with the value of effect **2** obtained from the complete factorial, 109.38.

Exercise 4.6. Use the runs of the complementary half-fraction in Table 4.2 to calculate the values of the l_1^*, l_3^* and l_4^* contrasts. Compare the results with the values given in Table 4.4 and also with the main effects calculated using the complete factorial.

If we join the two half-fractions, we will restore the original complete factorial. Taking a combination of the appropriate contrasts, we can recover the effect values without any confounding. For example, l_2 and l_2^* involve the same pair of effects, **2** and **134**. Their sum results in

$$l_2 + l_2^* = (\mathbf{2} + \mathbf{134}) + (\mathbf{2} - \mathbf{134}) = \mathbf{2} + \mathbf{2}.$$

The value of the main effect will therefore be half of the sum of the two contrasts:

$$\mathbf{2} = \frac{l_2 + l_2^*}{2} = \frac{114.75 + 104.00}{2} = 109.38.$$

In the same way, the **134** interaction will be given by half of the difference between l_2 and l_2^*:

$$\mathbf{134} = \frac{l_2 - l_2^*}{2} = \frac{114.75 - 104.00}{2} = 5.38.$$

Exercise 4.7. How would you combine the values of the contrasts to obtain the **1234** interaction effect? Do the corresponding calculations and compare the result with the value given in Table 4.3.

4.2. The concept of resolution

4.2.1. Resolution IV fractional factorial designs

The 2^{4-1} fractional design has an important characteristic. Its contrasts do not mix main effects with interaction effects involving two factors, but with interaction effects involving three factors, which in principle should be less significant. If these interaction effects are really negligible, the contrasts will furnish excellent approximations to the main effects calculated from the full factorial. We should have, for example, $l_2 \cong l_2^* \cong$ **2**. In general, we expect that $l_j \cong l_j^* \cong$ **J**, where **J** stands for any main effect.

As we should suspect already, something must be paid for not having to perform half of the experiments of the full design: the contrasts l_{ij} (or l_{ij}^*) are combinations of pairs of two-level interactions, and their interpretation becomes more difficult. Consider, for example, the value of l_{14} in Table 4.4, 26.75. According to the confounding patterns (Table 4.5), this value corresponds to the sum of interactions **14** and **23**. Should it be attributed mainly to **14**, to **23**, or divided in some way between the two interactions?

Strictly speaking, with only the results in Table 4.4, we have no way of knowing. However, since these same results indicate that factor **1** (the concentration of H_2SO_4) does not have a significant main effect, in contrast to factors **2** ([KI]) and **3** ([H_2O_2]), it is not unreasonable to suppose that the **23** interaction should be, in principle, more important than the **14** interaction. Consequently, the value of the l_{23} contrast (or l_{14}, which is the same) should be a good approximation to the interaction between factors [KI] and [H_2O_2].

This hypothesis is admittedly risky, but in this specific investigation it turns out to be valid. Using all runs of the full factorial, we find that **23** = 25.63 whereas the **14** value is only 1.13. Unfortunately, this does not always work. If only the runs corresponding to a half-fraction had been performed for a given design, and if it were necessary to distinguish between two binary interaction effects, we would have to execute the other eight runs to complete the factorial. Only thus could we estimate the effects separately.[33]

[33]In compensation, in designs used for screening a large number of variables we can live with confounded effects. In later sections we will see examples of such designs.

The 2^{4-1} design is an example of a fractional factorial design of resolution IV. In a factorial of resolution IV the main effects do not mix with two-factor interactions but these, in turn, are confounded with each other. The notation employed to represent the resolution of a design is a subscript written in Roman numerals. Here, for example, we write 2^{4-1}_{IV}.

The generating relations of a design determine its resolution. The number of factors composing the shortest term of these relations is, by definition, the **resolution** of the design. To define a half-fraction, one generating relation suffices. In our example this relation ($\mathbf{I} = \pm \mathbf{1234}$) contains four factors, and for this reason the resolution of the 2^{4-1} factorial is 4. In practice, most fractional designs have resolutions between 3 and 5.

Exercise 4.8. The results presented below were obtained in an investigation aimed at optimizing the yield of an organic reaction (R.S. Correa Neto, S.M.G. de Barros, M. Penso and D. Landini, *Optimization of an Alkylation under PTC Conditions*, 5th Brazilian Meeting on Organic Synthesis, Campinas, 1992).

Factors	(–)	(+)
1: Temperature	Room	Reflux
2: Base	K_2CO_3/NaOH	K_2CO_3
3: Solvent	CH_2Cl_2	CH_3CN
4: Catalyst	None	TEBA

Run	1	2	3	4	Yield (%)
1	–	–	–	–	0
2	+	–	–	+	70
3	–	+	–	+	65
4	+	+	–	–	0
5	–	–	+	+	100
6	+	–	+	–	85
7	–	+	+	–	50
8	+	+	+	+	95

What conclusions can you draw from these results?

4.2.2. Resolution V fractional factorial designs

We must confess that five factors were actually investigated in the Mo(VI) analytical signal study. The flow rate of the monosegmented system was also included in the design, in addition to the four factors already discussed. The 16 runs in Table 4.2 do not really correspond to a full 2^4 design but rather to a half-fraction of a 2^5 design, which is presented in Table 4.6. This fraction was constructed by making $\mathbf{5} = \mathbf{1234}$ or, what is the same thing, from the generating relation

$$\mathbf{I} = \mathbf{12345}. \tag{4.5}$$

Thus, it is a half-fraction of resolution V, which we denote by 2_V^{5-1}. The main effects are mixed only with four-factor interactions, while the two-factor interactions are mixed with three-factor ones. The full set of relations between the contrasts of this fractional factorial and the effects of the complete design are given in Table 4.7, together with the values of the contrasts calculated from the data in Table 4.6.

Assuming that the interaction effects involving three or more factors are negligible, we can conclude that only the main effects, **2**, **3** and **4** ([KI], $[H_2O_2]$ and t, respectively) and the **23** and **24** interactions are significant. (Perhaps the **5** and **34** effects are also significant, but in any case their values are much lower than those of the more important effects.)

Exercise 4.9. In a half-fractional factorial design of resolution VI, the main effects are confounded with what other effects? And the two-factor interactions?

Exercise 4.10. Explain why the results in runs 1 (+++++) and 9 (−+++−) in Table 4.6 are almost the same.

4.2.3. Inert variables and factorials embedded in fractions

Examining the values in Table 4.7, we conclude that all the contrasts involving factors **1** and **5** (sulfuric acid concentration and flow rate) are apparently negligible. In practice, this means that these variables do not affect the intensity of the analytical signal, at least at the levels studied in these experiments. They are **inert** variables and need not be considered further in this study. We may therefore remove the corresponding columns from Table 4.6, which leaves us with a complete 2^3 factorial in duplicate in the other three variables. This situation, which you can easily verify by using two pencils to hide columns **1** and **5** in Table 4.6, is illustrated in Fig. 4.2.

In the cube depicted in Fig. 4.2, the response varies much more between one vertex and another than within the same vertex. The largest values occur when all three factors are at their higher levels, that is, for the sign combination $(\mathbf{2\,3\,4}) = (+\,+\,+)$. Since the objective of the experiment is to increase the sensitivity of the analytical method, we can conclude that it would be advisable to carry out more experiments in the region around this vertex.

The appearance of embedded full factorials as a result of the inertness of one or more of the variables studied may occur in any factorial design. Fig. 4.3 illustrates the reason, using as an example the 2_{III}^{3-1} fractional factorial design with signs defined by the relation $\mathbf{3} = \mathbf{12}$. If we eliminate

Table 4.6
A 2_{V}^{5-1} fractional factorial design for the study of the catalytic response of Mo(VI)

Factor	Level	
	(−)	(+)
1 $[H_2SO_4]$ (mol L^{-1})	0.16	0.32
2 [KI] (mol L^{-1})	0.015	0.030
3 $[H_2O_2]$ (mol L^{-1})	0.0020	0.0040
4 Time (s)	90	130
5 Flow rate (mL min^{-1})	1.2	3.0

Order	**1** $[H_2SO_4]$	**2** [KI]	**3** $[H_2O_2]$	**4** Time	**5** Flow rate	Response
5	−	−	−	−	+	52
13	+	−	−	−	−	61
12	−	+	−	−	−	124
4	+	+	−	−	+	113
11	−	−	+	−	−	85
3	+	−	+	−	+	66
6	−	+	+	−	+	185
14	+	+	+	−	−	192
10	−	−	−	+	−	98
2	+	−	−	+	+	86
7	−	+	−	+	+	201
15	+	+	−	+	−	194
8	−	−	+	+	+	122
16	+	−	+	+	−	139
9	−	+	+	+	−	289
1	+	+	+	+	+	286

factor **3** from this design, we have a 2^2 complete factorial in the **1** and **2** variables. Geometrically, when we eliminate variable **3** we are removing its axis from the figure. The cube then collapses to a square situated in the plane defined by factors **1** and **2**, that is, the fractional factorial is projected on the **12** plane. It is easy to see that the same holds for the other axes. Eliminating any one variable leaves us with a 2^2 complete design in the remaining variables.

The two half-fractions of the 2^4 and 2^5 factorial designs are presented in Table 4.8. If we eliminate any variable from the 2_{IV}^{4-1} fractional design, we are left with a 2^3 full factorial in the other variables. If we eliminate two variables, we have two replicates of a 2^2 factorial in the other two variables. In the Mo(VI) example, we have already seen what happens

Table 4.7
Estimates of the contrasts of the 2_V^{5-1} fractional factorial design and their relations with the effects of the complete design. The values in boldface are the most significant

Relations between the columns	Contrasts	Estimates
1 = 2345	l_1 → **1+2345**	$l_1 = -2.38$
2 = 1345	l_2 → **2+1345**	$l_2 =$ **109.38**
3 = 1245	l_3 → **3+1245**	$l_3 =$ **54.38**
4 = 1235	l_4 → **4+1235**	$l_4 =$ **67.13**
5 = 1234	l_5 → **5+1234**	$l_5 = -8.88$
12 = 345	l_{12} → **12+345**	$L_{12} = -1.13$
13 = 245	l_{13} → **13+245**	$l_{13} = 2.88$
14 = 235	l_{14} → **14+235**	$l_{14} = 1.13$
15 = 234	l_{15} → **15+234**	$l_{15} = 0.13$
23 = 145	l_{23} → **23+145**	$l_{23} =$ **25.63**
24 = 135	l_{24} → **24+135**	$l_{24} =$ **21.88**
25 = 134	l_{25} → **25+134**	$l_{25} = 5.38$
34 = 125	l_{34} → **34+125**	$l_{34} = 9.88$
35 = 124	l_{35} → **35+124**	$l_{35} = -2.63$
45 = 123	l_{45} → **45+123**	$l_{45} = 2.63$
I = 12345	l_I → Average + $\frac{1}{2}$(**12345**)	$l_I =$ **143.31**

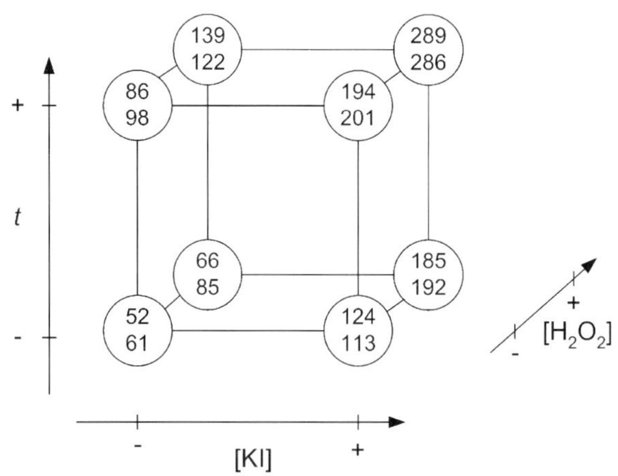

Fig. 4.2. 2^3 Factorial design in duplicate obtained from the 2^{5-1} fractional upon removal of the $[H_2SO_4]$ and flow rate variables.

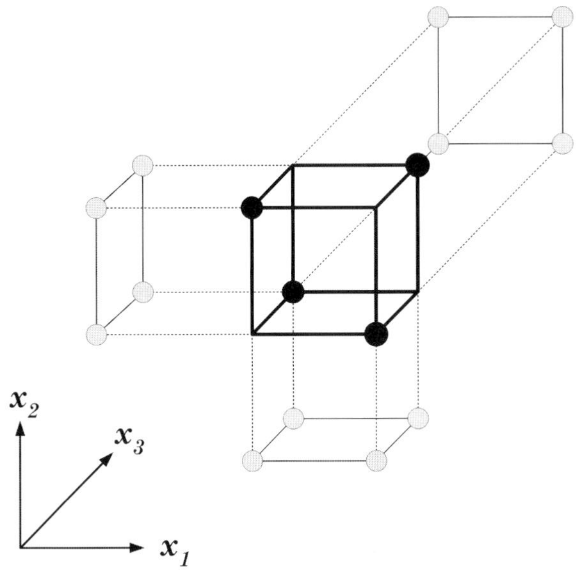

Fig. 4.3. Geometrical representation of the three complete 2^2 factorials embedded in a 2^{3-1} fraction.

when two variables are excluded from a 2_{V}^{5-1} fractional. What would happen if we eliminated three variables?

4.2.4. Half-fractions of maximum resolution

To construct the half-fractions discussed so far, we used the interaction effect of highest order to determine the column signs for one of the factors. In the first example, we started with a 2^3 factorial and used the **123** interaction to define the levels of the fourth variable, using $\mathbf{I} = \mathbf{1234}$. This resulted in a fractional factorial of resolution IV.

In the second example, we started with a 2^4 factorial and, by means of the generating relation $\mathbf{I} = \mathbf{12345}$, arrived at a resolution V half-fraction. This procedure, always using the highest order interaction of the starting factorial, is the most recommended, although we are free to choose any interaction to generate the fractional factorial. For example, we could define the signs of variable **5** in the 2^{5-1} half-fraction by the relation $\mathbf{5} = \mathbf{123}$. In this case the generating relation becomes $\mathbf{I} = \mathbf{1235}$, and consequently the resolution is reduced to 4. Since the half-fractions we have studied until now were constructed using the highest possible interaction, they also have the highest possible resolution, for the number of factors considered. For this reason, they are in principle better than the possible alternatives.

Table 4.8
2_{IV}^{4-1} and 2_V^{5-1} half-fractions

2_{IV}^{4-1}

I = 1234				**I = −1234**			
1	**2**	**3**	**4**	**1**	**2**	**3**	**4**
−	−	−	−	−	−	−	+
+	−	−	+	+	−	−	−
−	+	−	+	−	+	−	−
+	+	−	−	+	+	−	+
−	−	+	+	−	−	+	−
+	−	+	−	+	−	+	+
−	+	+	−	−	+	+	+
+	+	+	+	+	+	+	−

2_V^{5-1}

I = 12345					**I = −12345**				
1	**2**	**3**	**4**	**5**	**1**	**2**	**3**	**4**	**5**
−	−	−	−	+	−	−	−	−	−
+	−	−	−	−	+	−	−	−	+
−	+	−	−	−	−	+	−	−	+
+	+	−	−	+	+	+	−	−	−
−	−	+	−	−	−	−	+	−	+
+	−	+	−	+	+	−	+	−	−
−	+	+	−	+	−	+	+	−	−
+	+	+	−	−	+	+	+	−	+
−	−	−	+	−	−	−	−	+	+
+	−	−	+	+	+	−	−	+	−
−	+	−	+	+	−	+	−	+	−
+	+	−	+	−	+	+	−	+	+
−	−	+	+	+	−	−	+	+	−
+	−	+	+	−	+	−	+	+	+
−	+	+	+	−	−	+	+	+	+
+	+	+	+	+	+	+	+	+	−

In general, to construct a 2^{k-1} fractional factorial of maximum resolution, we must do the following:

1. Write a complete factorial design for k–1 variables.
2. Attribute to the remaining variable the signs of the **123**…**(*k*–1)** interaction or, alternatively, signs that are exactly opposite to them.

Exercise 4.11. Construct a 2^{5-1} fractional design using **5** = **124**. Determine, for this fraction, the relations between the contrasts involving one or two factors and the effects calculated from a complete factorial. Can you imagine a situation in which this design would be preferable, instead of a fraction of maximum resolution?

4.3. Screening variables

4.3.1. Resolution III fractional factorial designs

In the above example we concluded that only three, of the five starting variables, significantly affect the intensity of the analytical signal. Since a flow analytical system is relatively complex, it would have been difficult to predict which variables would have the most important effects on the response. Actually, in this study, the research workers were not even sure whether the levels chosen for each variable were the most appropriate for an initial investigation. In situations like this, where we do not know much about the behavior of the system under study, it is better to start performing only a fraction of the total number of runs of a complete factorial. Later, after analyzing the results of this fraction, we can decide whether it is worthwhile to execute the remaining runs. If the results indicate that only a few variables are important, we can introduce new variables, or change the levels of the ones already studied, in an effort to obtain better response values. On the other hand, if the initial results are found promising, we can carry out more runs in that region to complete the factorial or try to adjust a response surface to the results (a subject that will be treated in Chapter 6).

When we wish to investigate the effects of many variables, it is not a good policy to perform a complete factorial at the beginning. It is better to use a fractional factorial design to carry out a **screening**, that is, trying to separate the factors that deserve a more detailed study from those that are negligible. The factorial design can always be completed later, if the results indicate that this is worthwhile.

Until now, we have treated only half-fractions, for which we performed half of the runs required for a complete factorial. Depending on the number of factors, even this fraction could be too large. If this were the case, our next step would be to consider fractional factorial designs containing only a fourth of the total number of runs. With five variables, as in the Mo(VI) example, the design would have only eight runs and would correspond to a 2^{5-2} fractional factorial, that is, a one-quarter fraction. To construct its design matrix, we start with a 2^3 factorial based on any three of the five factors. Then we will need two generating relations to define the levels of the remaining two variables.

To construct the design shown in Table 4.9, we start with **4** = **123** and **5** = **12**, the equivalent of making **I** = **1234** and **125**. These in turn lead to a third generating relation (see Exercise 4.12). Since the shortest term in these three relations contains three factors, the design has resolution III, and its complete notation is 2^{5-2}_{III}. Clearly, this kind of design is more economical. On the other hand, it produces contrasts that confound main effects with two-factor interactions. This complicates the analysis of the results, but it is possible that some of these contrasts have values small enough to justify eliminating their corresponding variables from the design. Since this is always a possibility at the beginning of an investigation, performing a variable screening with a low-resolution fractional is not necessarily the bad idea that it might appear at first sight. The worst that can happen, in an unfavorable case, is having to perform the remaining runs to obtain a fraction of higher resolution or even the complete factorial.

Table 4.9
Analysis of a 2^{5-2}_{III} fractional factorial with partially simulated results, for the study of the catalytic response of Mo(VI)

Factor	Level	
	(−)	(+)
1 [H_2SO_4] (mol L^{-1})	0.16	0.32
2 [KI] (mol L^{-1})	0.015	0.030
3 [H_2O_2] (mol L^{-1})	0.0020	0.0040
4 Time (s)	90	130
5 Flow rate (mL min^{-1})	1.2	3.0

1 [H_2SO_4]	**2** [KI]	**3** [H_2O_2]	**4** Time	**5** Flow rate	Signal (× 1000)
−	−	−	−	+	52
+	−	−	+	−	92[a]
−	+	−	+	−	198[a]
+	+	−	−	+	113
−	−	+	+	+	122
+	−	+	−	−	76[a]
−	+	+	−	−	189[a]
+	+	+	+	+	286
Average = 141.00		$l_3 = 54.50$			
$l_1 = 1.50$		$l_4 = 67.00$			
$l_2 = 111.00$		$l_5 = 4.50$			

[a] Simulated results, based on the responses presented in Table 4.6. For example, result 92 for the (+ − − + −) run was obtained by taking the average of the 98 and 86 response values, observed in the (− − − + −) and (+ − − + +) runs of Table 4.6.

Four of the runs in Table 4.9 are identical to runs in Table 4.6. The responses for these runs are the same in both tables and represent real values. The other four runs have level combinations for which the experiments had not been performed. Their response values are simulations obtained from the experimental data in Table 4.6. The calculated contrasts are also shown in Table 4.9, where we can observe that their values are in excellent agreement with the estimates of the average and the main effects determined from the 2_{V}^{5-1} design (Table 4.7). Analyzing the results of the 2^{5-2} quarter-fraction, which would be obtained in the initial stage of the investigation, the research workers could decide if they should perform more runs to arrive at a half-fraction or even the 2^5 complete factorial, if they should introduce new factors in place of the **1** and **5** variables (which appear to have little influence on the response), or even if they would rather change the levels of the variables.[34]

Exercise 4.12. The confounded effects in a certain contrast are determined by the generating relations of the factorial and by all their possible products. For the 2^{4-1} and 2^{5-1} designs there was only one generating relation, and for this reason only two effects were confounded in each contrast. In the 2^{5-2} design there are two generators and we need to consider three equations: the two defining relations, $\mathbf{I} = \mathbf{1234}$ and $\mathbf{125}$, and their product, $(\mathbf{I})(\mathbf{I}) = \mathbf{I} = (\mathbf{1234})(\mathbf{125}) = \mathbf{345}$. Each effect will thus be confounded with three others. (a) Use these relations to show that main effect **1** is confounded with the **25**, **234** and **1345** interactions; (b) which interactions are confounded with main effect **5**?

4.3.2. Saturated designs

Optimization problems involving more than, say, half a dozen factors are not very common in academic research projects. Not only has the researcher ample liberty to define his research projects — and consequently choose simpler problems — but it is much easier to control the variables when the experiment is done in a university laboratory. In industry, the situation is exactly the opposite. The problem to be attacked is generally imposed by circumstances beyond the researcher's control and trying to solve it usually presents greater difficulties, including the tight deadlines dictated by industrial dynamics. Fractional factorial designs, which allow efficient screening of many variables, are particularly important for industrial laboratories.

In our previous examples, at the cost of fragmenting — and therefore confounding — a complete design, we were able to study more variables

[34]It must be stressed that this is a didactic example, where we made use of results from another design. In a real application, the research workers should actually perform the four runs whose response values were simulated here.

with fewer experiments. Since this does not involve any magic, obviously there must be a limit to this procedure. Given a certain number of runs, there must be a maximum number of factors that we can study with them. When this limit is reached, we say that the design is **saturated**. In this section, we are going to use a saturated design to analyze an empirical study performed outside the laboratory.

One of the authors, an amateur tennis player, decided to study the influence of several factors on the quality of his serves. For this, he chose seven factors that interested him and constructed the fractional design shown in Table 4.10. The factors selected were the technique used to execute the serve, the frequency of serving, the time, the type of court surface, the side of the court, the use (or not) of a shirt and the size of the racquet head. The design required the execution of eight runs. With them we can study seven factors at most, because 1 degree of freedom is necessarily spent on calculating the average. Since this is exactly equal to the number of selected variables, we have a saturated design.

Other factors, like wind and rain, can affect the accuracy of the serves. Controlling factors like these is obviously beyond the researcher's power, but precautions were taken to avoid their influence. When, upon executing the experiments, the atmospheric conditions became adverse, our researcher retired to the comforts of the clubhouse, putting off the runs until the weather improved. Each run consisted of a total of 100 serves. The number of successful serves in each run is given in the last column of Table 4.10.

The design was constructed from a complete factorial in the first three factors, using the generating relations $\mathbf{I} = \mathbf{124} = \mathbf{135} = \mathbf{236} = \mathbf{1237}$, to define the levels of the four remaining variables. This design is called saturated because all possible interactions of the factors included in the starting complete factorial — **12**, **13**, **23** and **123** in this case — were used to determine the levels of the other variables. This makes it impossible to define new variables whose levels are not exactly coincident with the levels of one of the factors already included in the factorial. Since the shortest term in the set of generating relations[35] contains three factors, the design is of resolution III, and its complete notation is 2_{III}^{7-4}.

The 2_{III}^{7-4} design corresponds to a one-sixteenth fraction of a complete 2^7 design, and has only 2^3 independent observations, which are the response values of the eight runs. Since the complete design has $2^7 = 128$ parameters (including the overall average), by using only eight observations to estimate their values we confound in each contrast a total of $128/8 = 16$ effects, which is precisely the ratio of the size of the complete design to the size of the fractional factorial. Each main effect will be confounded with 15 other effects. Of these, Table 4.11 shows only the

[35]And their possible combinations.

Table 4.10
A saturated fractional factorial design with eight runs, to evaluate how seven factors affect the accuracy of the serves of an amateur tennis player

Factor	Level	
	(–)	(+)
1 Technique	Normal	Slice
2 Frequency	Low	High
3 Time	Day	Night
4 Surface	Clay	Concrete
5 Side	Right	Left
6 Shirt	With	Without
7 Racquet	Medium	Large

Equivalences	**4 = 12**	**5 = 13**	**6 = 23**	**7 = 123**

Runs	**1**	**2**	**3**	**4**	**5**	**6**	**7**	% valid
1	–	–	–	+	+	+	–	56
2	+	–	–	–	–	+	+	66
3	–	+	–	–	+	–	+	51
4	+	+	–	+	–	–	–	52
5	–	–	+	+	–	–	+	54
6	+	–	+	–	+	–	–	70
7	–	+	+	–	–	+	–	42
8	+	+	+	+	+	+	+	64

two-factor interactions. We are assuming, for the time being, that the higher-order terms are negligible.

With the response values in Table 4.10, which represent the percent valid serves for each run, we can calculate the contrast values shown in Table 4.11. The value $l_1 = 12.25$, for example, is given by $\frac{1}{4}(-56 + 66 - 51 + 52 - 54 + 70 - 42 + 64)$. The rest are calculated in the same way, using the appropriate columns of signs.

Exercise 4.13. Use the data of Table 4.10 to calculate the value of the contrast corresponding to the main effect of the side of the tennis court.

Exercise 4.14. In the 2^{7-4} fractional factorial each main effect is confounded with 15 interactions. To discover which terms are confounded we must use, in addition to the four generating relations, their six binary products, their four ternary products and the product of all four relations. Determine which interactions are confounded with the main effect that represents the side of the tennis court.

Table 4.11
Contrasts calculated for the 2^{7-4}_{III} fractional factorial, and their expressions in terms of main and binary effects of a complete 2^7 factorial

Technique	$l_1 = 12.25 \rightarrow \mathbf{1} + \mathbf{24} + \mathbf{35} + \mathbf{67}$
Frequency	$l_2 = -9.25 \rightarrow \mathbf{2} + \mathbf{14} + \mathbf{36} + \mathbf{57}$
Time	$l_3 = 1.25 \rightarrow \mathbf{3} + \mathbf{15} + \mathbf{26} + \mathbf{47}$
Surface	$l_4 = -0.75 \rightarrow \mathbf{4} + \mathbf{12} + \mathbf{56} + \mathbf{37}$
Side	$l_5 = 6.75 \rightarrow \mathbf{5} + \mathbf{13} + \mathbf{46} + \mathbf{27}$
Shirt	$l_6 = 0.25 \rightarrow \mathbf{6} + \mathbf{23} + \mathbf{45} + \mathbf{17}$
Racquet	$l_7 = 3.75 \rightarrow \mathbf{7} + \mathbf{34} + \mathbf{25} + \mathbf{16}$

We can easily interpret the results of Table 4.11 if we assume that all two-factor interactions can also be neglected, as we have already done for the higher-order effects. Each contrast then simply represents a main effect, leading to the conclusion that the serving technique and frequency are the most important factors, of the seven studied in the design. Changing from the normal serve to the sliced one results in a 12.25% increase in valid serves, whereas an increase in the serving frequency reduces the accuracy by 9.25%.[36]

Two contrasts in Table 4.11, l_5 and l_7, corresponding, respectively, to the side of the tennis court and the size of the racquet head, have intermediate values. The side from where the serve is made seems to be important: on the left side the accuracy increases by about 7%. Taking into account that the tennis player is left-handed, this result seems quite natural. The use of a larger racquet head improves the accuracy by approximately 4%, and is also understandable, since the player is obviously an amateur. The other factors (time, shirt and court surface) do not appear to be important.

This approximate analysis neglected all the interaction effects. Could we be wrong? Who can tell if the effects we neglected are not the ones responsible for the high values of the contrasts?

Sometimes a contrast can present a high value not because of a main effect but because of a binary interaction involving two factors, both having important main effects. Consider for example factors **1** and **2**, which individually seem to be important. The **12** interaction is contained in the l_4 contrast, whose value is almost 0. If the **12** interaction were important, we would expect a higher value for the l_4 contrast. Since this is not the case, we conclude provisionally that this interaction must not be

[36] After some practice, the sliced serve is actually easier to control than the normal one. As for the other factor, serving at larger intervals allows the server to concentrate more intensely and therefore helps to increase accuracy.

important. A similar argument applies to the **15** and **17** interactions, which contribute to the l_3 and l_6 contrasts, respectively.

That leaves the **25**, **27** and **57** interactions, which participate in the most important contrasts. The **27** interaction, for example, contributes to the l_5 contrast, which we already interpreted as the main effect of the side of the court. If the **27** value is significant, we will have to change our interpretation of l_5. Perhaps the side of the court is not so important, and the 6.75 value really indicates that with a larger racquet it is possible to serve more frequently without sacrificing accuracy.

In an attempt to clarify this question, our athletic researcher performed a second design, constructed on purpose to separate (or, as sometimes said, **unconfound**) the value of main effect **5** from the sum **13**+**46**+**27**. This design, given in Table 4.12, is based on the same relations used in Table 4.10, with one exception. For factor **5**, which we wish to unconfound, we used the relation $\mathbf{5} = -\mathbf{13}$ instead of $\mathbf{5} = \mathbf{13}$. The two factorials are therefore identical except for the levels attributed to factor **5**, which have exactly opposite signs in the two tables. When we combine the results of the two designs, this characteristic allows us to isolate main effect **5** from the two-factor interactions.

Exercise 4.15. Each run in Tables 4.10 and 4.12 correspond to performing serves under the experimental conditions specified by the signs of the respective design matrices. Describe the experiment represented by run 4 in Table 4.10. In practice, what is the difference between this run and run 4 in Table 4.12?

The values of the contrasts calculated for the new fraction are given in Table 4.13, together with the relations between them and the effects of the complete design. Note that for these relations (a) all the binary interactions containing factor **5** have negative signs, and (b) all the binary interactions in the l_5 contrast have negative signs. The only contrast with a clearly significant value is the one corresponding to the serving technique (either normal or sliced). The contrasts for frequency and racquet head size now have absolute values a bit above 4%. Surprisingly, the contrast for using a shirt is a little higher than the contrast for the side of the court from which the serve is made. Perhaps this result can be attributed to the contribution of the **17** binary interaction.

To isolate main effect **5**, we combine the two contrasts in which it appears:

$$\mathbf{5} = \frac{l_5 + l_5^*}{2} = \frac{6.75 + 2.75}{2} = 4.75.$$

In the same way,

$$\mathbf{13} + \mathbf{46} + \mathbf{27} = \frac{l_5 - l_5^*}{2} = 2.00.$$

Table 4.12
A saturated fractional factorial design with eight runs that, together with the fractional of Table 4.10, permits separating the main effect for factor **5** from the two-factor interactions

Run	**1**	**2**	**3**	**4**	**5**	**6**	**7**	% valid
1	−	−	−	+	−	+	−	52
2	+	−	−	−	+	+	+	74
3	−	+	−	−	−	−	+	50
4	+	+	−	+	+	−	−	60
5	−	−	+	+	+	−	+	54
6	+	−	+	−	−	−	−	62
7	−	+	+	−	+	+	−	50
8	+	+	+	+	−	+	+	63

Note: The signs of factor 5 are opposite to those in Table 4.10. The other factors have the same signs.

Table 4.13
Calculated contrasts of the second 2^{7-4} fractional and their expressions in terms of main and binary interaction effects of the full 2^7 design

Technique	$l_1^* = 13.25 = \mathbf{1} + \mathbf{24} - \mathbf{35} + \mathbf{67}$
Frequency	$l_2^* = -4.75 = \mathbf{2} + \mathbf{14} + \mathbf{36} - \mathbf{57}$
Time	$l_3^* = -1.75 = \mathbf{3} - \mathbf{15} + \mathbf{26} + \mathbf{47}$
Surface	$l_4^* = -1.75 = \mathbf{4} + \mathbf{12} - \mathbf{56} + \mathbf{37}$
Side	$l_5^* = 2.75 = \mathbf{5} - \mathbf{13} - \mathbf{27} - \mathbf{46}$
Shirt	$l_6^* = 3.25 = \mathbf{6} + \mathbf{23} - \mathbf{45} + \mathbf{17}$
Racquet	$l_7^* = 4.25 = \mathbf{7} - \mathbf{25} + \mathbf{34} + \mathbf{16}$

We can conclude, then, that the main effect of the side of the court on serving accuracy is almost 5%, whereas the combined effect of the **13**, **46** and **27** interactions is 2%.

Table 4.14 shows the combinations for the other contrasts. Note that, besides isolating main effect **5**, we also were able to isolate all binary interactions involving this factor. The absolute values of these interactions are all inferior to 2.25. If we assume that the true values of all of them are 0, we can use the seven values of Table 4.14 that correspond to two-factor interactions to estimate an error for the value of a contrast:

$$s_c^2 = \frac{(-0.50)^2 + (-2.25)^2 + (1.50)^2 + (0.50)^2 + (2.00)^2 + (-1.50)^2 + (-0.25)^2}{7} = 2.02,$$

which leads to a standard error of 1.42.

With this error estimate, we can finally conclude that only the contrasts containing the main effects of technique (**1**), frequency (**2**), courtside (**5**) and size of racquet head (**7**) have significant values, at the 95% confidence level. Other fractions could be run, if we wished to isolate other effects. In all, 16 different 2_{III}^{7-4} designs exist, defined by the possible combinations of the relations $\mathbf{4} = \pm\mathbf{12}$, $\mathbf{5} = \pm\mathbf{13}$, $\mathbf{6} = \pm\mathbf{23}$ and $\mathbf{7} = \pm\mathbf{123}$.

If, for example, we wish to know if the size of the racquet head (factor **7**) is really important, we can perform another design keeping the first three generating relations as they are and making $\mathbf{7} = -\mathbf{123}$. Combining the results of this design with those of the first fractional, we will obtain isolated estimates of the main effect for racquet size and all its binary interactions. Usually, however, this strategy for isolating effects is applied first to the contrasts with the largest absolute values. In this example the first candidate for isolation is factor **1**, whose contrasts are $l_1 = 12.25$ and $l_1^* = 13.25$. Running a new fraction with all signs in column **1** reversed and combining its results with those of the first fraction, we would have a better estimate of the main effect of this factor (presumably the most important of all), and also estimates of all binary interactions in which it participates.

4.3.3. How to construct resolution III fractional factorial designs

Saturating a 2^{m} complete design, we can obtain fractional factorial designs of resolution III for up to $2^{\mathrm{m}}–1$ variables. To do this we use, as we have seen for $m = 3$, generating relations derived from all possible interactions of the m initial factors. Starting, for example, with a 2^4 factorial, we can employ the following 11 relations: ***5*** = **12**, ***6*** = **13**, ***7*** = **14**, ***8*** = **23**, ***9*** = **24**, ***10*** = **34**, ***11*** = **123**, ***12*** = **124**, ***13*** = **134**, ***14*** = **234** and ***15*** = **1234.** (Here, to avoid mistaking a factor for an effect, we are italicizing the new factors *5–15*.) The resulting fractional factorial design will have 16 runs and it will be possible to study the effects of 15 ($2^4–1$) variables. Its notation will be 2_{III}^{15-11}.

The same procedure can be extended to any number of starting factors. It is important to note, however, that using a saturated design does not oblige us to study 7, 15 or, in general, 2^m-1 factors. We can study any number of factors, as long as it is not larger than the maximum number allowed by the fractional design. The columns of signs that do not correspond to real variables can be used to calculate estimates of the standard error of the contrasts, as we shall see in Section 4A.9.

4.3.4. How to construct a 2_{IV}^{8-4} fraction from a 2_{III}^{7-4} fraction

Resolution IV designs are easily constructed from saturated designs of resolution III. For example, starting from our first 2_{III}^{7-4} fractional, we can construct the 2_{IV}^{8-4} design presented in Table 4.15. To do this, we start by adding to the initial 2_{III}^{7-4} design a column of positive signs for factor **8**.

Table 4.14
Estimates of main effect **5** and all binary interaction effects involving this factor

	Contrast
1	$\frac{l_1 + l_1^*}{2} = \frac{12.25 + 13.25}{2} = 12.75 = \mathbf{1 + 24 + 67}$
	$\frac{l_1 - l_1^*}{2} = \frac{12.25 - 13.25}{2} = -0.50 = \mathbf{35}$
2	$\frac{l_2 + l_2^*}{2} = \frac{-9.25 - 4.75}{2} = -7.00 = \mathbf{2 + 14 + 36}$
	$\frac{l_2 - l_2^*}{2} = \frac{-9.25 + 4.75}{2} = -2.25 = \mathbf{57}$
3	$\frac{l_3 + l_3^*}{2} = \frac{1.25 - 1.75}{2} = -0.25 = \mathbf{3 + 26 + 47}$
	$\frac{l_3 - l_3^*}{2} = \frac{1.25 + 1.75}{2} = 1.50 = \mathbf{15}$
4	$\frac{l_4 + l_4^*}{2} = \frac{-0.75 - 1.75}{2} = -1.25 = \mathbf{4 + 12 + 37}$
	$\frac{l_4 - l_4^*}{2} = \frac{-0.75 + 1.75}{2} = 0.50 = \mathbf{56}$
5	$\frac{l_5 + l_5^*}{2} = \frac{6.75 + 2.75}{2} = 4.75 = \mathbf{5}$
	$\frac{l_5 - l_5^*}{2} = \frac{6.75 - 2.75}{2} = 2.00 = \mathbf{13 + 46 + 27}$
6	$\frac{l_6 + l_6^*}{2} = \frac{0.25 + 3.25}{2} = 1.75 = \mathbf{6 + 23 + 17}$
	$\frac{l_6 - l_6^*}{2} = \frac{0.25 - 3.25}{2} = -1.50 = \mathbf{45}$
7	$\frac{l_7 + l_7^*}{2} = \frac{3.75 + 4.25}{2} = 4.00 = \mathbf{7 + 34 + 16}$
	$\frac{l_7 - l_7^*}{2} = \frac{3.75 - 4.25}{2} = -0.25 = \mathbf{25}$

Since a 2^{8-4} design must have 16 runs, we need eight more rows. To obtain the signs for these rows, we choose signs opposite to those of the first eight runs, line by line. The ninth run will be the same as the first one, except that all signs are reversed, the 10th one will have signs opposite to the second, and so on until the 16th row, which will have only negative signs since it is obtained from the eighth row.

With this design we will have to perform twice as many runs, but the additional work is compensated by a gain in resolution. Since the new design is defined by the relations **I** = **1248** = **1358** = **2368** = **1237**, its resolution becomes 4.

Exercise 4.16. How are the generating relations of Table 4.15 derived? Why don't they change when the signs of the runs are changed, as was done for runs 9–16?

Exercise 4.17. Use the generating relations given in Table 4.15 and find the three-factor interactions confounded with factor **1**.

With resolution IV designs, we can completely separate all the main effects from the two-factor interactions, as shown in Table 4.16. The contrasts corresponding to the actual columns of the design estimate the main effects of the eight variables, while the contrasts defined by products of two columns estimate combinations of two-factor interactions.[37]

4.3.5. Saturated Plackett–Burman designs

We have just discussed designs with which we can study the influence of up to 7, 15, 31,…,2^m–1 factors using 8, 16, 32,…,2^m experimental runs. Another class of fractional factorial designs employs a total of 12, 20, 24, 28,… runs to simultaneously investigate up to 11, 19, 23, 27,… factors. With these designs, proposed by R.L. Plackett and J.P. Burman, it is possible to estimate all $k = n-1$ main effects (where n represents the number of runs) with minimum variance (Plackett and Burman, 1946). Table 4.17 presents a Plackett–Burman design for $n = 12$.

Plackett–Burman designs have one characteristic in common with the other fractional factorial designs we have studied up to this point. The $n/2$ positive signs of any column always correspond, in the other columns, to $n/4$ positive signs and $n/4$ negative signs. The same happens for the negative signs. This means that the columns are all orthogonal, and this symmetry allows the main effect of each factor to be determined individually, assuming that all interaction effects are negligible.

With a saturated design of n runs we can study up to n–1 factors, but it is advisable to choose a smaller number. The columns not used to set the factor levels then play the roles of inert factors and can be used to estimate the error associated with the contrasts. For Plackett–Burman

[37]We are of course neglecting all higher-order interactions.

Table 4.15
A resolution IV fractional design, 2_{IV}^{8-4}, defined by
$\mathbf{I} = \mathbf{1248} = \mathbf{1358} = \mathbf{2368} = \mathbf{1237}$

Run	**1**	**2**	**3**	**4**	**5**	**6**	**7**	**8**
1	−	−	−	+	+	+	−	+
2	+	−	−	−	−	+	+	+
3	−	+	−	−	+	−	+	+
4	+	+	−	+	−	−	−	+
5	−	−	+	+	−	−	+	+
6	+	−	+	−	+	−	−	+
7	−	+	+	−	−	+	−	+
8	+	+	+	+	+	+	+	+
9	+	+	+	−	−	−	+	−
10	−	+	+	+	+	−	−	−
11	+	−	+	+	−	+	−	−
12	−	−	+	−	+	+	+	−
13	+	+	−	−	+	+	−	−
14	−	+	−	+	−	+	+	−
15	+	−	−	+	+	−	+	−
16	−	−	−	−	−	−	−	−

Table 4.16
Contrasts of the 2_{IV}^{8-4} fraction as a function of the main and binary interactions of the 2^8 complete factorial, neglecting interactions involving more than two factors

$l_1 = \mathbf{1}$	$l_{12} = \mathbf{12} + \mathbf{37} + \mathbf{48} + \mathbf{56}$
$l_2 = \mathbf{2}$	$l_{13} = \mathbf{13} + \mathbf{27} + \mathbf{46} + \mathbf{58}$
$l_3 = \mathbf{3}$	$l_{14} = \mathbf{14} + \mathbf{28} + \mathbf{36} + \mathbf{57}$
$l_4 = \mathbf{4}$	$l_{15} = \mathbf{15} + \mathbf{26} + \mathbf{38} + \mathbf{47}$
$l_5 = \mathbf{5}$	$l_{16} = \mathbf{16} + \mathbf{25} + \mathbf{34} + \mathbf{78}$
$l_6 = \mathbf{6}$	$l_{17} = \mathbf{17} + \mathbf{23} + \mathbf{68} + \mathbf{45}$
$l_7 = \mathbf{7}$	$l_{18} = \mathbf{18} + \mathbf{24} + \mathbf{35} + \mathbf{67}$
$l_8 = \mathbf{8}$	

designs, it is recommended that the number of factors studied not be larger than n–4. With the design of Table 4.17, for example, we should investigate eight factors at most. The three remaining degrees of freedom[38] can be used to estimate the errors of the values calculated for the main effects.

[38]One degree of freedom is used for calculating the average (first column of the design matrix).

Table 4.17
The Plackett–Burman saturated fractional design for studying 11 variables with 12 runs

Runs	**I**	**1**	**2**	**3**	**4**	**5**	**6**	**7**	**8**	**9**	**10**	**11**
1	+	+	+	−	+	+	+	−	−	−	+	−
2	+	+	−	+	+	+	−	−	−	+	−	+
3	+	−	+	+	+	−	−	−	+	−	+	+
4	+	+	+	+	−	−	−	+	−	+	+	−
5	+	+	+	−	−	−	+	−	+	+	−	+
6	+	+	−	−	−	+	−	+	+	−	+	+
7	+	−	−	−	+	−	+	+	−	+	+	+
8	+	−	−	+	−	+	+	−	+	+	+	−
9	+	−	+	−	+	+	−	+	+	+	−	−
10	+	+	−	+	+	−	+	+	+	−	−	−
11	+	−	+	+	−	+	+	+	−	−	−	+
12	+	−	−	−	−	−	−	−	−	−	−	−

A disadvantage of Plackett–Burman designs is that the relations between the calculated contrasts and the effects of a complete factorial are quite complex. This makes it difficult to choose the additional runs necessary to unconfound the effects.

4.3.6. Taguchi techniques for quality engineering

In post-war Japan, Genichi Taguchi, an engineer with the Nippon Telephone & Telegraph Co., set himself the task of helping Japanese engineers to develop quality products, in spite of the unfavorable conditions prevailing at that time: inferior raw material, outdated equipment and — even worse — lack of qualified personnel. To face these problems, Taguchi came up with a strategy based on experimental multivariate designs, constructed with the objective of developing products and processes that

- were insensitive to environmental conditions;
- were insensitive to component variations;
- had minimum variations about a target value.

With respect to traditional thinking, the Taguchi approach, which later became quite popular in the West, introduced two new ideas:

- Any deviation from the target value is undesirable, even if the product meets the specification limits.

- During product design one must take into account not only the factors that can be controlled in the manufacturing process *but also* those difficult or impossible to control but that might affect the response, like small fluctuations in component materials, equipment degradation or changes in the way the customer uses the product.

Consider a cake mixture, manufactured, let us say, with four ingredients: wheat flour, sugar, eggs and margarine. When the cook makes the cake, he has to add milk, adjust the temperature of the oven and control the baking time. These factors also affect the final result, but are beyond the manufacturer's control, no matter how explicit be the instructions printed on the mixture package.

Taguchi gave the name of **parameters** to the first factors, those that can be controlled during the manufacture of the cake mixture. The others are sources of **noise**. In the Taguchi approach, the latter should also be included in the design and development of the product. For this he recommended using orthogonal factorial designs, similar to those we have seen in this chapter.

Two types of design should be constructed: an **inner array**, involving only the parameters, and an **outer array**, based on the noise sources. These two arrangements are then **crossed**, that is, runs are performed at all their possible combinations. In the cake mixture, for example, if we consider only two levels for each of the seven factors mentioned above, the Taguchi procedure results in the scheme shown in Table 4.18.

For Taguchi, the response should be as close to target as possible, but also should be **robust** — that is, very insensitive — to the influence of noise. This means that we must take into account not only the response values of the inner array but also the variations due to the noise factors. Two runs in the table, the second and the eighth, produce average response values exactly at the target value (80). Nevertheless, the second run should be preferred, because it has a standard deviation of only 1.83, compared to 4.97 for the eighth run.

Actually, in Taguchi analysis, we should choose the best run by analyzing a signal-to-noise ratio, whose form depends on the objective of the experiment. For this example, for which the objective is to arrive at a certain nominal value.[39] Taguchi recommends maximizing the relation

$$SN_T = 10\ \log\frac{\bar{y}^2}{s^2},$$

whose values appear in the last column of Table 4.18. By this criterion, the second run would also be chosen. Now the eighth run, which is the only other on target, would wind up two places before the last, beating out

[39]Instead of maximizing or minimizing a response.

Table 4.18
Taguchi design for the development of a cake mixture

F	M	S	E	Time/milk/temperature				$\bar{y}$	s	SN_T
				− − +	− + −	+ − −	+ + +			
−	−	−	−	85	96	97	92	92.5	5.45	24.6
−	−	−	+	82	81	78	79	80	1.83	32.8
−	+	+	−	75	80	70	73	74.5	4.20	25.0
−	+	+	+	66	75	83	70	73.5	7.33	20.0
+	−	+	−	84	91	95	90	90	4.55	25.9
+	−	+	+	78	72	80	69	74.8	5.12	23.3
+	+	−	−	86	85	90	91	88	2.94	29.5
+	+	−	+	86	82	77	75	80	4.97	24.1

Note: The L8 internal array for four parameters (**F**lour, **M**argarine, **S**ugar and **E**ggs) is crossed with an L4 external array for three environmental factors. The meaning of SN_T is explained in the text. The target value for $\bar{y}$ (a texture measure) is 80.

only the fourth and sixth runs. This is a consequence of Taguchi's emphasis on robustness of the response relative to noise. It is also one of the reasons for the criticisms suffered by Taguchi's approach.

The strategy to improve quality advocated by Taguchi is intrinsically multivariate, and formally does not bring any new ideas. His designs involving two levels, for example, have the same structure as the factorial designs discussed in this chapter and the last. In Taguchi's methodology, as we have seen, these designs should be performed to discover the combination of factor levels that produce responses with the smallest variation and closest to the desired objective.

Taguchi suggests experiments at two levels, defined by orthogonal design arrays designated as L4, L8, L12, L16 and L32, where the number indicates the total number of runs for each design. The L4 array is a 2^{3-1} fractional factorial design, in which the levels of the third variable are defined by the **I** = **123** generating relation. The L8 design is equivalent to the 2^{7-4} factorial presented in Table 4.10, except it is defined by the relations **I** = −**124** = −**135** = −**236** = **1237**. In the Taguchi design matrices, the columns are arranged in a different order than the standard one adopted in this book. However, since any factor can be attributed to any column of signs, the two classes of designs are identical.

The Taguchi design for 12 runs, L12, is quite different from the Plackett–Burman saturated design to study 11 factors in 12 runs (Table 4.17). Even so, the two designs are orthogonal and should present similar results, if all interaction effects are negligible. If this does not happen, the interpretations of the contrasts can be different, because the relations between the contrasts and the main and interaction effects are different for the two designs. Just as for Plackett–Burman designs, these relations are quite complex, making it difficult to unconfound the effects by expanding the design.

Taguchi also proposed designs with three and four levels, which could be used to identify non-linear tendencies in the relations between responses and factors, but these designs are unable to estimate two-factor interactions.

Everyone praises the Taguchi philosophy of attempting to develop products that are robust to noise, but his methods of design and analysis have been widely criticized. Some authors argue that the same results could be obtained in a more efficient way using other designs (see, for example, Myers and Montgomery, 2002, Chapter 10). The crossed designs proposed by Taguchi not only demand a large number of runs but also ignore interactions between controlled factors. In the presence of significant interactions, the interpretation of the results is not clear, and so we lose an opportunity to better understand how the system works, which would be of great help for attacking future problems.

An alternative procedure is to substitute the Taguchi arrays by fractional factorial designs in which the noise and controlled factors are treated in the same way. For example, instead of using the crossed L8 $\times$ L4 arrangement of Table 4.18, which requires 32 runs, we could use a 2^{7-3} fractional design **combining** all seven factors, which requires only 16 runs and does not confound the main effects with the two-factor interactions.

The use of signal-to-noise ratios has also been criticized. In combining $\bar{y}$ and s^2 into a single numerical value, we are mixing proximity to target with fluctuations caused by noise, and for this reason many researchers prefer to analyze separately the results for the responses and their variances. We have seen the difference that can result from using one approach or the other, in the analysis of the results in Table 4.18.

Enthusiasts of the Taguchi philosophy rebut the criticisms arguing that the methods work. There are at least two reasons for this. The most important, perhaps, is that engineers, like Taguchi himself, draw on the knowledge they already have about the system to choose the most appropriate design.[40] The combination of specialized technical information with statistical methodology (though imperfect) is a powerful tool for dealing with research and development problems. The other reason is that the Taguchi methodology was applied mostly in industries that had not yet used multivariate designs. These, as we already know, usually work better than univariate methods, even if they are not used in the best way. The industries were quick to perceive the difference.

[40]And an excellent practice it is, by the way.

Taguchi methods, and the other fractional factorials that we studied, have in common the idea of performing multivariate studies by means of orthogonal designs. Even if the researcher does not choose the ideal one, just by using them she will have more chances of succeeding than if she uses traditional univariate methods.

4A. Applications

4A.1. Adsorption on organofunctionalized silicas

In a preliminary experiment to study the adsorption of Cu(II) on organofunctionalized silica surfaces (obtained when Si–OH groups on the silica surface bind with alcoxysilanes), the fractional factorial design whose data are presented in Table 4A.1 was carried out (Cestari et al., 1996).

The normal plot shows how contrast **3** (+3.58) is clearly separated from the rest (Fig. 4A.1). We could have anticipated this result, because the last four response values, which correspond to the higher level of factor **3**, have larger values than the first four. But that is not the whole story. The other contrasts, although much smaller in absolute value, are all negative, suggesting that they represent a systematic behavior rather than pure error. This suspicion is confirmed by the cubic graph (Fig. 4A.2). Factors **2** (salt) and **4** (silica amount), both tend to diminish the response when they are changed from their lower levels to their higher ones. This occurs in three of the four individual effects for both factors. The two exceptions involve the response value for the (**1 2 3 4**) = (– – – –) run, which is very low (0.39). In any case, Fig. 4A.2 makes it very clear that, if we want to increase the quantity of Cu(II) adsorbed by the silica, we should not only set factor **3** at its higher level (i.e., use ethanol as a solvent), but also fix factor **4** at its lower level — that is, use less silica. Under these conditions, the type of salt does not appear to make any difference. The second most significant contrast is the **34** interaction (= **12** = –1.50). Can you see why, from an inspection of Fig. 4A.2?

4A.2. Calcium oxalate thermogravimetry

A 2^{5-2}_{III} fractional factorial design was used to estimate the effects of nitrogen flow, sample weight, heating rate, crucible type and baseline correction on the thermal decomposition of calcium monohydrate oxalate, monitored by thermogravimetry (Mathias et al., 1999). One of the responses studied was the average position of a certain peak in the thermogravimetric curve (Table 4A.2).

Table 4A.1
Experimental data

Factors	(–)	(+)
1 Silica type	Sil-et-1	Sil-et-2
2 Salt	$CuCl_2$	$Cu(C_2H_3O_2)_2$
3 Solvent	Water	Ethanol
4 Silica amount (mg)	100	200

Run	**1**	**2**	**3**	**4**	y (mol g^{-1}) ($\times 10^4$)
1	–	–	–	–	0.39
2	+	–	–	+	1.74
3	–	+	–	+	1.37
4	+	+	–	–	1.68
5	–	–	+	+	4.66
6	+	–	+	–	6.12
7	–	+	+	–	6.09
8	+	+	+	+	2.61

Contrasts

1 = 234	**2 = 134**	**3 = 124**	**4 = 123**	**12 = 34**	**13 = 24**	**14 = 23**
–0.09	–0.29	3.58	–0.98	–1.50	–0.92	–0.75

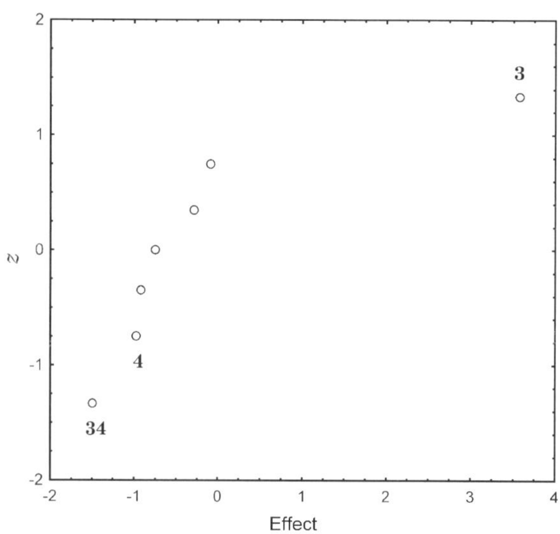

Fig. 4A.1. Normal plot of the contrasts.

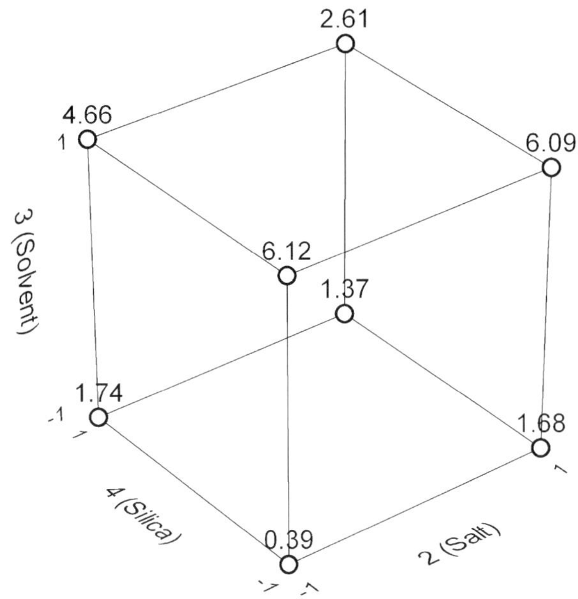

Fig. 4A.2. Cubic plot of the response values of Table 4A.1.

Here, since we are dealing with a one-quarter fraction, the effects are confounded in groups of four, and interpreting the contrasts becomes a bit more complex. We must keep in mind that the column labels in Table 4A.2 actually stand for the sums of four effects. The most significant value, 83.43, is in fact the sum of four effects, **3+45+124+1235**. Also, we cannot ignore that seven of the eight contrasts have positive values, and that the only negative contrast is precisely the least significant of all (Fig. 4A.3). Even so, it is possible to draw some conclusions. Since the most significant contrasts are **3**, **2**, **5** and **23**, a cubic graph of the response values as a function of the **2**, **3** and **5** factors (Fig. 4A.4) helps us in the interpretation, which is simpler than in the preceding example. All the individual contrasts are now positive, along the three axes in the figure. Making the reservation that we are talking of contrasts that are actually sums of four effects, we can perceive some evidence of interaction between the factors. For example, when factor **2** (mass) is at its lower level, the average effect of factor **3** (heating rate) is 64.0. When factor **2** is at its higher level, that is, when we use a heavier sample, the average effect of the heating rate increases to 103.0. Since the nitrogen flow rate and the crucible type do not affect peak position, we can use a smaller flow rate (15 mL min^{-1}) and an Al_2O_3 crucible, which is less expensive than a platinum one. The researchers had already foreseen positive effects from sample weight and heating rate, because of the delays observed on reaching thermal equilibrium.

Table 4A.2
Experimental data

Factors	–	+
1 Nitrogen flow rate ($mL\,min^{-1}$)	30	15
2 Sample mass (g)	4	10
3 Heating rate ($^{\circ}C\,min^{-1}$)	10	50
4 Crucible type	Al_2O_3	Pt
5 Baseline correction	Without	With

Run	**1**	**2**	**3**	**4**	**5**	Average position (°C)
1	–	–	–	–	+	726.4
2	+	–	–	+	–	695.4
3	–	+	–	+	–	734.7
4	+	+	–	–	+	738.4
5	–	–	+	+	+	780.8
6	+	–	+	–	–	768.9
7	–	+	+	–	–	822.8
8	+	+	+	+	+	856.1

Contrasts

1	**2**	**3**	**4**	**5**	**13**	**23**
−1.48	45.13	83.43	2.63	19.98	12.18	19.48

4A.3. Chromatographic analysis of gases

The performance of electrical transformers can be monitored by chromatographic analysis of the gases dissolved in the mineral oil contained in the isolating system. In a study of the effects of five factors on chromatographic results for the analysis of ethylene, a fractional factorial was executed by two students (M.M. da Silva Jr. and C.A.G. da Câmara) as an out-of-class assignment in a chemometrics course (Table 4A.3).

Of the five main effects, the one for factor **3** (equilibrium time) is clearly secondary. As for the other four, we cannot discard the possibility that some of their values are mainly due to interactions. For example, the effect of stirring (**4**) is confounded with interaction **12** (sample volume and temperature) while the effect of pressurization (**5**) is confounded with the interaction of temperature and time (**23**). Actually,

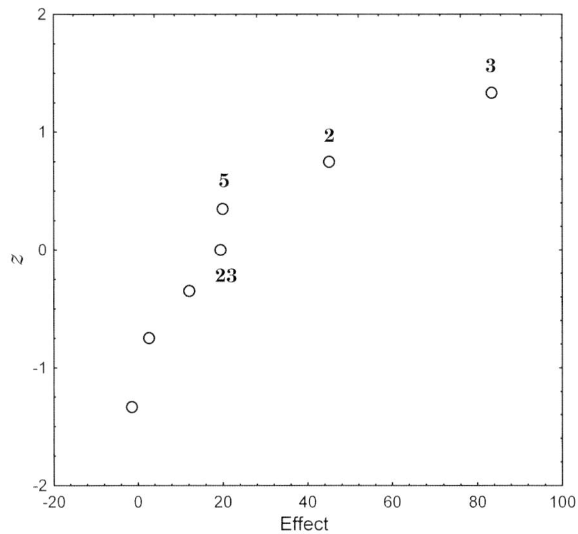

Fig. 4A.3. Normal plot of the effects.

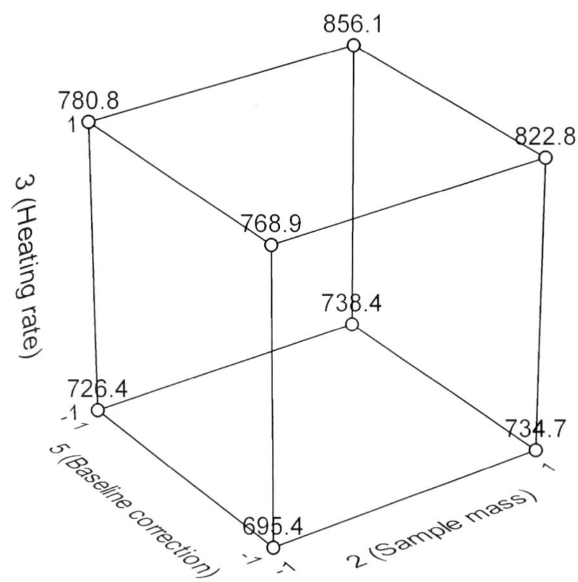

Fig. 4A.4. Cubic plot of the response values of Table 4A.2.

these were the relations used to define the signs of columns **4** and **5** in this design.

Fig. 4A.5 shows the responses as a function of the factor levels whose main effects seem to be the most important. Some patterns can

Table 4A.3
Experimental data

Factors	(–)	(+)
1 Sample volume (mL)	10	19
2 Sample temperature (°C)	50	80
3 Sample equilibrium time (min)	5	20
4 Stirring	Without	With
5 Sample pressurization (psi)	3	11

Run	**1**	**2**	**3**	**4**	**5**	Signal (au[a])
1	–	–	–	+	+	49
2	+	–	–	–	+	21
3	–	+	–	–	–	15
4	+	+	–	+	–	1
5	–	–	+	+	–	42
6	+	–	+	–	–	2
7	–	+	+	–	+	25
8	+	+	+	+	+	32

Contrasts

1	**2**	**3**	**4**	**5**	**13**	**15**
–18.75	–10.25	3.75	15.25	16.75	2.25	8.25

[a] au = arbitrary units.

be detected,[41] but we must not forget that this design is highly confounded and that factor **2**, which does not appear in the graph, also has a relatively high contrast. Since the objective of the experiment was to increase the analytical signal, the best result of the fractional factorial has sample volume and temperature at their lower levels (10 mL and 50 °C), with sample pressurization (**5**) at 11 psi, under stirring (**4**). Because equilibrium time (**3**) does not seem to have an important effect, the shorter time can be used without compromising the results.

4A.4. Mn-porphyrin catalytic response

Students in a chemometrics course performed a fractional factorial design to investigate how the catalytic response of $MnM_{2\text{-}Br}PTTP$

[41]Which ones?

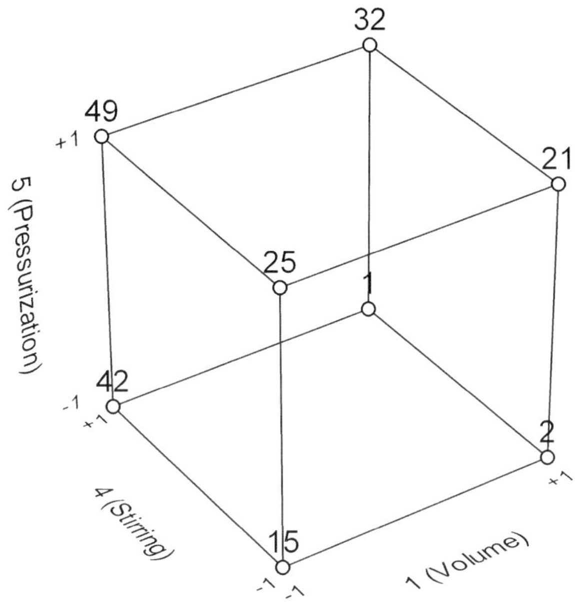

Fig. 4A.5. Responses of Table 4A.3 as a function of factors **1**, **4** and **5**.

Mn-porphyrin is affected by seven factors. The results obtained are in Table 4A.4. The response is the ratio between the amount of cyclohexanol produced in the porphyrin-catalyzed reaction and a fixed quantity of an internal standard previously added to the reaction medium.

Two factors, time (**3**) and the presence of imidazole (**6**), do not seem to be important, which transforms the design into a 2^{5-2} fraction. Of the remaining five factors, two — the catalyst concentration and the ϕIO/MnP ratio — act in the direction of increasing cyclohexanol content, while the other three (stirring method, solvent nature and especially the temperature) tend to decrease it. Fig. 4A.6 shows how the two apparently most important factors affect the response. Should we suspect a possible interaction between them? Which of the other three important factors (**4**, **5** and **7**) could you associate with the difference between the response values obtained in the pairs of experiments where (**1 2**) = (– –) and (**1 2**) = (+ –)?

4A.5. Oxide drainage in the steel industry

The principal objective of a project executed by students in a chemometrics course was to optimize a procedure to decrease the drainage time of oxides used in the steel industry. Four factors were

Table 4A.4
Experimental data

Factors	(−)	(+)
1 Stirring	Magnetic	Ultrasound
2 Temperature (°C)	0	Room
3 Time (min)	30	90
4 [Catalyst] (M)	10^{-4}	10^{-3}
5 ϕIO/MnP ratio	90	15
6 Imidazole	Absent	Present
7 Solvent	Dichloromethane	Acetonitrile

Run	**1**	**2**	**3**	**4**	**5**	**6**	**7**	% cyclohexanol
1	−	−	−	+	+	+	−	34.3
2	+	−	−	−	−	+	+	5.6
3	−	+	−	−	+	−	+	3.6
4	+	+	−	+	−	−	−	2.9
5	−	−	+	+	−	−	+	19.8
6	+	−	+	−	+	−	−	19.6
7	−	+	+	−	−	+	−	4.4
8	+	+	+	+	+	+	+	3.85

Contrasts

1	**2**	**3**	**4**	**5**	**6**	**7**
−7.54	−16.14	0.31	6.91	7.16	0.56	−7.09

studied with a 2^{4-1} fractional factorial design: percent amount of additive, type of solvent used for homogenizing, presence of a catalyst and time spent in an oven (Table 4A.5).

The values calculated for the contrasts strongly indicate that there are only two important factors (**1** and **4**, i.e., solvent and time), and that the interaction between them is also important. Treating factors **2** and **3** as inert ones, we obtain a 2^2 full factorial design in duplicate (Table 4A.6), from which we can extract an estimate of pure error and consequently a measure of the significance of the effects.

The pooled variance of the duplicate runs is 44.91. The standard deviation of a response is the square root of this value, 6.70. The variance of an effect is half that of the pooled variance, 22.45, and therefore its standard error is 4.74. Since the estimate of the pooled variance has 4

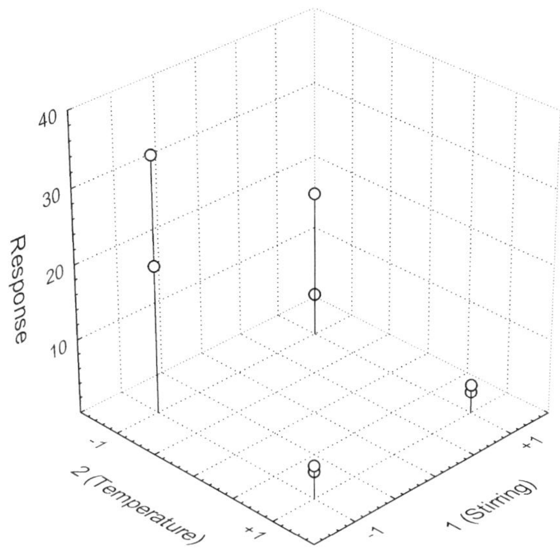

Fig. 4A.6. Responses of Table 4A.4 as a function of factors **1** and **2**.

degrees of freedom, the 95% confidence interval is given by

$$\pm(t_4 \times 4.74) = \pm(2.776 \times 4.74) = \pm 13.2,$$

confirming that only three contrasts are significant, just as the simple inspection of the calculated values had indicated. The practical significance of the calculated values can be seen in Fig. 4A.7. When the solvent is hexane, increasing oven time considerably increases the response. When it is alcohol, this increase does not make a difference. The objective of the experiment was to minimize the response value — the drainage time of the oxide. Our analysis shows this can be accomplished in more than one way, and it is interesting to note that the two smallest response values obtained in the eight experiments are those of runs 7 and 2, which are opposite in signs for both factors **1** and **4**. In short, to lower the drainage time what we must *not* do is use hexane as a solvent or leave the oxide in the oven for a long time.

4A.6. Violacein production by bacteria

In a project for a chemometrics course, a 2^{15-11} fractional factorial design was used to study the production of the pigment violacein by *Chromobacterium violaceum* (Mendes et al., 2001). Violacein is produced in relatively large quantities by this bacterium, and has several pharmacological applications, for example as an agent against other bacteria, mycobacteria, some tumors and even parasites of the

Table 4A.5
Experimental data

Factors	(−)	(+)
1 Solvent	Hexane	Alcohol
2 *Additive (%)*	1	2
3 Catalyst	Without	With
4 Oven time (min)	5	10

Run	**1**	**2**	**3**	**4**	Drainage time (s)
1	−	−	−	−	32.5
2	+	−	−	+	26.0
3	−	+	−	+	76.0
4	+	+	−	−	38.5
5	−	−	+	+	74.0
6	+	−	+	−	35.5
7	−	+	+	−	23.0
8	+	+	+	+	42.0

Contrasts

1 = 234	**2 = 134**	**3 = 124**	**4 = 123**	**12 = 34**	**13 = 24**	**14 = 23**
−15.88	2.88	0.38	22.13	6.63	6.13	−25.13

Table 4A.6
2^2 factorial design in duplicate for factors **1** and **4**

1	**4**	Drainage	time (s)	Average	Variance
−	−	32.5	23.0	27.8	45.13
+	−	38.5	35.5	37.0	4.5
−	+	76.0	74.0	75.0	2.0
+	+	26.0	42.0	34.0	128.0

Trypanosoma genus, which cause Chagas disease. The results obtained are reproduced in Table 4A.7.

Since the signs of columns 12–15 in the table do not correspond to any real factors, we can use the values of their calculated contrasts as estimates of standard errors of the effects (assuming, of course, that all interactions are negligible). We can obtain a pooled estimate of the

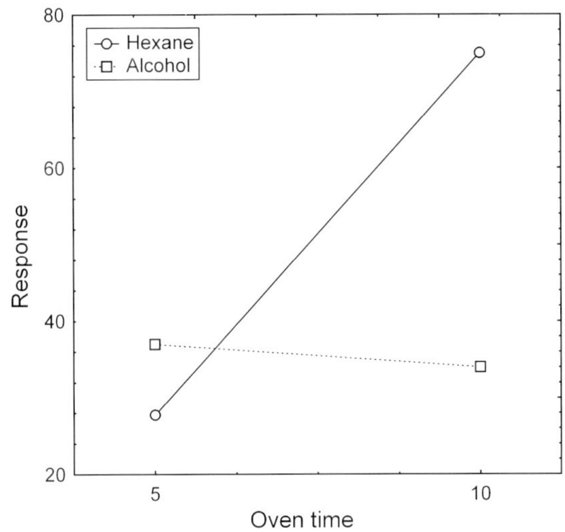

Fig. 4A.7. Average responses of Table 4A.5, as a function of factors **1** and **4**.

variance of an effect, with 4 degrees of freedom, from the expression

$$\hat{V}(effect) = \frac{(0.039)^2 + (0.024)^2 + (-0.029)^2 + (0.031)^2}{4} = 97.5 \times 10^{-4}.$$

The standard error of an effect is estimated by the square root of this value, 0.031. At the 95% confidence level, this corresponds to an interval of $\pm 2.776 \times 0.031 = \pm 0.087$, indicating that the effects of factors **9** (peptone), **1** (glucose) and **7** (temperature), in this order, are the most significant. Increasing the peptone concentration and the temperature tends to bolster violacein production, while increasing the glucose concentration tends to reduce it. All of this becomes clear on the cubic graph of the response averages as a function of these three factors (Fig. 4A.8). If the only response of interest were the production of violacein, these three factors would be strong candidates for closer study.

4A.7. Polyester resin cure

The industrial production of an unsaturated polyester resin is controlled by several factors, whose levels confer on it certain properties that are important to other industries that use the resin as raw material for making finished products. In the curing stage, a catalyst is added to accelerate gel formation by the resin. The time when the gel starts to form, after the catalyst is added, is an important characteristic known as gel time.

A.D. Liba and C.E. Luchini, graduate students in chemical engineering, decided to investigate (as an assignment for a chemometrics course) how

Table 4A.7
Experimental data

Factors	(–)	(+)
1 Carbon source (D-glucose) (%)	0.25	1
2 Yeast extract (%)	0.1	0.6
3 DL methionine (g L^{-1})	0.01	0.1
4 Vitamin B_{12}	0.012 ng mL^{-1}	0.01 μg mL^{-1}
5 L-tryptophan (%)	0.02	0.1
6 Stirring (rpm)	150	250
7 Temperature (°C)	28	33
8 pH (phosphate buffer)	6.8	7.8
9 Bacteriological peptone (%)	0.2	0.8
10 Inoculum (with 10 h and ABS = 0.26) (%)	1	5
11 $ZnSO_4$ (mM)	0.05	0.1

Run	**1**	**2**	**3**	**4**	**5**	**6**	**7**	**8**	**9**	**10**	**11**	**12**	**13**	**14**	**15**	Response
1	–	–	–	–	+	+	+	+	+	+	–	–	–	–	+	0.39
2	+	–	–	–	–	–	–	+	+	+	+	+	+	–	–	0.19
3	–	+	–	–	–	+	+	–	–	+	+	+	–	+	–	0.26
4	+	+	–	–	+	–	–	–	–	+	–	–	+	+	+	0.14
5	–	–	+	–	+	–	+	–	+	–	+	–	+	+	–	0.26
6	+	–	+	–	–	+	–	–	+	–	–	+	–	+	+	0.15
7	–	+	+	–	–	–	+	+	–	–	–	+	+	–	+	0.31
8	+	+	+	–	+	+	–	+	–	–	+	–	–	–	–	0.00
9	–	–	–	+	+	+	–	+	–	–	–	+	+	+	–	0.09
10	+	–	–	+	–	–	+	+	–	–	+	–	–	+	+	0.00
11	–	+	–	+	–	+	–	–	+	–	+	–	+	–	+	0.24
12	+	+	–	+	+	–	+	–	+	–	–	+	–	–	–	0.26
13	–	–	+	+	+	–	–	–	–	+	+	+	–	–	+	0.17
14	+	–	+	+	–	+	+	–	–	+	–	–	+	–	–	0.14
15	–	+	+	+	–	–	–	+	+	+	–	–	–	+	–	0.26
16	+	+	+	+	+	+	+	+	+	+	+	+	+	+	+	0.31

Contrasts

1	**2**	**3**	**4**	**5**	**6**	**7**	**8**
–0.99	0.049	0.04	–0.029	0.009	–0.001	0.086	–0.009
9	**10**	**11**	**12**	**13**	**14**	**15**	
0.119	0.069	–0.039	0.039	0.024	–0.029	0.031	

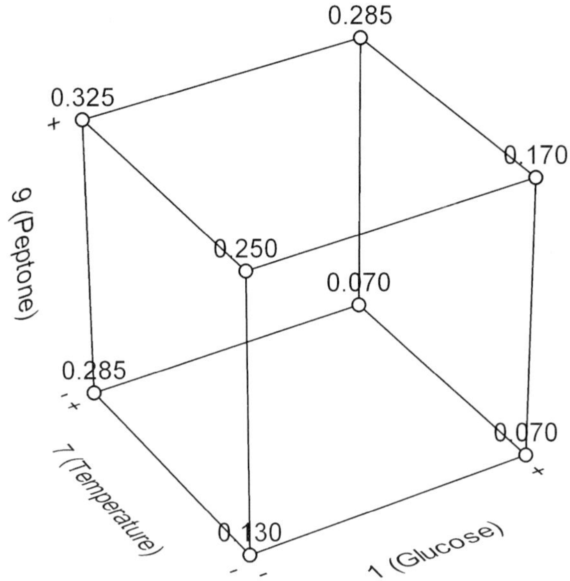

Fig. 4A.8. Average responses from Table 4A.7, as a function of factors **9**, **7** and **1**.

the concentrations of five additives used in the resin production process influenced gel time. To do this they used the 2^{5-1} fractional factorial design whose data are presented in Table 4A.8. The gel was produced with the catalyst methyl ethyl ketone, added at a constant proportion of 1% of resin weight.

The normal plot of the effects (Fig. 4A.9) is very easy to interpret, and shows that gel time is totally controlled by three factors (**5** = dimethylaniline, **4** = copper octanoate and **1** = hydroquinone, in this order of importance). Factors **4** and **5** contribute to lower the gel time, whereas factor **1** tends to increase it. There is no evidence of significant interactions. This is welcome, because it means the factors can be varied independently of one another, at the customer's convenience. The numerical values on the right side of the plot are the cumulative probabilities corresponding to the z values on the left.

Another interesting point worth mentioning in this application also serves to remind us of a crucial question in any design. Even though changing the levels of the factors directly affects the intensity of the response, we are not always at liberty to change them as we please. This is particularly true of an experiment performed in industry,[42] where the characteristics of the process usually impose several restrictions. All the levels of the factors in this experiment are concentrations, measured in

[42]Such as this one.

Table 4A.8
Experimental data. All levels are in ppm.

Factors	(–)	(+)
1 Hydroquinone (HQ)	190	210
2 Benzoquinone (BQ)	20	30
3 Copper octanoate (Cu)	180	360
4 Cobalt octanoate (Co)	900	1800
5 Dimethylaniline (DMA)	270	540

Run	Order	**1**	**2**	**3**	**4**	**5**	Gel time (min)
1	11	–	–	–	–	+	14.02
2	4	+	–	–	–	–	29.42
3	1	–	+	–	–	–	26.07
4	12	+	+	–	–	+	17.58
5	2	–	–	+	–	–	25.18
6	13	+	–	+	–	+	17.03
7	14	–	+	+	–	+	15.24
8	16	+	+	+	–	–	33.54
9	3	–	–	–	+	–	18.30
10	9	+	–	–	+	+	12.17
11	10	–	+	–	+	+	10.57
12	5	+	+	–	+	–	22.20
13	15	–	–	+	+	+	10.19
14	8	+	–	+	+	–	23.52
15	6	–	+	+	+	–	21.14
16	7	+	+	+	+	+	13.10

Contrasts

1	**2**	**3**	**4**	**5**	**12**	**13**	**14**
3.48	1.20	1.08	−5.86	−11.18	−0.13	0.38	−0.78
15	**23**	**24**	**25**	**34**	**35**	**45**	
−1.01	0.57	−0.49	−0.43	0.10	−0.77	1.40	

the same units, permitting us to make direct comparisons of the quantities used in the 16 runs. For the last three factors, the higher-level concentration is twice the one at the lower level. Gel time is not affected by factor **3**, but factors **4** and **5** are precisely those exhibiting the largest effects. It is interesting that the remaining significant effect, that of hydroquinone (**1**), is the one corresponding to the smallest concentration change between the two levels (only about 11%). If it were possible to

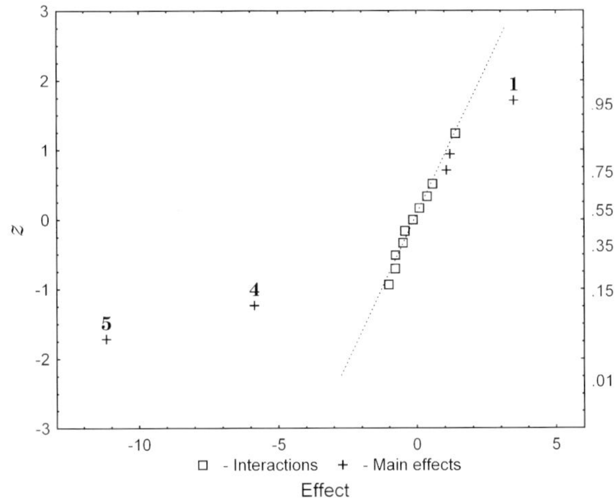

Fig. 4A.9. Normal plot of the effects.

vary the hydroquinone levels in the same proportion as those of the other two important factors (subjected, of course, to all the possible constraints we have just mentioned), perhaps its effect would become the most significant of all. Fig. 4A.10 illustrates this possibility, and facilitates visualization of the experimental results.

4A.8. Screening design for earplug production

In the initial phase of an optimization study (Sato, 2002), the researchers executed a screening design to evaluate the effects of seven factors on three responses considered critical for determining the quality of earplugs sold as hearing protectors by its manufacturer. The company had two production lines and the researchers used the smaller one for their study. Since they had to limit the number of experiments to minimize the time the production line would be out of routine operation, they decided to perform a 2^{7-4}_{III} fractional factorial design, which only requires eight experiments. The factors, their levels and the three response values obtained from these experiments are presented in Table 4A.9.

For the apparent density, the contrasts calculated for the seven factors (in $\mathrm{g\,cm^{-3}}$) and their expressions in terms of confounded factorial effects are as follows:

$l_{\mathrm{A}} = 0.0015 = \mathbf{A} + \mathbf{BD} + \mathbf{CE} + \mathbf{FG}$, $l_{\mathrm{B}} = 0.0020 = \mathbf{B} + \mathbf{AD} + \mathbf{CF} + \mathbf{EG}$,
$l_{\mathrm{C}} = -0.0025 = \mathbf{C} + \mathbf{AE} + \mathbf{BF} + \mathbf{DG}$, $l_{\mathrm{D}} = -0.0435 = \mathbf{D} + \mathbf{AB} + \mathbf{EF} + \mathbf{CG}$,
$l_{\mathrm{E}} = -0.0110 = \mathbf{E} + \mathbf{AC} + \mathbf{DF} + \mathbf{BG}$, $l_{\mathrm{F}} = -0.0005 = \mathbf{F} + \mathbf{BC} + \mathbf{DE} + \mathbf{AG}$,
$l_{\mathrm{G}} = 0.0000 = \mathbf{G} + \mathbf{CD} + \mathbf{BE} + \mathbf{AF}$.

The corresponding normal probability plot is given in Fig. 4A.11.

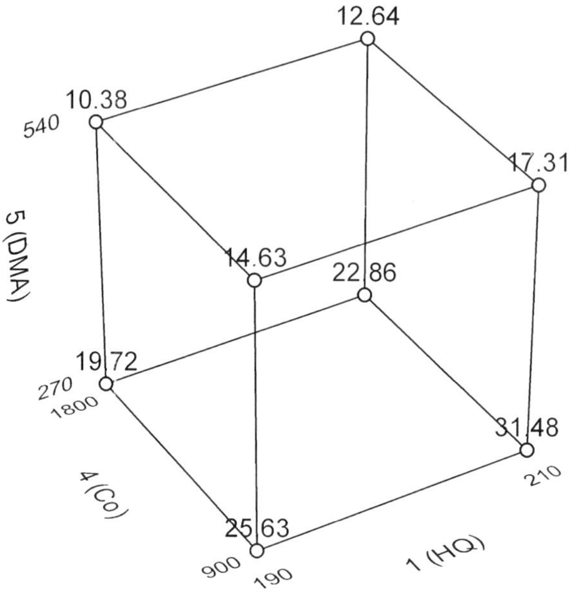

Fig. 4A.10. Averages response values from Table 4A.8, as a function of factors **1**, **4** and **5**.

The contrast values for factors **A**, **B**, **C**, **F** and **G** fit quite well to a straight line through the origin, and therefore can be interpreted as resulting from experimental error. The values of the other two contrasts, which include the main effects **D** and **E**, lie far from the line, especially the value of **D**. It appears, then, that the reticulate solution flow rate (**D**) and the mold temperature (**E**) do affect the apparent density values. The **E** contrast can be interpreted as the mold temperature effect if one assumes that the two-factor interactions in the definition of contrast l_E are negligible. If a similar assumption is made for l_D, it can be interpreted as the main effect of the reticulate solution flow rate.

The mold temperature effect is negative, resulting in lower apparent density values, which is desirable for the final product. Since operational constraints did not permit decreasing this temperature below 56 °C, the temperature was held at this value for further experimentation. The reticulate solution flow rate, on the other hand, was varied in subsequent phases of the optimization study (see case studies in Sections 3A.9 and 6A.4).

4A.9. Plackett–Burman designs for screening factors

Factorial design and response surface techniques were used in combination with modeling and simulation to design and optimize an

Table 4A.9
Factors, levels and response values observed for the 2^{7-4} screening design used to study earplug properties

Factors	(–)	(+)
A = Polyol solution flow rate ($g\,min^{-1}$)	446	454
B = Prepolymer solution flow rate ($g\,min^{-1}$)	318	324
C = Catalyst/extensor solution flow rate ($g\,min^{-1}$)	32	36
D = Reticulate solution flow rate ($g\,min^{-1}$)	15	23
E = Mold temperature (°C)	56	73
F = Curing temperature (°C)	93	149
G = Curing time (h)	2.83	5.58

Responses

AD = Apparent density ($g\,cm^{-3}$)
EF = Equilibrium force (lb)[a]
RT = Retention time (s)

Run	A	B	C	D	E	F	G	*AD*	*EF*	*RT*
1	–	–	–	+	+	+	–	0.122	0.089	8.910
2	+	–	–	–	–	+	+	0.178	0.275	0.226
3	–	+	–	–	+	–	+	0.168	0.331	0.255
4	+	+	–	+	–	–	–	0.137	0.116	6.720
5	–	–	+	+	–	–	+	0.131	0.082	10.150
6	+	–	+	–	+	–	–	0.165	0.234	1.470
7	–	+	+	–	–	+	–	0.176	0.363	0.260
8	+	+	+	+	+	+	+	0.123	0.091	9.500

[a] 1 lb = 4.448N.

industrial bioprocess, a multistage alcoholic fermentation process composed of four ideal continuously stirred tank reactors linked in series, with cell recycling (Kalil et al., 2000). Operational conditions for maximum yield and productivity were investigated as a function of the levels of 10 factors, the temperature ($\mathbf{T}_i$) and residence time ($\mathbf{rt}_i$) of each reactor (i = 1, 2, 3 and 4), cell recycling concentration (**c**) and the fraction of fresh medium (**f**) fed into the second fermentation stage. A half-fraction of a 20-run Plackett–Burman design was used to screen for the most important factors affecting yield and productivity. This design and the yield and productivity results obtained in the experiments are given in Table 4A.10.

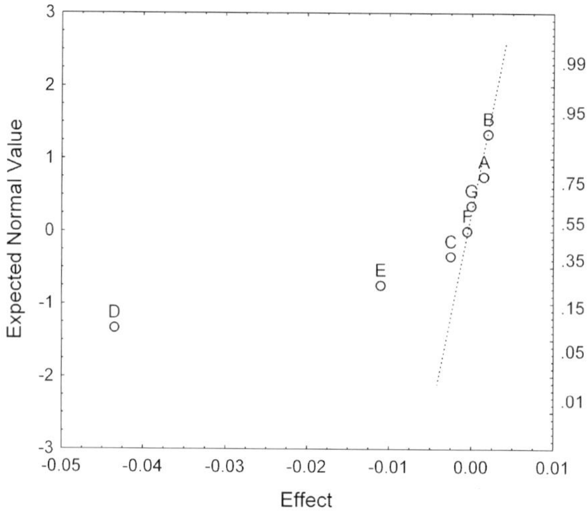

Fig. 4A.11. Normal plot of the contrasts for the apparent density data.

Table 4A.10
Yield and productivity results for the Plackett–Burman design based on different levels of reactor temperatures and retention times, cell cycling concentration and fraction of fresh medium fed into the second fermentation stage

Run	$\mathbf{T}_1$	$\mathbf{T}_2$	$\mathbf{T}_3$	$\mathbf{T}_4$	$\mathbf{rt}_1$	$\mathbf{rt}_2$	$\mathbf{rt}_3$	$\mathbf{rt}_4$	$\mathbf{c}$	$\mathbf{f}$	$\mathbf{d}_1$	$\mathbf{d}_2$	$\mathbf{d}_3$	$\mathbf{d}_4$	$\mathbf{d}_5$	$\mathbf{d}_6$	$\mathbf{d}_7$	$\mathbf{d}_8$	$\mathbf{d}_9$	Yield	Productivity
1	+	+	−	−	+	+	+	+	−	+	−	+	−	−	−	−	+	+	−	86.46	6.19
2	+	−	−	+	+	+	+	−	+	−	+	−	−	−	−	+	+	−	+	85.63	7.64
3	−	−	+	+	+	+	−	+	−	+	−	−	−	−	+	+	−	+	+	81.19	6.53
4	−	+	+	+	+	−	+	−	+	−	−	−	−	+	+	−	+	+	−	72.54	7.52
5	+	+	+	+	−	+	−	+	−	−	−	−	+	+	−	+	+	−	−	73.26	6.74
6	+	+	+	−	+	−	+	−	−	−	−	+	+	−	+	+	−	−	+	82.40	7.58
7	+	+	−	+	−	+	−	−	−	−	+	+	−	+	+	−	−	+	+	79.84	8.57
8	+	−	+	−	+	−	−	−	−	+	+	−	+	+	−	−	+	+	+	83.13	8.92
9	−	+	−	+	−	−	−	−	+	+	−	+	+	−	−	+	+	+	+	75.00	10.58
10	+	−	+	−	−	−	−	+	+	−	+	+	−	−	+	+	+	+	−	84.88	9.82

Natural variation in the simulation procedure was evaluated from the effect values calculated for the dummy factors ($\mathbf{d}_i$, $i = 1, 2, \ldots, 9$), just as done to estimate the standard effect error in case study in Section 4A.6. The effect values are listed in Table 4A.11.

For the yield we find that

$$\hat{V}(\text{effect}) = \frac{(2.06)^2 + (-0.31)^2 + \cdots + (0.82)^2}{9} = 1.35.$$

Table 4A.11
Effect values for yield and productivity obtained from the Plackett–Burman screening design

Effect	$\mathbf{T}_1$	$\mathbf{T}_2$	$\mathbf{T}_3$	$\mathbf{T}_4$	$\mathbf{rt}_1$	$\mathbf{rt}_2$	$\mathbf{rt}_3$	$\mathbf{rt}_4$	**C**	**F**
Yield	−1.06	−4.25	−4.23	−6.99	0.96	1.97	1.77	0.63	−2.31	−2.09
Productivity	0.06	−0.33	−0.62	−0.69	−0.99	−1.06	−1.21	−1.20	0.41	−0.22

Effect	$\mathbf{d}_1$	$\mathbf{d}_2$	$\mathbf{d}_3$	$\mathbf{d}_4$	$\mathbf{d}_5$	$\mathbf{d}_6$	$\mathbf{d}_7$	$\mathbf{d}_8$	$\mathbf{d}_9$
Yield	2.06	−0.31	−0.33	−1.38	0.11	1.62	−0.02	1.57	0.82
Productivity	0.06	0.00	−0.15	−0.35	−0.15	0.12	0.09	0.18	−0.02

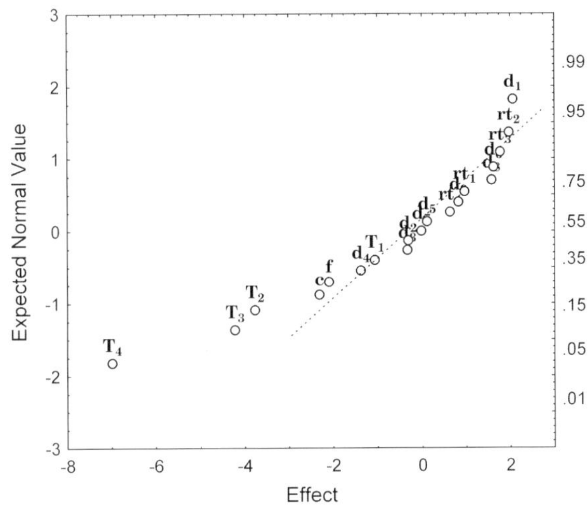

Fig. 4A.12. Normal plot of the effects on yield determined from the Plackett–Burman design.

The square root of this value, 1.16, provides an estimate of the uncertainty in the yield effect values. At the 95% confidence level we have limits of $\pm(1.16)(2.16) = \pm 2.62$. At this level of confidence only the $\mathbf{T}_2$, $\mathbf{T}_3$ and $\mathbf{T}_4$ effects are significant. Fig. 4A.12 shows the corresponding normal probability plot. The values of these effects are represented by the points farthest from a straight line passing through the origin in the normal plot. In contrast, all the $\mathbf{d}_i$ dummy factor effects lie closer to this line, despite a slight curvature on the right side of the plot.

For productivity, an uncertainty of ± 0.16 in the effect values is determined from the results for the dummy variables,

$$\hat{V}(\text{effect}) = \frac{(0.064)^2 + (0.004)^2 + \cdots + (-0.24)^2}{9} = \frac{0.2268}{9} = 0.02520.$$

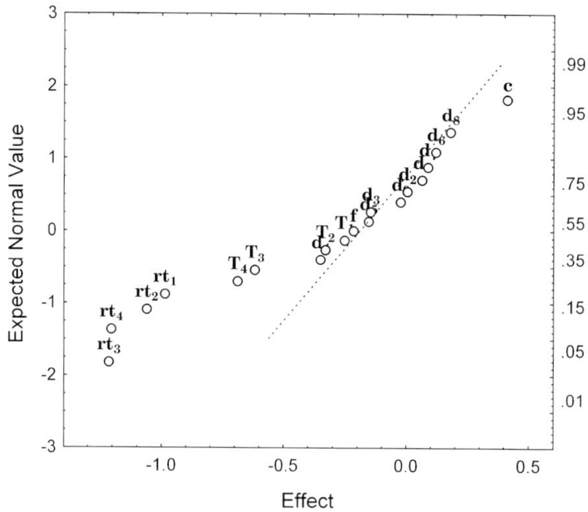

Fig. 4A.13. Normal plot of the effects on productivity determined from the Plackett–Burman design.

The 95% confidence interval is now $\pm(0.16)(2.26) = \pm 0.36$. The significant effects at this confidence level are those involving $\mathbf{T}_3$ and $\mathbf{T}_4$, the four retention times and the cell recycling concentration. Fig. 4A.13 shows the normal plot graph of the effects on productivity. Clearly, all the significant effect values lie off the line passing close to the origin, whereas the dummy effect values fit this line quite well.

CHAPTER 5
Empirical model-building

In the experimental designs we saw in the preceding chapters, each factor was studied at only two levels. Because of this economy, we had to be content with a limited form of the function describing the effects of the factors on the response. Consider, for example, the variation in reaction yield with temperature discussed in Chapter 3. The average yields observed with catalyst A are 59% at 40 °C, and 90% at 60 °C, as given in Table 3.1. We can see in Fig. 5.1a that these two pairs of values are compatible with an infinite number of functions. In Chapter 3, we fitted a model with a linear part and an interaction term to the response values, but we had no guarantee that this was the correct model. To clarify this situation, more information is needed.

If we make, say, three more measurements at intermediate temperatures and find that the plot of the five points is similar to Fig. 5.1b, we gain more confidence in the linear model. A graph like Fig. 5.1c, on the other hand, would be taken as evidence that the linear model is not the best one.

These considerations also serve to remind us that two-level designs are appropriate only in the initial stages of an investigation. To improve the description of the response surface, we have to perform a larger number of experiments.

5.1. A model for $y = f(X)$

Table 5.1 contains the yields obtained from a new design, in which the reaction was run at five equally spaced temperatures between 40 and 60 °C, in the presence of catalyst A. From the graph of these values (Fig. 5.2), the linear model really appears to be the one that best describes

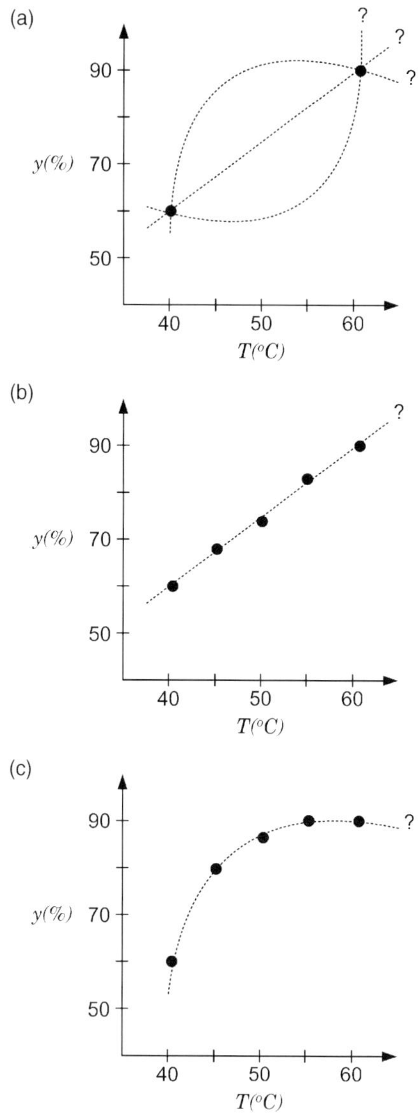

Fig. 5.1. (a) Two data points. Many functions can pass through both of them. (b) Pattern of points suggesting a linear function. (c) Pattern of points for which a linear model would be inadequate.

the variation of the yield with temperature. Recalling that each observation is affected by random error, we can represent this model by the equation

$$y_i = \beta_0 + \beta_1 T_i + \varepsilon_i, \quad (5.1a)$$

Table 5.1
Reaction yield as a function of temperature, in the 40–60°C range, with catalyst A

Temperature (°C)	40	45	50	55	60
Yield (%)	60	70	77	86	91

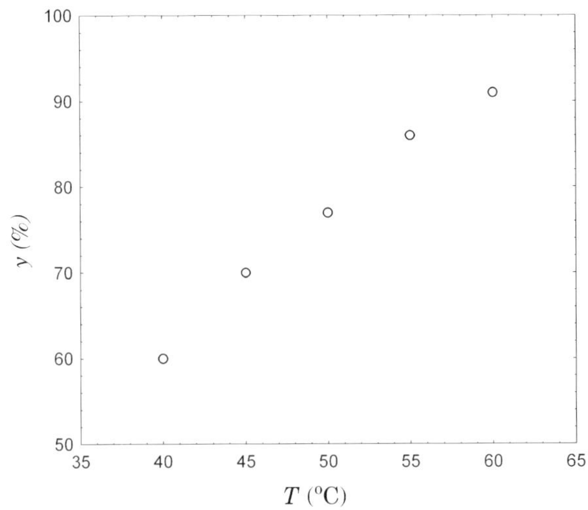

Fig. 5.2. Reaction yield as a function of temperature. Data from Table 5.1.

where y_i is the yield corresponding to temperature T_i and ε_i is the random error associated with the experimental determination of its value.[43] β_0 and β_1 are the model's parameters.

To determine the values of β_0 and β_1 we should fit Eq. (5.1a) to the five pairs of values (y_i, T_i) in Table 5.1. That is, we should solve a system of five equations,

$$\begin{aligned} y_1 &= \beta_0 + \beta_1 T_1 + \varepsilon_1 \\ y_2 &= \beta_0 + \beta_1 T_2 + \varepsilon_2 \\ &\vdots \\ y_5 &= \beta_0 + \beta_1 T_5 + \varepsilon_5, \end{aligned}$$

where each equation contains a pair of values (y_i, T_i) = (yield, temperature), and whose unknowns are β_0 and β_1. This system can be represented

[43]By convention, we use lower case letters to represent random variables and upper case ones for controlled variables.

in a more compact form by a single matrix equation,

$$\mathbf{y} = \mathbf{X}\boldsymbol{\beta} + \boldsymbol{\varepsilon}, \tag{5.1b}$$

where

$$\mathbf{y} = \begin{bmatrix} y_1 \\ y_2 \\ \vdots \\ y_5 \end{bmatrix} \quad \mathbf{X} = \begin{bmatrix} 1 & T_1 \\ 1 & T_2 \\ \vdots & \vdots \\ 1 & T_5 \end{bmatrix} \quad \boldsymbol{\varepsilon} = \begin{bmatrix} \varepsilon_1 \\ \varepsilon_2 \\ \vdots \\ \varepsilon_5 \end{bmatrix} \quad \boldsymbol{\beta} = \begin{bmatrix} \beta_0 \\ \beta_1 \end{bmatrix}.$$

Matrix Eq. (5.1b) has the important advantage of remaining valid in general, for any number of observations or model parameters. All that needs to be done, as the model changes, is to augment the matrices correspondingly, as we will see later.

You can verify, using a ruler, that it is not possible to draw a straight line passing simultaneously through all five points in Fig. 5.2. Any line we choose will miss at least some of the observations, as illustrated, with a bit of exaggeration, in Fig. 5.3. The deviations can be positive or negative, depending on whether the observed yields lie above or below the chosen line. Clearly, the best line is the one that passes closest to the experimental points, since it is impossible to pass through all of them.

In practical terms, "to pass closest to" means to minimize the distance of the points to the line, that is, minimize the total length of the vertical segments in Fig. 5.3. The traditional way of doing this is to locate the line in such a way that the sum of the squares of the residuals — the vertical distances — is minimal. For this reason, the method is called a

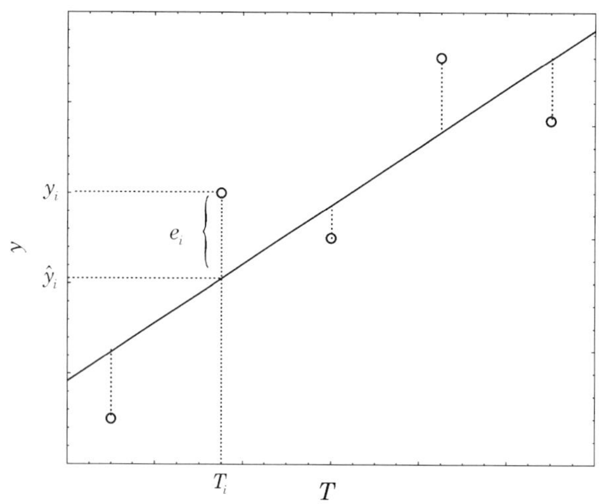

Fig. 5.3. Residuals left by a linear model. A residual is the difference between an observed value and its estimate obtained from the chosen model: $e_i = y_i - \hat{y}_i$.

least-squares fit. It is also known as **regression analysis**, a termed used for the first time in this context by Sir Francis Galton, one of the pioneers of statistics, in a paper entitled Regression towards Mediocrity in Hereditary Stature (Galton, 1886).

If, at temperature T_i, the observed yield is y_i and the yield predicted by the regression line is $\hat{y}_i$ (see Fig. 5.3), the residual left by the model is

$$e_i = y_i - \hat{y}_i, \tag{5.2}$$

where $\hat{y}_i = b_0 + b_1 T_i$, and b_0 and b_1 are the coefficients defining the intercept and the slope of the fitted line, that is, the estimators of the population parameters β_0 and β_1. Using matrices, we can write

$$\hat{\mathbf{y}} = \mathbf{Xb}, \tag{5.3}$$

where $\hat{\mathbf{y}}$ and $\mathbf{b}$ are column vectors containing, respectively, the predicted values of the model and the estimators of the parameters:

$$\hat{\mathbf{y}} = \begin{bmatrix} \hat{y}_1 \\ \hat{y}_2 \\ \vdots \\ \hat{y}_5 \end{bmatrix} \quad \text{and} \quad \mathbf{b} = \begin{bmatrix} b_0 \\ b_1 \end{bmatrix}.$$

Since the y_i values are known before making the regression, the residuals will depend only on the values we choose for b_0 and b_1. For the least-squares fit, these values are the ones that make the $\sum e_i^2$ sum as small as possible.[44]

To minimize the value of $\sum e_i^2$, it is necessary that its derivatives relative to b_0 and b_1 be both zero (Box et al., 1978, Chapter 14; Pimentel and Neto, 1996):

$$\frac{\partial(\sum e_i^2)}{\partial b_0} = 0, \tag{5.4a}$$

$$\frac{\partial(\sum e_i^2)}{\partial b_1} = 0. \tag{5.4b}$$

To generalize this derivation, we will represent the independent variable, which in this example is the temperature, by X. The regression equation then becomes $\hat{y}_i = b_0 + b_1 X_i$. Substituting this expression into Eq. (5.2), taking the square, and doing the sum, we have

$$\sum e_i^2 = \sum (y_i - \hat{y}_i)^2 = \sum (y_i - b_0 - b_1 X_i)^2.$$

[44]In general, sums are performed over the index i, from $i = 1$ to n, the total number of observations. We will only write indices below the summation sign if needed to avoid confusion.

Taking the derivatives and setting them equal to zero, we arrive at the expressions

$$\frac{\partial(\sum e_i^2)}{\partial b_0} = -2\sum(y_i - b_0 - b_1 X_i) = 0, \tag{5.5a}$$

$$\frac{\partial(\sum e_i^2)}{\partial b_1} = -2\sum X_i(y_i - b_0 - b_1 X_i) = 0. \tag{5.5b}$$

Canceling the -2 factor and separating out the sums, we wind up with a system of two linear equations in b_0 and b_1, called **normal equations**:

$$nb_0 + b_1\sum X_i = \sum y_i, \tag{5.6a}$$

$$b_0\sum X_i + b_1\sum X_i^2 = \sum X_i y_i. \tag{5.6b}$$

Isolating b_0 in (5.6a), we obtain

$$b_0 = \frac{\sum y_i - b_1\sum X_i}{n}$$

or

$$b_0 = \bar{y} - b_1\bar{X}. \tag{5.7}$$

Substituting the first of these expressions into (5.6b), we can write

$$\frac{\left(\sum y_i - b_1\sum X_i\right)\sum X_i}{n} + b_1\sum X_i^2 = \sum X_i y_i,$$

and from it

$$\frac{\sum X_i\sum y_i}{n} + b_1\left(\sum X_i^2 + \frac{\left(\sum X_i\right)^2}{n}\right) = \sum X_i y_i.$$

Isolating b_1, we finally have

$$b_1 = \frac{\sum X_i y_i - \frac{\sum X_i \sum y_i}{n}}{\sum X_i^2 - \frac{\left(\sum X_i\right)^2}{n}}. \tag{5.8}$$

This equation can be put in a form easier to remember, using the residuals about the averages of X and y (Exercise 5.1):

$$b_1 = \frac{\sum(X_i - \bar{X})(y_i - \bar{y})}{\sum(X_i - \bar{X})^2}, \tag{5.9}$$

or, in a more compact notation,

$$b_1 = \frac{S_{xy}}{S_{xx}}. \tag{5.10}$$

Exercise 5.1. Rewrite Eq. (5.9) to show that it is equivalent to Eq. (5.8).

Exercise 5.2. Show that

$$\mathbf{X}^t\mathbf{X} = \begin{bmatrix} n & \sum X_i \\ \sum X_i & \sum X_i^2 \end{bmatrix} \quad \text{and} \quad \mathbf{X}^t\mathbf{y} = \begin{bmatrix} \sum y_i \\ \sum X_i y_i \end{bmatrix}.$$

Exercise 5.3. With the notation introduced in Eq. (5.10), how would we represent the sample standard deviation of variable y?

We can calculate the values of b_0 and b_1 solving a single matrix equation. With the results of Exercise 5.2, the normal Eqs. (5.6a) and (5.6b) reduce to

$$\mathbf{X}^t\mathbf{X}\mathbf{b} = \mathbf{X}^t\mathbf{y}, \tag{5.11}$$

a result that you can easily prove, by explicitly writing the matrices. To solve this equation, we must multiply the left-hand side by the inverse of $\mathbf{X}_t\mathbf{X}$, to isolate vector **b**, whose elements are the estimators we seek:

$$\begin{aligned} (\mathbf{X}^t\mathbf{X})^{-1}(\mathbf{X}^t\mathbf{X})\mathbf{b} &= (\mathbf{X}^t\mathbf{X})^{-1}\mathbf{X}^t\mathbf{y}, \\ \mathbf{I}\mathbf{b} &= (\mathbf{X}^t\mathbf{X})^{-1}\mathbf{X}^t\mathbf{y}, \\ \mathbf{b} &= (\mathbf{X}^t\mathbf{X})^{-1}\mathbf{X}^t\mathbf{y}. \end{aligned} \tag{5.12}$$

This equation is very important and worth remembering. Augmenting the **X** and **y** matrices adequately, we have the general solution for a least-squares model fit, for any number of observations or parameters needed to specify the model. A solution only exists, however, if

(a) The $(\mathbf{X}^t\mathbf{X})^{-1}$ matrix can be calculated, that is, the $\mathbf{X}^t\mathbf{X}$ matrix is non-singular.
(b) The models are linear in their parameters, which means that they cannot contain terms like b_0^2 or $b_0 b_1$. This restriction, however, is not as drastic as it might first appear. With a little thought we can write quite complex equations that are nevertheless linear in their parameters (Exercise 5.4).

Using the data of Table 5.1, we can write

$$\mathbf{y} = \begin{bmatrix} 60 \\ 70 \\ 77 \\ 86 \\ 91 \end{bmatrix} \quad \text{and} \quad \mathbf{X} = \begin{bmatrix} 1 & 40 \\ 1 & 45 \\ 1 & 50 \\ 1 & 55 \\ 1 & 60 \end{bmatrix},$$

and from them

$$\mathbf{X}^t\mathbf{X} = \begin{bmatrix} 5 & 250 \\ 250 & 12,750 \end{bmatrix} \quad \text{and} \quad \mathbf{X}^t\mathbf{y} = \begin{bmatrix} 384 \\ 19,590 \end{bmatrix}.$$

Substituting these matrices into Eq. (5.11), we arrive at the system of linear equations

$$5b_0 + 250b_1 = 384,$$

$$250b_0 + 12,750b_1 = 19,590,$$

whose solution is

$$b_0 = -1.200 \quad \text{and} \quad b_1 = 1.560.$$

Choosing instead the matrix solution, we have

$$(\mathbf{X}^t\mathbf{X})^{-1} = \begin{bmatrix} 10.2 & -0.2 \\ -0.2 & 0.004 \end{bmatrix}$$

and therefore, according to Eq. (5.12),

$$\mathbf{b} = \begin{bmatrix} 10.2 & -0.2 \\ -0.2 & 0.004 \end{bmatrix} \times \begin{bmatrix} 384 \\ 19,590 \end{bmatrix} = \begin{bmatrix} -1.200 \\ 1.560 \end{bmatrix}.$$

With a larger number of experimental results or with more complicated models, the solutions are numerically more tedious, but can be easily calculated on a microcomputer, or even a scientific calculator.

Equation $\hat{y}_i = b_0 + b_1X_i$ gives us an estimate of the response obtained when the independent variable takes on the value X_i. With the values of b_0 and b_1 that we have just calculated, we can write

$$\hat{y}_i = -1.200 + 1.560X_i. \tag{5.13}$$

Substituting the X values (the temperatures) into this equation, we obtain the predicted yields (the $\hat{y}$ values). Using matrix equation (5.3), we can calculate all the predictions in a single operation:

$$\hat{\mathbf{y}} = \mathbf{Xb} = \begin{bmatrix} 1 & 40 \\ 1 & 45 \\ 1 & 50 \\ 1 & 55 \\ 1 & 60 \end{bmatrix} \times \begin{bmatrix} -1.200 \\ 1.560 \end{bmatrix} = \begin{bmatrix} 61.2 \\ 69.0 \\ 76.8 \\ 84.6 \\ 92.4 \end{bmatrix}.$$

Relative to the yields actually observed, these predictions leave the residuals

$$\mathbf{e} = \mathbf{y} - \hat{\mathbf{y}} = \begin{bmatrix} 60 \\ 70 \\ 77 \\ 86 \\ 91 \end{bmatrix} - \begin{bmatrix} 61.2 \\ 69.0 \\ 76.8 \\ 84.6 \\ 92.4 \end{bmatrix} = \begin{bmatrix} -1.2 \\ 1.0 \\ 0.2 \\ 1.4 \\ -1.4 \end{bmatrix}.$$

Fig. 5.4 shows how the fitted line is placed in relation to the observed points, providing visual confirmation that the linear model is indeed an excellent representation of the data in Table 5.1.

The values predicted by the regression line are given by $\hat{y}_i = b_0 + b_1 X_i$. Substituting the value of b_0, given by Eq. (5.7), into this equation, we have

$$\begin{aligned} &\hat{y}_i = \bar{y} - b_1\bar{X} + b_1 X_i, \\ \text{or } &\hat{y}_i = \bar{y} + b_1(X_i - \bar{X}). \end{aligned} \tag{5.14}$$

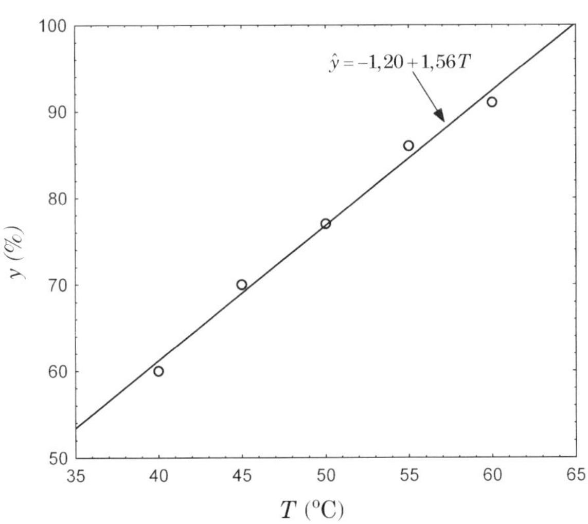

Fig. 5.4. Least-squares line for the data in Table 5.1.

When $X_i = \bar{X}$, the second term becomes zero and we obtain $\hat{y}_i = \bar{y}$. This shows that

- The least-squares line always passes through the average point of all observations, that is, the $(\bar{X}, \bar{y})$ point.
  From this fact we can also conclude that
- The sum of all the residuals left by the model is zero:

$$\sum e_i = \sum (y_i - \hat{y}_i)$$
$$= \sum \left[y_i - \bar{y} - b_1(X_i - \bar{X}) \right]$$
$$= \sum (y_i - \bar{y}) - b_1 \sum (X_i - \bar{X})$$
$$= 0 - b_1 \times 0 = 0,$$

since, as we have already seen in Chapter 2, the sum of the residual values of any variable relative to its average is always zero (Eq. (2.3)). It is for this reason, in fact, that we need to square the residuals before minimizing their sum.

Exercise 5.4. Distinguish, in the list below, the models that are linear in their parameters from those that are not.

(a) The function $y = b_0 \sin X + b_1 \cos(b_2 X)$
(b) Boyle's law, $pV = constant$
(c) The virial equation, $p\bar{V}/RT = 1 + B'P + C'P^2 + D'P^3 + \cdots$
(d) The model for the response of a 2^k factorial design
(e) A first-order reaction rate law, $[A] = [A]_0 e^{-kt}$

Exercise 5.5. Calibration is a very important procedure in instrumental analytical chemistry. In calibration, one attempts to find a quantitative relation between the response of an instrument and the concentration of a chemical species one wishes to determine. Normally, this relation is determined by means of the least-squares method. Consider the matrices

$$\mathbf{c} = \begin{bmatrix} 1 & 0.50 \\ 1 & 0.50 \\ 1 & 1.00 \\ 1 & 1.00 \\ 1 & 1.50 \\ 1 & 1.50 \\ 1 & 2.00 \\ 1 & 2.00 \\ 1 & 2.50 \\ 1 & 2.50 \\ 1 & 3.00 \\ 1 & 3.00 \end{bmatrix} \quad \text{and} \quad \mathbf{A} = \begin{bmatrix} 0.0937 \\ 0.0916 \\ 0.1828 \\ 0.1865 \\ 0.2782 \\ 0.2732 \\ 0.3776 \\ 0.3702 \\ 0.4562 \\ 0.4505 \\ 0.5593 \\ 0.5499 \end{bmatrix}$$

which were obtained in an experiment performed to construct a calibration curve. The elements of the second column of matrix **c** are concentrations of analytical standards of iron (in mg L^{-1}). The elements of vector **A** are the absorbances of these standards, determined by UV–visible spectroscopy (M. F. Pimentel, master's dissertation, UFPE, 1992). Fit a linear model to these data,

$$A_i = \beta_0 + \beta_1 c_i + \varepsilon_i,$$

knowing that

$$(\mathbf{c}^t\mathbf{c})^{-1} = \begin{bmatrix} 0.4333 & -0.2000 \\ -0.2000 & 0.1143 \end{bmatrix} \quad \text{and} \quad \mathbf{c}^t\mathbf{A} = \begin{bmatrix} 3.8696 \\ 8.3790 \end{bmatrix}.$$

Exercise 5.6. The fit of a statistical model to the results of a factorial design, which we discussed in Chapter 3, can also be done by the least-squares method, that is, by solving matrix equation $\mathbf{b} = (\mathbf{X}^t\mathbf{X})^{-1}\mathbf{X}^t\mathbf{y}$. Consider Eq. (3.11), corresponding to the 2^2 factorial

$$\hat{y}(x_1, x_2) = b_0 + b_1x_1 + b_2x_2 + b_{12}x_1x_2.$$

Choosing the elements of **X** in agreement with the signs in the table of contrast coefficients we can write

$$\mathbf{X} = \begin{bmatrix} 1 & -1 & -1 & 1 \\ 1 & 1 & -1 & -1 \\ 1 & -1 & 1 & -1 \\ 1 & 1 & 1 & 1 \end{bmatrix},$$

as we saw in Eq. (3.7). Calculate $\mathbf{X}^t\mathbf{X}$ and invert it to obtain $(\mathbf{X}^t\mathbf{X})^{-1}$. The inversion does not introduce problems, because $\mathbf{X}^t\mathbf{X}$ is an orthogonal matrix. Use the $\mathbf{X}^t\mathbf{y}$ matrix given on p. 97 and finally determine the **b** vector. Compare your results with the values given on p. 97.

5.2. The analysis of variance

Examining the residuals is fundamental for evaluating the fit of any model. In principle, the residuals should be small. If a certain model leaves residuals of considerable size, perhaps it can be improved. In an ideal model, all of the predicted values would exactly coincide with the observed response values, and there would be no residuals. Ideal models, however, are very hard to come by in a real world.

The method usually employed for a quantitative evaluation of the quality of fit of a model is the **analysis of variance** (or simply ANOVA). To perform an ANOVA, we start with an algebraic decomposition of the deviations of the observed response values relative to the overall average value.[45] As shown in Fig. 5.5, the deviation of an individual response from

[45]Some authors prefer the expression *grand average*. In this book, we use both expressions interchangeably.

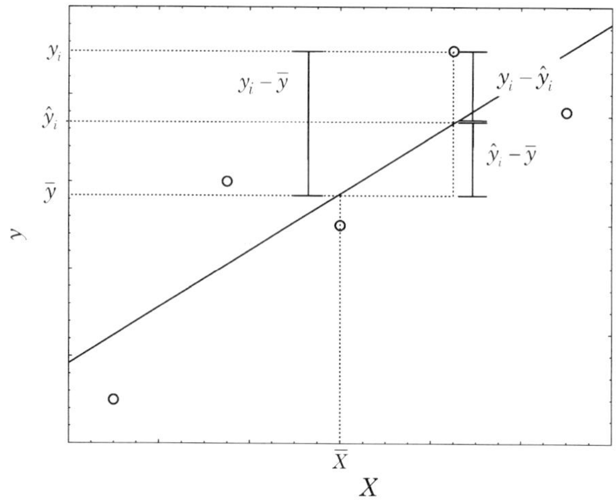

Fig. 5.5. Decomposition of the deviation of a response value from the grand average, $(y_i - \bar{y})$, as a sum of two contributions $(y_i - \hat{y}_i)$ and $(\hat{y}_i - \bar{y})$.

the average of all the observed values, $(y_i - \bar{y})$, can be decomposed into two parts:

$$(y_i - \bar{y}) = (\hat{y}_i - \bar{y}) + (y_i - \hat{y}_i). \tag{5.15}$$

The first part, $(\hat{y}_i - \bar{y})$, represents the deviation of the prediction made by the model for the point in question, $\hat{y}_i$, from the grand average, $\bar{y}$. The second contribution is the residual left by the model at this point, that is, the difference between the observed value and the corresponding prediction. In a well-adjusted model, we expect this second difference to be relatively small.[46] In terms of Eq. (5.15), this is equivalent to saying that the $(y_i - \bar{y})$ deviation should be approximately equal to the $(\hat{y}_i - \bar{y})$ deviation. To say that the deviations of the values predicted by the model are similar to those of the observed values (both relative to the overall average, $\bar{y}$) is the same as saying that the predictions are in good agreement with the observations.

The next step in an ANOVA is to express this comparison of deviations in quantitative terms. To do this, we square both sides of Eq. (5.15) and then take their sum over all points:

$$\begin{aligned}\sum (y_i - \bar{y})^2 &= \sum \left[(\hat{y}_i - \bar{y}) + (y_i - \hat{y}_i)\right]^2 \\ &= \sum (\hat{y}_i - \bar{y})^2 + 2\sum (\hat{y}_i - \bar{y})(y_i - \hat{y}_i) + \sum (y_i - \hat{y}_i)^2.\end{aligned}$$

[46]How small it can legitimately be, however, depends on the relative size of the random errors affecting the response values, else there would be model overfitting. We will return to this point later.

It can be demonstrated (Exercise 5.7) that the sum of products $(\hat{y}_i - \bar{y})(y_i - \hat{y}_i)$ is equal to zero, and consequently

$$\sum(y_i - \bar{y})^2 = \sum(\hat{y}_i - \bar{y})^2 + \sum(y_i - \hat{y}_i)^2. \tag{5.16}$$

These sums of squares of the deviations are often denoted by the acronym SS. Using this notation, we can read Eq. (5.16) as

$$\text{Total } SS = \text{Regression } SS + \text{Residual } SS.$$

In a more compact notation, we can write[47]

$$SS_T = SS_R + SS_r.$$

In other words, the regression equation accounts for part of the dispersion of the observations about the overall average $\bar{y}$, and the rest is left to the residuals. Evidently, the larger the fraction described by the regression, the better the model fit. This can be quantified by means of the ratio

$$R^2 = \frac{SS_R}{SS_T} = \frac{\sum(\hat{y}_i - \bar{y})^2}{\sum(y_i - \bar{y})^2}. \tag{5.17}$$

R^2 is called the **coefficient of determination** of the model. The maximum value of R^2 is 1, and will occur only if there are no residuals, that is, if all the variation about the average is explained by the regression. The closer the value of R^2 is to 1, the better the model will fit the observed values.

Exercise 5.7. Substitute $\hat{y}_i = \bar{y} + b_1 \sum(X_i - \bar{X})$ in $\sum(\hat{y}_i - \bar{y})(y_i - \hat{y}_i)$ and show that this sum is equal to zero.

A certain number of degrees of freedom is associated with each sum of squares, indicating how many independent values involving the n observations $y_1, y_2, \ldots, y_n$ are necessary to determine it. For the sum of squares of the n deviations about the average, the number of degrees of freedom is not n but $(n-1)$, because $\sum(y_i - \bar{y})$ is zero and this uses one degree of freedom, as we saw in Chapter 2.

To arrive at the number of degrees of freedom of SS_R we start from Eq. (5.14) and find that the regression sum of squares is given by

$$\sum(\hat{y}_i - \bar{y})^2 = b_1^2 \sum(X_i - \bar{X})^2. \tag{5.18}$$

Since the X_i variables are not random, the $\sum(X_i - \bar{X})^2$ term is fixed *a priori* by the design matrix used. The value of $\sum(\hat{y}_i - \bar{y})^2$ is therefore completely determined by a single number, the value of b_1. This value in turn is a random variable, because it depends on the experimentally determined responses. The result $b_1 = 1.56$ is valid only for the results of Table 5.1. If

[47]The total sum of squares is sometimes called the *corrected* total sum of squares, to emphasize that its values refer to deviations from the grand average.

we perform another series of identical experiments, carried out at the same temperatures, the experimental error will lead to different yields, with which we would calculate another value for b_1. This single new value again will be sufficient to determine a new value for the $\sum(\hat{y}_i - \bar{y})^2$.

These considerations show that the regression sum of squares has only one degree of freedom. Since the number of degrees of freedom of the total sum of squares is $(n-1)$, the residual sum of squares must have $(n-2)$ degrees of freedom, to be consistent with Eq. (5.16):

$$\nu_T = \nu_R + \nu_r,$$
$$(n-1) = 1 + (n-2).$$

The right side of this equation reflects the fact that our model contains only two parameters, β_0 and β_1. In the general case of a model with p parameters, the number of degrees of freedom of the residual sum of squares is given by the difference between the number of response values and the number of estimated parameters: $\upsilon_r = (n-p)$. Since $\upsilon_T = (n-1)$ always, the number of degrees of freedom of the regression sum of squares has to be equal to the number of parameters minus one: $\nu_R = (p-1)$.

The results of this section for the special case of a model with only two parameters are summarized in Table 5.2, which is called an ANOVA table. Dividing the sums of squares by their respective number of degrees of freedom, we obtain the mean squares (MS's) shown in the last column of the table.

So far, we have only made an algebraic decomposition of the total sum of squares. Soon we will see that, under certain assumptions, we can give the average sums of squares an interpretation that will let us use them to perform some statistical tests and to calculate confidence intervals.

In our example, with the response values of Table 5.1 and the predictions given by Eq. (5.13), the ANOVA results in the numerical values presented in Table 5.3. Substituting into Eq. (5.17) the values calculated for SS_R and SS_T, we obtain

$$R^2 = \frac{608.4}{614.8} = 0.9896,$$

meaning that 98.96% of the total variation about the average is explained by the regression. Only 1.04% is left for the residuals.

The residual sum of squares, SS_r, represents the part of the variation of the response values (about the average) that the model could not account for. Dividing it by ν_r, we obtain the residual mean square, which is an estimate, with $n-2$ degrees of freedom, of the variance of the points about the regression equation, that is, about the fitted model. The estimate can be interpreted as an approximate measure of the (squared) average error that would occur if we used the regression equation to predict the response y_i corresponding to a given X_i value. In our example,

Table 5.2
ANOVA for the fit of a linear model with two parameters

Source of variation	Sum of squares	Degree of freedom	Mean square
Regression	$\sum(\hat{y}_i - \bar{y})^2$	1	$MS_R = SS_R$
Residual	$\sum(y_i - \hat{y}_i)^2$	$n-2$	$MS_r = \frac{SS_r}{n-2} = s^2$
Total	$\sum(y_i - \bar{y})^2$	$n-1$	

Table 5.3
ANOVA for the fit of a linear model to the data of Table 5.1

Source of variation	Sum of squares	Degree of freedom	Mean square
Regression	608.4	1	608.4
Residual	6.4	3	6.4/3 = 2.13
Total	614.8	4	
% explained variation: 98.96			

$S^2 = 2.13$, with 3 degrees of freedom, as shown in the next to last line of Table 5.3.

Exercise 5.8. Start from Eq. (5.9) and show that $b_1 = \frac{\sum(X_i - \bar{X})\, y_i}{S_{xx}}$.

5.3. Confidence intervals

To postulate our model (Eq. (5.1)), we assumed that each observation y_i consists of a systematic part, $\beta_0 + \beta_1 X_i$, and a random part, ε_i. Let us assume now that the model is correct, that is, that the systematic part is indeed an accurate description of the actual relation between the y and X variables. In this case, if we performed a large number of repeated experiments at the same value of X_i we should observe a distribution of the y_i individual values around the response population value, $\beta_0 + \beta_1 X_i$. This value, therefore, is the mean of the observed responses at X_i, if the linear model is correct. Since there is no measure strictly without error, the response values of repeated experiments will always fluctuate, carrying uncertainty into the determination of the parameter values and the predictions made from the model, even if the model is adequate. In this section we will show how, by making a few assumptions about the behavior of the errors ε_i, we can quantify this uncertainty and determine confidence intervals for all estimated values.

If the true mean value of y_i is $\beta_0 + \beta_1 X_i$, we expect that repeated observations at the same point X_i will be distributed symmetrically about $\beta_0 + \beta_1 X_i$, with positive deviations as frequent as negative ones, in such a way that the errors will average out to zero. At a given X_i, the errors in the y_i values will be distributed with a certain variance σ_i^2, which in principle also varies with X_i. To proceed with our deductions we also need to make the following hypotheses:

1. The error variance is constant for all the investigated range of X, and is equal to a certain value σ^2. This hypothesis is often made by stating that the observed responses are **homoscedastic**.
2. The errors for responses observed at different values of the independent variable are uncorrelated, that is, Cov $(\varepsilon_i, \varepsilon_j) = 0$, if $i \neq j$. Since the sole random contribution to y_i is the error ε_i, we can also conclude that $V(y_i) = \sigma^2$ and Cov $(y_i, y_j) = 0$, if $i \neq j$.
3. The errors follow a normal distribution. This is a good approximation for most experiments where the response is a continuous variable, thanks to the central limit theorem (Section 2.3) and the efforts researchers make to eliminate systematic errors from their experiments.

These three hypotheses about the behavior of the random errors can be summarized by the following expressions:

$$\varepsilon_i \approx N(0, \sigma^2) \quad \text{and} \quad \mathrm{Cov}(\varepsilon_i, \varepsilon_j) = 0, \tag{5.19}$$

or, equivalently,

$$y_i \approx N(\beta_0 + \beta_1 X_i, \sigma^2) \quad \text{and} \quad \mathrm{Cov}(y_i, y_j) = 0. \tag{5.19a}$$

With these assumptions we can finally determine confidence intervals for the results of the model.

We have already seen (Exercise 5.8) that the slope of the regression line can be given by

$$b_1 = \frac{\sum (X_i - \bar{X}) y_i}{S_{xx}},$$

which is a linear combination of the random variables y_i, with $(X_i - \bar{X})/S_{xx}$ coefficients:

$$b_1 = \frac{(X_1 - \bar{X})}{S_{xx}} y_1 + \cdots + \frac{(X_n - \bar{X})}{S_{xx}} y_n.$$

Since by hypothesis the y_i's, besides not being correlated, have constant variance, we can apply Eq. (2.14) and write

$$V(b_1) = \left(\frac{X_1 - \bar{X}}{S_{xx}}\right)^2 V(y_1) + \cdots + \left(\frac{X_n - \bar{X}}{S_{xx}}\right)^2 V(y_n)$$
$$= \left[\sum \left(\frac{X_i - \bar{X}}{S_{xx}}\right)^2\right] \sigma^2,$$
$$= \frac{\sigma^2}{(S_{xx})^2} \sum (X_i - \bar{X})^2$$

or, because $\sum(X_i - \bar{X})^2 = S_{xx}$,

$$V(b_1) = \frac{\sigma^2}{S_{xx}}. \qquad (5.20)$$

Assuming now that the value of s^2, the residual variance about the regression model, is a good estimate of σ^2, we can obtain an estimate of the standard error of b_1 taking the square root of Eq. (5.20) and substituting σ by s:

$$\text{standard error of } b_1 = \frac{s}{\sqrt{S_{xx}}}. \qquad (5.21)$$

Since we are also assuming that the errors are normally distributed, we can use Student's distribution to test the significance of the value estimated for b_1. We follow the procedure discussed in Chapter 2 and define confidence intervals by means of

$$b_1 \pm t_{n-2} \times (\text{standard error of } b_1). \qquad (5.22)$$

Note that the number of degrees of freedom of the t-distribution point is $n-2$, which is the number of degrees of freedom of the s^2 estimate, and consequently of the standard error.

With the numerical values of our example, the standard error of b_1 becomes

$$\frac{s}{\sqrt{S_{xx}}} = \sqrt{\frac{2.13}{250}} = 0.0923,$$

which corresponds to the interval

$$1.560 \pm 3.182 \times 0.0923 = (1.266, 1.854),$$

at the 95% confidence level. Since its two limits are positive, the b_1 value we calculated is significantly different from zero, confirming the existence of a relation between the observed yields and the reaction temperatures.

Other confidence intervals are obtained in the same way. Starting from the algebraic expression for b_0 (Eq. 5.7) and following the same procedure used for b_1, we arrive at

$$V(b_0) = \frac{\sum X_i^2}{nS_{xx}} \sigma^2, \tag{5.23}$$

and from this, at the expression for the standard error of b_0:

$$\text{standard error of } b_0 = s\sqrt{\frac{\sum X_i^2}{nS_{xx}}}. \tag{5.24}$$

The standard error calculated from the data in our example is 4.665, resulting in the interval

$$\begin{aligned} & b_0 \pm t_{n-2} \times (\text{standard error of } b_0) \\ & = -1.200 \pm 3.182 \times 4.665 = (-16.044, 13.644). \end{aligned} \tag{5.25}$$

This means there is a 95% probability that the true value of the parameter β_0 lies between -16.044 and 13.644. Since these two limiting values have opposite signs, and since no value in a confidence interval is more probable than any other, the true value of β_0 could be zero. In other words, the value $b_0 = -1.200$ is not statistically significant and there is not enough evidence for maintaining the β_0 term in the model. All the same, it is customarily kept in the equation, for the sake of mathematical hierarchy.

Note that the values of the standard errors are very different. The standard error of b_0, 4.665, is about 50 times larger than the standard error of b_1, which is 0.0923. The reason for this is that the standard error of b_0 is affected by the sum of the squared X_i values, as shown in Eq. (5.24). If the calculations had been based in temperature values closer to zero, the two errors would have more similar values.

Exercise 5.9. Let a and c be two linear combinations of random variables y_i,

$$\begin{aligned} a &= a_1 y_1 + \cdots + a_n y_n \\ c &= c_1 y_1 + \cdots + c_n y_n. \end{aligned}$$

If $V(y_i) = \sigma^2$ and $\mathrm{Cov}(y_i, y_j) = 0$, then $\mathrm{Cov}(a, c) = (a_1 c_1 + \cdots + a_n c_n)\sigma^2$. Use the result of Exercise 5.8 and this last expression to show that the covariance of $\bar{y}$ and b_1 is zero.

When we develop a regression model, our objective is to obtain a relation that allows us to predict the response corresponding to any specific value of the independent variable, which we shall label X_a. In the model we are adopting, this relation is given by Eq. (5.14),

$$\hat{y}_a = \bar{y} + b_1(X_a - \bar{X}),$$

where the estimate $\hat{y}_a$ is a linear combination of the two random variables $\bar{y}$ and b_1. Its variance will be given by (see Eq. (2.14))

$$V(\hat{y}_a) = V(\bar{y}) + (X_a - \bar{X})^2 V(b_1) + 2(X_a - \bar{X})\text{Cov}(\bar{y}, b_1).$$

Since the covariance of $\bar{y}$ and b_1 is zero (Exercise 5.9), this expression reduces to

$$V(\hat{y}_a) = V(\bar{y}) + (X_a - \bar{X})^2 V(b_1).$$

Substituting the expressions for the variances of the average (Eq. (2.15a)) and b_1 (Eq. (5.20)), we obtain

$$V(\hat{y}_a) = \sigma^2 \left[\frac{1}{n} + \frac{(X_a - \bar{X})^2}{S_{xx}} \right]. \tag{5.26}$$

Substituting again σ^2 by s^2 and taking the square root, we finally obtain the standard error of the estimate $\hat{y}_a$:

$$\text{standard error of } \hat{y}_a = s\sqrt{\frac{1}{n} + \frac{(X_a - \bar{X})^2}{S_{xx}}}. \tag{5.27}$$

When $X_a = \bar{X}$, the second term under the root sign becomes zero and the standard error of $\hat{y}_a$ takes on its minimum value. As we move away from this point, in either direction, the uncertainties in the predicted values will increase. The farther we are from $X_a = \bar{X}$, the more uncertain will be the predictions of the model.

As always, we use the standard error to define confidence intervals about the predicted value:

$$\hat{y}_a \pm t_{n-2} \times (\text{standard error of } \hat{y}_a).$$

Equation (5.27) shows that the limits of the interval vary with the position along the X-axis. Geometrically, they determine hyperbolas above and below the regression line.

Equation (5.27) refers to the error in the estimate of the population mean of the response values at X_a. The individual response values distribute themselves about this mean (which is $\beta_0 + \beta_1 X_a$) with variance σ^2, as given by Eq. (5.19a). If we wish to refer to the prediction of *a single observed value* obtained at the X_a value, we need to include this variance. Thus, we write

$$\begin{bmatrix} \text{standard prediction error} \\ \text{for one observation} \end{bmatrix} = s\sqrt{1 + \frac{1}{n} + \frac{(X_a - \bar{X})^2}{S_{xx}}}, \tag{5.28}$$

which is an extension of Eq. (5.27). To obtain it, we simply added σ^2 to the variance of $\hat{y}_a$.

Likewise, if we are interested in predicting the average of q observations, we have

$$\begin{bmatrix} \text{standard prediction error} \\ \text{for the average of } q \text{ observations} \end{bmatrix} = s\sqrt{\frac{1}{q}+\frac{1}{n}+\frac{(X_a-\bar{X})^2}{S_{xx}}}. \tag{5.29}$$

Using these errors, we can determine confidence intervals based on the t distribution, as we did for the other estimates. We shall leave this as an exercise for interested readers. These intervals — and all confidence intervals, for that matter — should be interpreted exactly as the intervals for the population mean were interpreted in Section 2.3.

Matrices can also be used to calculate the uncertainties in the parameter estimates. To do this, we have to define first the **covariance matrix** of b_0 and b_1:

$$\mathbf{V}(\mathbf{b}) = \begin{bmatrix} V(b_0) & \mathrm{Cov}(b_0, b_1) \\ \mathrm{Cov}(b_0, b_1) & V(b_1) \end{bmatrix}.$$

It can be shown that this matrix is given simply by

$$\mathbf{V}(\mathbf{b}) = (\mathbf{X}^t\mathbf{X})^{-1}\sigma^2. \tag{5.30}$$

This is another very important equation that applies to the least-squares fit of any linear model,[48] under the assumptions that we made at the beginning of this section. This equation is also worth remembering.

For our example, we use the $(\mathbf{X}^t\mathbf{X})^{-1}$matrix calculated in Section 5.1 and substitute the population variance, σ^2, by its estimate $s^2 = 2.13$:

$$\mathbf{V}(\mathbf{b}) = \begin{bmatrix} 10.2 & -0.2 \\ -0.2 & 0.004 \end{bmatrix} \times 2.13 = \begin{bmatrix} 21.73 & -0.43 \\ -0.43 & 8.52 \times 10^{-3} \end{bmatrix}.$$

Taking the square roots of the elements on the main diagonal, we arrive at the standard errors of b_0 and b_1.

5.4. Statistical significance of the regression model

Now that we added the assumption that the errors follow a normal distribution to our hypotheses, we can return to the ANOVA and use the mean square values to test if the regression equation is statistically significant. When $\beta_1 = 0$, that is, when there is no relation between X and y, it can be demonstrated that the ratio of the MS_R and MS_r mean squares

[48]Linear in its parameters, as we discussed in Section 5.1.

follows an F distribution:

$$\frac{MS_R}{MS_r} \approx F_{1,n-2}, \tag{5.31}$$

where 1 and $n-2$ are the numbers of degrees of freedom of the regression mean square and the residual mean square, respectively. Since Eq. (5.31) is only valid for $\beta_1 = 0$, we can test this null hypothesis comparing the value calculated for MS_R/MS_r with the tabulated $F_{1,n-2}$ value, at the desired confidence level. If we find that $MS_R/MS_r > F_{1,n-2}$, we must reject the possibility that $\beta_1 = 0$. We would then have sufficient statistical evidence to believe in the existence of a relation between the y and X variables, and the larger the MS_R/MS_r value the higher our confidence in this conclusion.

For our example we need the value of $F_{1,3}$, which can be found in Table A.4, at the intersection of the column for $\nu_1 = 1$ with the row for $\nu_2 = 3$. At the 95% confidence level, this value is 10.13. Our regression will be statistically significant if we find that $MS_R/MS_r > 10.13$. Otherwise, we will have no reason to doubt that the value of β_1 is really zero and consequently there is no relation between the variables.

Using the values of Table 5.3, $MS_R/MS_r = 608.4/2.13 = 285.6$, which shows that our equation is highly significant. However, a regression found to be significant using the F-test is not always useful to make predictions. It might happen that the range spanned by the factors is not large enough to result in calculated effect values that can be clearly distinguished from experimental error. A useful rule of thumb is to consider a regression adequate for prediction if the MS_R/MS_r value is, at least, about 10 times larger than the F distribution point with the appropriate number of degrees of freedom, at the chosen confidence level (Box and Wetz, 1973; Box and Draper, 1987). In this example we would like to have an MS_R/MS_r value larger than 101. This condition is easily met by the results of our example.

5.5. A new model for $y = f(X)$

Heartened by the results obtained so far, our chemist decides to extend the temperature range and performs four more experiments, at 30, 35, 65 and 70 °C. The new yields appear in Table 5.4, together with the values previously obtained. Using Eq. (5.12) to fit the linear model to the nine pairs of values of this new data set, we obtain

$$\hat{y} = -7.33 + 1.52X. \tag{5.32}$$

Table 5.5 shows the ANOVA for this new fit. The percent variation explained by the model is now 80.63%. This value is reasonably high, but

Table 5.4
Reaction yield as a function of temperature, in the 30–70°C range, with catalyst A

Temperature (°C)	30	35	40	45	50	55	60	65	70
Yield (%)	24	40	60	70	77	86	91	86	84

Table 5.5
ANOVA for the fit of a linear model to the data of Table 5.4

Source of variation	Sum of squares	Degree of freedom	Mean square
Regression	3465.6	1	3465.6
Residual	832.4	7	118.9
Total	4298.0	8	
% explained variation: 80.63			

much less impressive than the 98.96% figure from the preceding example, which was limited to the 40–60 °C range.

The reason for the deterioration of the fit becomes clear when we place the points for the nine yields and the regression line on the same graph (Fig. 5.6a). The points have a curved pattern, indicating that a linear model is no longer able to represent them. This impression is confirmed by the plot of the residuals (Fig. 5.6b). Instead of being distributed in a random fashion, the residuals clearly suggest a geometrical pattern, with the positive values concentrated in the center of the plot while the negative ones lie close to the borders.

The value of $MS_R/MS_r = 29.14$, whereas $F_{1,7} = 5.59$, at the 95% level. This indicates that we have a significant regression,[49] but the use of an F-test pre-supposes that the residuals have a normal distribution, and we have just seen that this is not the case. We can only use an *F-test* if there is no evidence of abnormality in the distribution of the residuals.

Since the linear model has proved unsatisfactory, let us augment it, adding a quadratic term. We will try to model the influence of temperature on the yield by the equation

$$y_i = \beta_0 + \beta_1 X_i + \beta_2 X_i^2 + \varepsilon_i, \tag{5.33}$$

where X_i represents the temperature at the ith level. The fit of the new model to the observed values is also done by means of Eq. (5.12), except that the matrices must be correspondingly augmented to refer to Eq.

[49]Note that the number of degrees of freedom of the residual mean square changed, because of the larger number of points.

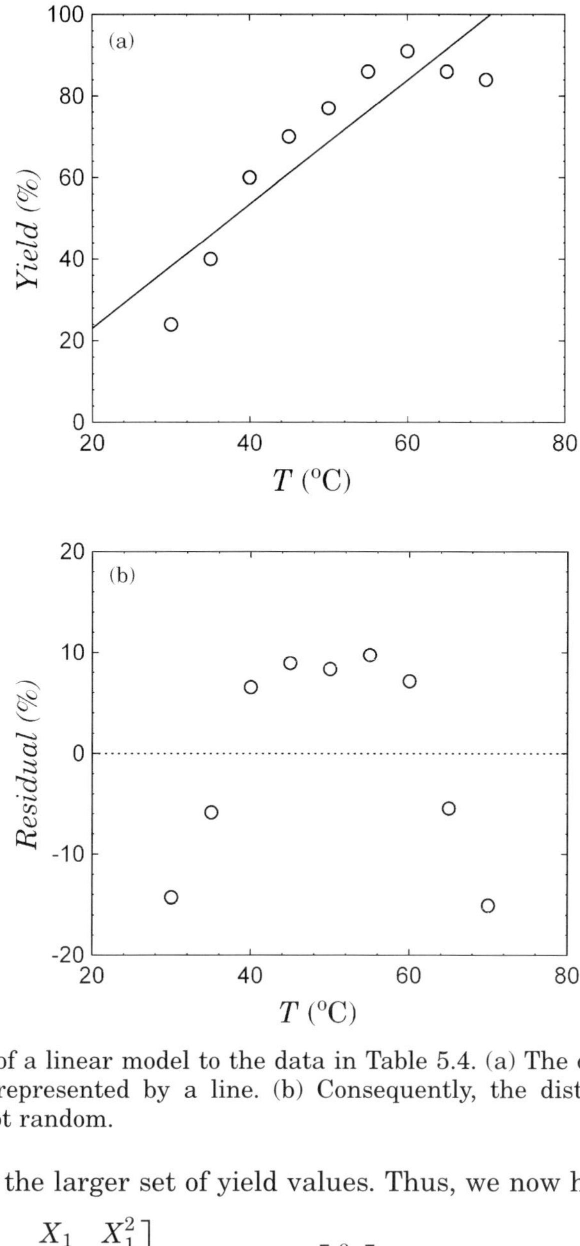

Fig. 5.6. Fit of a linear model to the data in Table 5.4. (a) The observed values are not well represented by a line. (b) Consequently, the distribution of the residuals is not random.

(5.33) and to the larger set of yield values. Thus, we now have

$$\mathbf{X} = \begin{bmatrix} 1 & X_1 & X_1^2 \\ 1 & X_2 & X_2^2 \\ \cdots & \cdots & \cdots \\ 1 & X_9 & X_9^2 \end{bmatrix} \quad \text{and} \quad \boldsymbol{\beta} = \begin{bmatrix} \beta_0 \\ \beta_1 \\ \beta_2 \end{bmatrix},$$

besides, of course

$$\mathbf{y} = \begin{bmatrix} y_1 \\ y_2 \\ \vdots \\ y_9 \end{bmatrix}.$$

Substituting the appropriate values into Eq. (5.12), we obtain

$$\mathbf{b} = \begin{bmatrix} -158.24 \\ 7.99 \\ -0.065 \end{bmatrix},$$

which means our quadratic model estimates the yields by the equation

$$\hat{y} = -158.24 + 7.99T - 0.065T^2. \tag{5.34}$$

Fig. 5.7a shows a graph of this expression, together with the observed values. The fit is much better than the one in Fig. 5.6a, and this is reflected in the graph of the new residuals. As we can see in Fig. 5.7b, it is no longer incompatible with a random distribution, and this authorizes us to use an F-test. Also, the residuals are now much smaller than those left by the linear model.

The ANOVA (Table 5.6) confirms the superiority of the quadratic model. The new model accounts for 99.37% of the total variance, against only 80.63% for the linear model. The value of MS_R/MS_r increases to 471.4, from 29.14 for the linear model. Since introducing the β_2 parameter in the model transfers 1 degree of freedom from the residual mean square to the regression mean square, the new MS_R/MS_r value should be compared with $F_{2,6}$ (which is 5.14 at the 95% level), rather than with $F_{1,7}$. In any case, from these results we may conclude that now we have a highly significant fit.

This comparison of linear and quadratic models is a good occasion to remember that empirical models are *local* models, that is, models that can only be applied to specific regions. This characteristic makes extrapolation a very risky endeavor. We only have to remember that the linear model was shown to be quite satisfactory for the first set of values, but the small extension of the temperature range made it necessary to employ a quadratic model, even though the data in Table 5.1 are all contained in Table 5.4. Even this second model should not be extrapolated, and we do not have to go far to appreciate this. If we let, for example, $T = 20\,^\circ\mathrm{C}$ in Eq. (5.34), which represents only $10\,^\circ\mathrm{C}$ less than the lowest temperature investigated, we obtain $\hat{y} = -24.44\%$, an absurd value, since negative yields are meaningless. In short, we should never trust extrapolations. If, even so, we dare make them, they should always be tested with further experiments, called for this reason **confirmatory experiments**.

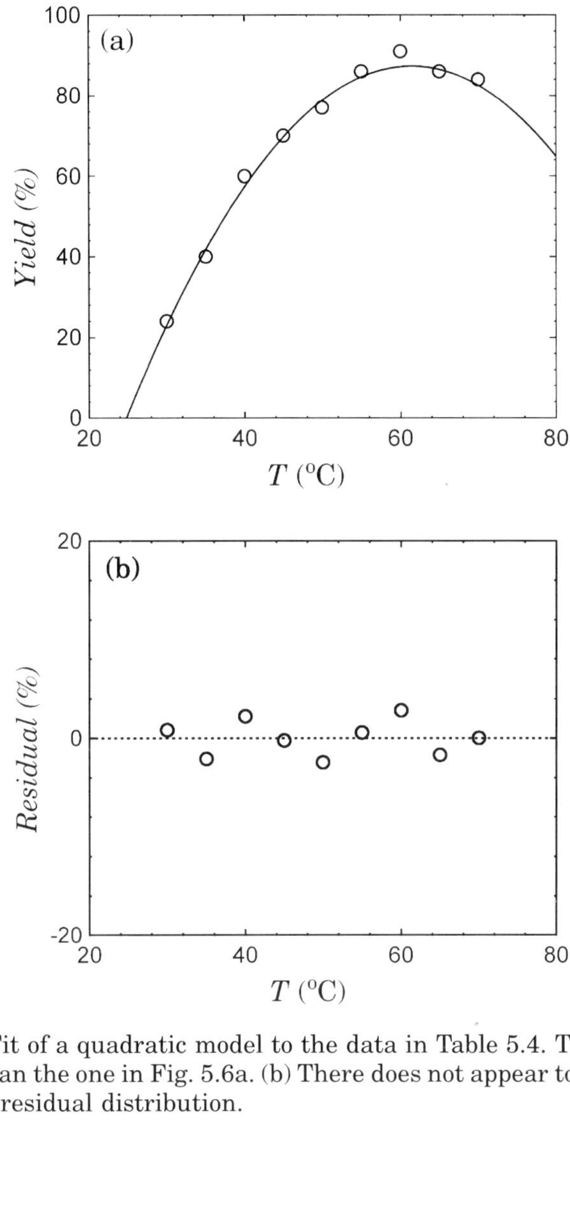

Fig. 5.7. (a) Fit of a quadratic model to the data in Table 5.4. The agreement is much better than the one in Fig. 5.6a. (b) There does not appear to be a systematic pattern in the residual distribution.

5.6. Lack of fit and pure error

So far we have based the evaluation of our models on the appearance of the residual graph. If the distribution of the residuals does not show any systematic structure, that is, if it seems random, we consider the model satisfactory. This approach is no doubt quite subjective but we should not

Table 5.6
ANOVA for the fit of a quadratic model to the data of Table 5.4

Source of variation	Sum of squares	Degree of freedom	Mean square
Regression	4270.8	2	2135.4
Residual	27.2	6	4.53
Total	4298.0	8	
% explained variation: 99.37			

Table 5.7
Reaction yield as a function of temperature, in the 30–70°C range, with catalyst A. The runs are in duplicate

Temperature (°C)	30	35	40	45	50	55	60	65	70
Yield (%)	24	40	60	70	77	86	91	86	84
	20	43	57	72	80	89	88	89	80

underestimate it, since we have no alternative in the absence of some measure of random error. Moreover, examining the distribution of the residuals always helps us to see if there is anything wrong with the model, and may indicate how to improve it if needed. A careful examination of the residual plots should be considered mandatory in any statistical analysis.

If our experiments furnish replicated response values, we can use them to obtain an estimate of random error. With this estimate, we will have a quantitative criterion to judge whether the chosen model is a good representation of the observations, or if we need to improve it. To show how this is done, we will present a numerical example, based on duplicates of the runs performed in the 30–70 °C range.

Suppose that the runs in Table 5.4 have been performed in duplicate, and that our data are the 18 yields given in Table 5.7. For each X value there are now two different values of y. It is evident that no model will be able to pass through these two points at the same time. There will always be residuals no matter what model is chosen. We can attribute them, at least in part, to random errors.

We will see that when replicate values are available the residual sum of squares left by the model can be decomposed into two parts: one caused by random errors, and the other owing to lack of fit of the model. The second part can be reduced by improving the model. The other part cannot.

Let us set our numerical example aside for a moment and consider the general case where n_i replicate response values have been determined for each value of X_i. To identify the replicates we will use a second subscript, j, so that a given response value is now represented by y_{ij}, meaning the jth value observed at the ith level. The total number of response values in the experiment will be equal to the sum of all replicates: $n = \sum n_i$.

At each level i the model leaves n_i residuals, one for each replicate. Summing the squares of all of them over all replicates and all levels, we will obtain the residual sum of squares. We can then write, assuming there are m different levels of the X variable, the following expressions:

- Residual sum of squares at the ith level : $(SS_r)_i = \sum_j^{n_i} (y_{ij} - \hat{y}_i)^2$;
- Residual sum of squares: $SS_r = \sum_i^m (SS_r)_i = \sum_i^m \sum_j^{n_i} (y_{ij} - \hat{y}_i)^2$.

Each single residual can be algebraically decomposed into the difference of two terms:

$$(y_{ij} - \hat{y}_i) = (y_{ij} - \bar{y}_i) - (\hat{y}_i - \bar{y}_i), \tag{5.35}$$

where $\bar{y}_i$ is the average of the response values observed at the ith level. Squaring this equation and summing over all the observations, we will have the residual sum of squares, SS_r, on the left side, as we have just seen. As happened in the decomposition of the total sum of squares (Eq. (5.16) and Exercise 5.7), the summation over the cross terms is zero, leaving us with the sums of the squares of the two terms on the right side. We can then write

$$\sum_i^m \sum_j^{n_i} (y_{ij} - \hat{y}_i)^2 = \sum_i^m \sum_j^{n_i} (y_{ij} - \bar{y}_i)^2 + \sum_i^m \sum_j^{n_i} (\hat{y}_i - \bar{y}_i)^2. \tag{5.36}$$

The first sum on the right-hand side has nothing to do with the model and therefore does not depend on the estimates $\hat{y}_i$, reflecting only the dispersion, at each level i, of the replicate response values y_{ij} about their own averages $\bar{y}_i$. This term, which will give us a measure of the random error, is called the **pure error sum of squares** (SS_{pe}). The second sum, on the contrary, depends on the model. Its value will be affected by how much the estimates at the different levels, $\hat{y}_i$, deviate from the corresponding average response values, $\bar{y}_i$. This term furnishes a measure of the lack of fit of the model to the observed responses, being called for this reason the **lack of fit sum of squares**, SS_{lof}. With this terminology, Eq. (5.36) can be read as

$$\begin{bmatrix} \text{Residual sum} \\ \text{of squares} \end{bmatrix} = [\text{Pure error } SS] + [\text{Lack of fit } SS]$$

or

$$SS_r = SS_{pe} + SS_{lof}. \tag{5.36a}$$

Dividing each of these sums of squares by their respective numbers of degrees of freedom we obtain three mean squares, whose values we can compare to evaluate the model's possible lack of fit.

The residuals $(y_{ij} - \bar{y}_i)$ that are combined to give SS_{pe} have $n_i - 1$ degrees of freedom at each level i. Taking the sum over all levels we obtain the number of degrees of freedom of the pure error sum of squares:

$$\nu_{pe} = \sum (n_i - 1) = (n - m),$$

where n is the total number of observations and m is the number of levels of variable X.

We have already seen that the number of degrees of freedom of the residual sum of squares is the total number of observed values minus the number of parameters in the model, $\nu_r = (n - p)$. Subtracting from this the number of degrees of freedom corresponding to SS_{pe} we will have the number of degrees of freedom for lack of fit:

$$\nu_{lof} = (n - p) - (n - m) = (m - p).$$

Note that ν_{lof} is given by the difference between the number of levels of the independent variable and the number of parameters in the model. This has an implication of great practical importance. We will only be able to test for lack of fit if the number of levels in our experimental design is larger than the number of parameters in the model we are trying to fit. For example, if the model is a straight line (which is defined by two parameters) we must have at least three levels for the X variable, so that ν_{lof} is larger than zero. If we attempt to fit a line to response values determined at only two levels,[50] it will necessarily pass through the averages of the response values at each level. This annihilates SS_{lof} in Eq. (5.36a) and reduces the residual sum of squares to a pure error sum of squares, making it impossible to discover any lack of fit.

With the partitioning of the residual sum of squares into contributions from lack of fit and pure error, the ANOVA table gains two new lines and becomes the complete version (Table 5.8). The pure error mean square,

$$MS_{pe} = \frac{\sum_i^m \sum_j^{n_i} (y_{ij} - \bar{y}_i)^2}{n - m},$$

does not depend on the model and is an estimate of the variance σ^2 that we postulated for the responses, be the model adequate or not. The lack of

[50]And with the same number of replicates at both levels.

Table 5.8
ANOVA table for the least-squares fit of a model that is linear in its parameters

Source of variation	Sum of squares	Degree of freedom	Mean square
Regression	$SS_R = \sum_i^m \sum_j^{n_i} (\hat{y}_i - \bar{y})^2$	$p-1$	$MS_R = \frac{SS_R}{p-1}$
Residual	$SS_r = \sum_i^m \sum_j^{n_i} (y_{ij} - \hat{y}_i)^2$	$n-p$	$MS_r = \frac{SS_r}{n-p}$
Lack of fit	$SS_{\text{lof}} = \sum_i^m \sum_j^{n_i} (\hat{y}_i - \bar{y}_i)^2$	$m-p$	$MS_{\text{lof}} = \frac{SS_{\text{lof}}}{m-p}$
Pure error	$SS_{\text{pe}} = \sum_i^m \sum_j^{n_i} (y_{ij} - \bar{y}_i)^2$	$n-m$	$MS_{\text{pe}} = \frac{SS_{\text{pe}}}{n-m}$
Total	$SS_T = \sum_i^m \sum_j^{n_i} (y_{ij} - \bar{y})^2$	$n-1$	

% explained variation: $\frac{SS_R}{SS_T}$

Maximum % explainable variation: $\frac{SS_T - SS_{pe}}{SS_T}$

Note: $n_i =$ number of replicates at the ith level; $m =$ number of distinct levels of the independent variables; $n = \Sigma n_i =$ total number of observations; $p =$ number of parameters in the model.

fit mean square,

$$MS_{\text{lof}} = \frac{\sum_i^m \sum_j^{n_i} (\hat{y}_i - \bar{y}_i)^2}{m - p},$$

also estimates σ^2 if the model is adequate, that is, if there is no lack of fit. If there is, the MS_{lof} value estimates σ^2 *plus* a contribution from lack of fit. We can then apply an F-test on the $MS_{\text{lof}}/MS_{\text{pe}}$ ratio to evaluate whether our model provides a good fit or not to the observations. The larger the $MS_{\text{lof}}/MS_{\text{pe}}$ value obtained from the ANOVA the larger the model's lack of fit, and vice versa.

We return now to the data in duplicate in Table 5.7. We already know that the linear model is inadequate for this temperature range but we are going to adjust it anyway, to show how the F-test for lack of fit works. Once again, we start by using matrix Eq. (5.12) to find the regression equation, taking care to match the corresponding elements of the **X** and **y** matrices, which now have 18 rows instead of 9. We then write, from Table 5.7,

$$\mathbf{X} = \begin{bmatrix} 1 & 30 \\ 1 & 30 \\ 1 & 35 \\ 1 & 35 \\ \cdots & \cdots \\ 1 & 70 \\ 1 & 70 \end{bmatrix} \quad \text{and} \quad \mathbf{y} = \begin{bmatrix} 24 \\ 20 \\ 40 \\ 43 \\ \cdots \\ 80 \\ 84 \end{bmatrix}.$$

The regression line determined from these matrices is given by

$$\hat{y} = -7.42 + 1.52X.$$

Table 5.9 presents the ANOVA for this new fit. The percent variation explained by the regression, that is, the ratio of the regression sum of squares to the total sum of squares, is $SS_R/SS_T = 77.79\%$. This value, however, must not be compared with 100%, because we must discount the contribution arising from pure error.

Since no model can reproduce the pure error sum of squares, the maximum explainable variation is the total sum of squares minus SS_{pe}. In our case, $SS_T - SS_{pe} = 8930.00 - 45.00 = 8885.00$, which corresponds to $8885.00/8930.00 = 99.50\%$ of the total sum of squares. This percentage is close to 100%, because the pure error contribution is relatively small, but it is with this new value that we should compare the variation explained by the regression, 77.79%. The model inadequacy appears clearly in the two first graphs of Fig. 5.8. Once again the residuals are distributed in a curved pattern.

The value of the MS_R/MS_r ratio for the linear model is 56.03. Were it not for the lack of fit, this value would indicate a significant regression when compared to $F_{1,16} = 4.49$ (at the 95% level). The lack of fit is confirmed by the high value of MS_{lof}/MS_{pe},

$$\frac{MS_{lof}}{MS_{pe}} = \frac{276.94}{5.00} = 55.39,$$

which is much larger than $F_{7,9} = 3.29$.

Since the linear model presents lack of fit, we proceed to adjust a quadratic model, which is given by Eq. (5.33). Constructing the appropriate matrices and substituting them into Eq. (5.12) we arrive at

$$\mathbf{b} = \begin{bmatrix} -172.42 \\ 8.59 \\ -0.071 \end{bmatrix},$$

or

$$\hat{y} = -172.42 + 8.59X - 0.071X^2.$$

Table 5.9
ANOVA for the fit of a linear model to the data of Table 5.7

Source of variation	Sum of squares	Degree of freedom	Mean square
Regression	6946.41	1	6946.41
Residual	1983.59	16	123.97
Lack of fit	1938.59	7	276.94
Pure error	45.00	9	5.00
Total	8930.00	17	

% explained variation: 77.79
Maximum % explainable variation: 99.50

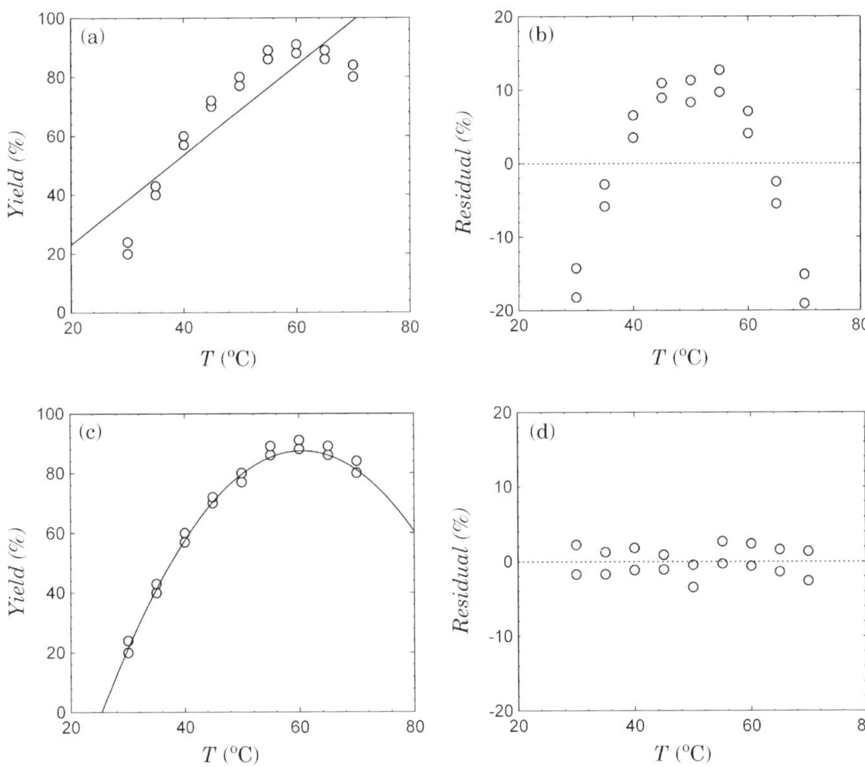

Fig. 5.8. (a, b) Fit of a linear model to the data of Table 5.7. The observed values are not well represented by a straight line. The distribution of the residuals is not random. (c, d) The agreement is much better with the quadratic model. The residuals now appear randomly distributed.

Table 5.10
ANOVA for the fit of a quadratic model to the data of Table 5.7

Source of variation	Sum of squares	Degree of freedom	Mean square
Regression	8871.61	2	4435.80
Residual	58.40	15	3.89
Lack of fit	13.39	6	2.23
Pure error	45.00	9	5.00
Total	8930.00	17	

% explained variation: 99.35
Maximum % explainable variation: 99.50

The plots of the quadratic model (Fig. 5.8c, d) show a much better fit than those of the linear model, as we expected. The improvement is numerically confirmed by the ANOVA values (Table 5.10). The difference is striking: the value of the MS_R/MS_r ratio increases to $4435.80/3.89 = 1140.3$, while the $MS_{\text{lof}}/MS_{\text{pe}}$ ratio, which was 55.39 for the linear fit, is now reduced to only 0.45, a statistically insignificant value. There is no evidence of lack of fit, and therefore we may calculate confidence intervals for the model parameters.

Since there is no lack of fit, both MS_{lof} and MS_{pe} estimate σ^2. We can take advantage of this fact to obtain a variance estimate with a larger number of degrees of freedom, summing SS_{lof} and SS_{pe} and dividing the total by $(\nu_{\text{lof}} + \nu_{\text{pe}})$. With this operation, we again calculate the residual mean square, which now becomes a legitimate estimate of the pure error.

Substituting σ^2 in Eq. (5.30) by the value of the residual mean square, $s^2 = 3.89$, we arrive at estimates of the parameter variances. Then, taking the square root, we obtain their standard errors. With them we can write the final result of our fit as

$$\hat{y} = \underset{(\pm 7.65)}{-172.42} + \underset{(\pm 0.32)}{8.59X} - \underset{(\pm 0.003)}{0.071X^2}.$$

The values in parentheses are the standard errors of the parameters. Since they are much smaller than the estimates of the regression coefficients, we conclude that all three parameters are statistically significant. If a more rigorous analysis is necessary, we can perform a t test on each one.[51]

[51]With how many degrees of freedom?

Exercise 5.10. This is an incomplete version of an ANOVA table for the fitted model of Exercise 5.5. Complete this table and verify if the linear model is satisfactory, or if there is evidence of lack of fit.

Source of variation	Sum of squares	Degree of freedom	Mean square
Regression	2.95146×10^{-1}	?	?
Residual	?	?	?
Lack of fit	?	?	?
Pure error	1.09355×10^{-4}	?	?
Total	2.95425×10^{-1}	11	
% explained variation: ?			
Maximum % explainable variation: ?			

Exercise 5.11. The result of Exercise 5.10 should show that there is no evidence of lack of fit in the model of Exercise 5.5. Use the residual mean square as an estimate of the variance of the observations and determine the standard errors of the coefficient estimates. Are they statistically significant at the 95% confidence level?

5.7. Correlation and regression

It is quite common to see, in the literature, the results of a regression analysis being discussed in terms of the correlation of the dependent variable with the independent variable. Strictly, this does not make sense, because correlation is defined for *pairs* of random variables, whereas in regression only the dependent variable is supposed to be random. However, if we set aside this conceptual detail, some algebraic relations between correlation and regression do exist that are worthwhile discussing, if only to clarify their true meanings and limitations.

Imagine that both X and y are random variables and consequently there is a legitimate correlation coefficient between them, defined by

$$r(X,y) = \frac{\sum((X_i - \bar{X})/s_x)((y_i - \bar{y})/s_y)}{n-1},$$

as we saw in Section 2.4. With the simplified notation we introduced in Eq. (5.10) (see also Exercise 5.3), we can rewrite this expression as follows:

$$r(X,y) = \frac{S_{xy}}{\sqrt{S_{xx}S_{yy}}}. \tag{5.37}$$

In the same notation, the β_1 estimate is given as

$$b_1 = \frac{S_{xy}}{S_{xx}}. \tag{5.10}$$

Combining the two expressions, we obtain a relation between the slope of the regression line, b_1, and the correlation coefficient between the two variables, $r(X,y)$:

$$b_1 = r(X,y)\sqrt{\frac{S_{yy}}{S_{xx}}}, \tag{5.38}$$

or

$$b_1 = r(X,y)\frac{s_y}{s_x}, \tag{5.39}$$

where s_y and s_x are the standard deviations of the y and X variables, respectively. Even so, the meanings of b_1 and $r(X, y)$ are still intrinsically different. The correlation coefficient, as we know, is a measure of the linear association between the X and y variables, both assumed random. The value of the slope, b_1, represents the variation in $\hat{y}$ per unit variation in X, or more succinctly, the $\mathrm{d}\hat{y}/\mathrm{d}X$ derivative.

For a linear model, we can also establish a relation between the percent explained variation (or coefficient of determination),

$$R^2 = \frac{SS_R}{SS_T} = \frac{\sum(\hat{y}_i - \bar{y})^2}{\sum(y_i - \bar{y})^2},$$

and the correlation coefficient $r(X, y)$. To do this, we use Eq. (5.18) and rewrite R^2 as

$$R^2 = b_1^2 \frac{\sum(X_i - \bar{X})^2}{\sum(y_i - \bar{y})^2} = b_1^2 \frac{S_{xx}}{S_{yy}}.$$

Now, using Eq. (5.38) to substitute b_1^2, we arrive at

$$R^2 = r^2(X,y)\frac{S_{yy}}{S_{xx}}\frac{S_{xx}}{S_{yy}}$$

or, simplifying,

$$R^2 = r^2(X,y). \tag{5.40}$$

This equality shows that, when we adopt the $y_i = \beta_0 + \beta_1 X_i + \varepsilon_i$ model, the percent variation explained by the regression is also a measure of the linear association between X and y. A common error, perhaps induced by Eq. (5.40), is to interpret the value of R, the square root of R^2 with the appropriate algebraic sign, as the correlation coefficient between X and y in any regression. As we have just seen, this is only valid for the fit of a straight line. Besides, in least-squares modeling X is not even a random variable. In fact, the value of R can be interpreted as a correlation coefficient, but not between X and y. It can be demonstrated that in any circumstance, for any linear regression whatever, R is the correlation coefficient between the observed responses *and the values predicted* by the fitted model:

$$R = r(y, \hat{y}). \tag{5.41}$$

This relation is legitimate, since both the observed responses and the predicted values are random variables. This R value, which is called the multiple correlation coefficient, is never negative. It is the largest correlation that a linear combination of independent variables, in the form specified by the model, could have with the observed y values.

5A. Applications

5A.1. The spring of air

In 1661, Robert Boyle presented a communication to the Royal Society describing his discovery of the relation, later known as Boyle's law,[52] between the pressure and the volume of a given mass of air. Boyle's original data, which were published in 1662, in the second edition of his *New Experiments Physio-Mechanical, Touching the Spring of Air and its Effects*, are presented in Table 5A.1. We are going to fit a few polynomial models to them and evaluate these fits by means of the ANOVA and the inspection of residual plots.

The fits are performed in the usual manner, by least squares (Eq. (5.12)) and produce the following results:

	Fitted model	MQ_R/MQ_r	R^2 (%)
Linear	$p_i = \underset{(\pm 0.170)}{4.125} - \underset{(\pm 0.006)}{0.0742}\,V_i$	166.66	87.68
Quadratic	$p_i = \underset{(\pm 0.185)}{6.183} - \underset{(\pm 0.014)}{0.238}\,V_i + \underset{(\pm 0.00023)}{0.00279}\,V_i^2$	651.50	98.34
Cubic	$p_i = \underset{(\pm 0.187)}{8.253} - \underset{(\pm 0.022)}{0.491}\,V_i + \underset{(\pm 0.0008)}{0.012}\,V_i^2 - \underset{(\pm 0.000009)}{0.000105}\,V_i^3$	3241.45	99.78

All of the terms in these models are significant, as we can see from the relatively small sizes of their respective standard errors. Each additional term eliminates 1 degree of freedom from the residuals but this does not worsen the models — judging by the ANOVA results, they are progressively improving. The cubic model, for example, has $MS_R/MS_r =$ 3214.15, and explains 99.78% of the total variation. With these numbers, we would hardly bother about finding a better model. When we look at the

[52]Or Mariotte's law, as they say in France. The Frenchman E. Mariotte discovered the same relation, independently of Boyle, and even noted that it was only valid if the temperature remained constant.

Table 5A.1
Experimental data from Boyle's experiments. The pressures are in atmospheres. The volume unit is arbitrary

p	1.000	1.049	1.097	1.150	1.212	1.270	1.350	1.429	1.517
V	48	46	44	42	40	38	36	34	32
p	1.616	1.727	1.865	2.019	2.105	2.199	2.302	2.427	2.545
V	30	28	26	24	23	22	21	20	19
p	2.674	2.841	3.017	3.195	3.449	3.702	4.036		
V	18	17	16	15	14	13	12		

residual graphs, though, we see another story (Fig. 5A.1). The residuals do become smaller, but it is evident that they form systematic patterns that change with the complexity of the model. For the linear model, the residuals have a pattern that resembles a parabola. When we include the quadratic term, the new residuals suggest a cubic function. Then, on adding a cubic term, the residuals take on the appearance of a quartic function.

What we are really doing is adjusting a power series to a set of data that could be much better represented by a simpler function. As we know from elementary chemistry, Boyle's law is given by the expression $pV = k$, or $p = k(1/V)$, where k is a constant that depends on the temperature and the quantity of gas. Using the inverse volume as an independent variable instead of the volume itself, Boyle's law becomes the expression of a straight line passing through the origin. Fig. 5A.2 shows this line and the residuals it leaves after fitted Boyle's data.

Changing the independent variable from V to $1/V$ is, for obvious reasons, an example of what is known as a **linearizing transformation**. This is an important lesson. Before thinking of adding new terms and complicating a model, we should always reflect if there is not some variable transformation that might produce a more satisfactory fit. In any case, the residual plot is always an excellent indication of the success of a particular model.

As for the ANOVA, the transformation also provides excellent results. The adjusted equation is now

$$p_i = \underset{(\pm 0.04270)}{0.065} + \underset{(\pm 0.088)}{48.18}\frac{1}{V_i}, \quad \text{with} \quad MS_R/MS_r = 299,273.$$

and $R_2 = 99.992\%$.

In other words, these results are much better than those for the three polynomial model fits with the volume as the independent variable. Also,

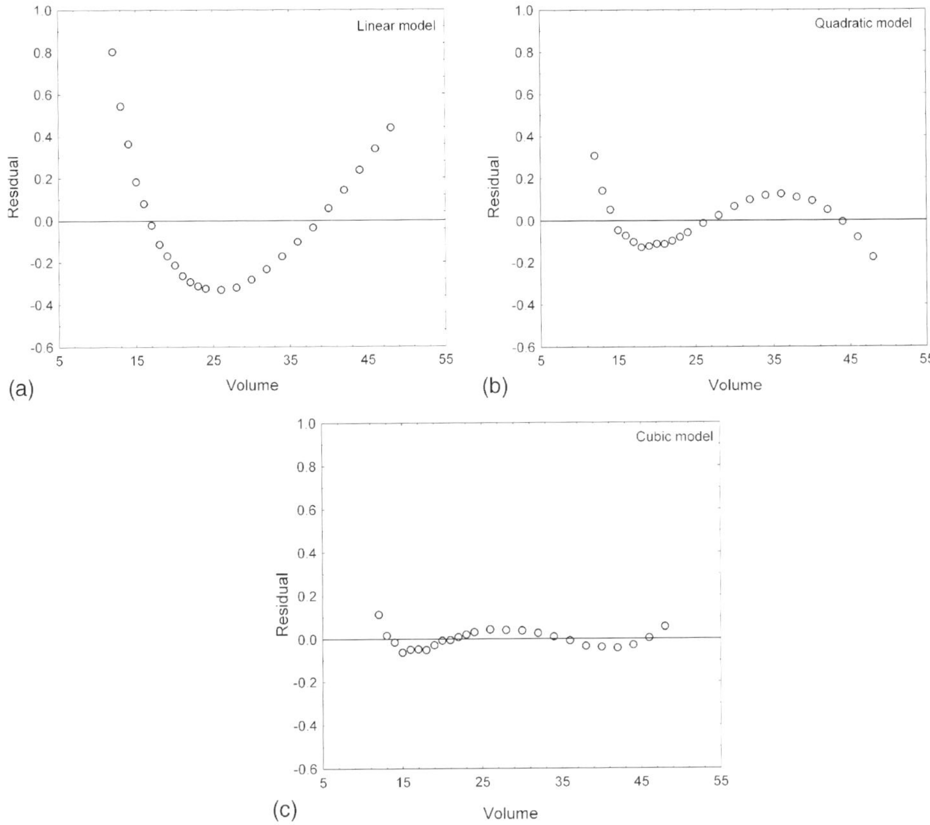

Fig. 5A.1. Plots of the residuals left by the three polynomial models fitted to Boyle's data.

the constant term is no longer statistically significant, and Boyle's law is reproduced perfectly.[53]

5A.2. Chromatographic calibration

To construct a calibration curve, G.M.F. Silva and J.F. Pinto measured the height of a chromatographic peak (A) for six different concentrations of benzatone (C). Various replicate determinations were made for each concentration, resulting in a total of 21 runs (Table 5A.2). Judging from the plot of the heights against concentrations, shown in Fig. 5A.3, a straight line would be a good model for these data. Doing a least-squares

[53]Did you notice that when we use 1/V the points become more distant from one another as we move from left to right on the graph? Why does this happen?

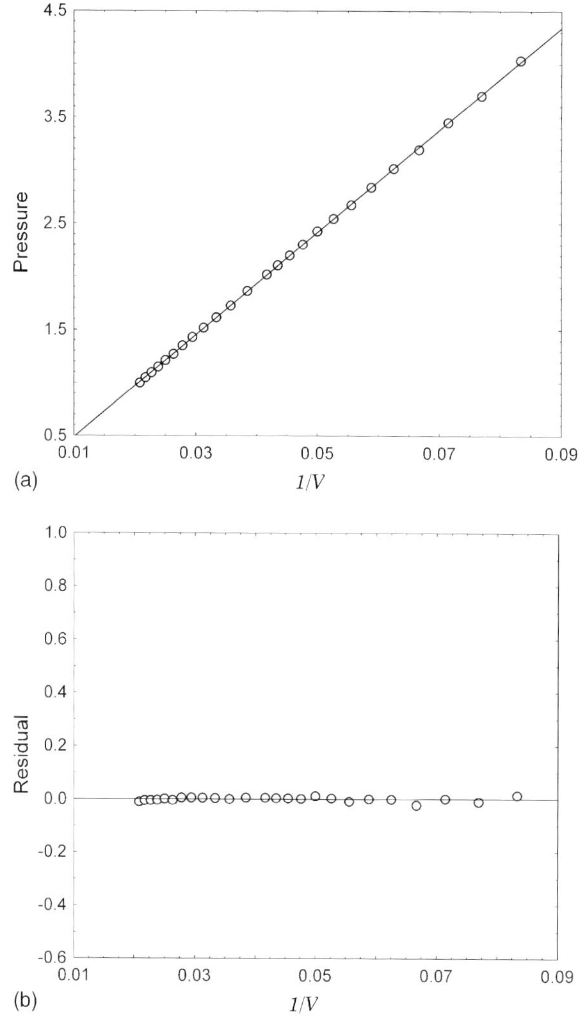

Fig. 5A.2. Fit of Boyle's data by a linear model involving the inverse volume.

fit, we arrive at the equation

$$\hat{A} = \underset{(\pm 0.0930)}{-0.1906} + \underset{(\pm 0.1244)}{15.3488C} \quad \text{with} \quad R^2 = 99.87\%.$$

At first sight, this equation appears to be excellent, but the truth is that it presents lack of fit. Using the equations given in Table 5.8, we obtain $MS_{\text{lof}} = MS_{\text{pe}} = 6.11$, a value larger than $F_{4,\,15} = 3.06$ (95% confidence level). The plot of the residuals against the peak heights predicted by the linear model confirms this finding (Fig. 5A.4). In an attempt to eliminate the lack of fit, we then add a quadratic term, and arrive at a model with three parameters, given by

$$\hat{A} = \underset{(\pm 0.0694)}{-0.0010} + \underset{(\pm 0.3116)}{13.7456C} + \underset{(\pm 0.2333)}{1.2418C^2} \quad \text{with} \quad R^2 = 99.95\%.$$

This model may not appear much of an improvement, but its residual plot (Fig. 5A.4) shows that it is in fact superior, because it does not show lack of fit. Its lack of fit ratio is $MS_{\mathrm{lof}}/MS_{\mathrm{pe}} = 0.142$, a value that, being less than one, dispenses with the F-test. Since there is no lack of fit, the standard errors appearing in the second equation were calculated using the residual mean square rather than the pure error mean square. Also, it is worthwhile noting that the constant term, which is slightly significant in the linear model, is no longer so in the quadratic model. In other words, applying the second model to a zero concentration solution predicts the absence of a chromatographic peak for benzatone, just as it should.

Table 5A.2
Chromatographic calibration data for benzatone

Concentration (mg L^{-1})	Peak height (cm)				
0.0133	0.1836	0.1787	0.1837	0.1806	0.1861
0.0665	0.9373	0.9177	0.9224		
0.3325	4.6227	4.7812	4.6256		
0.6650	9.6905	9.9405	9.5754		
0.9975	14.7607	15.0113	14.9641		
1.3300	21.0033	20.2700	20.5719	20.0915	

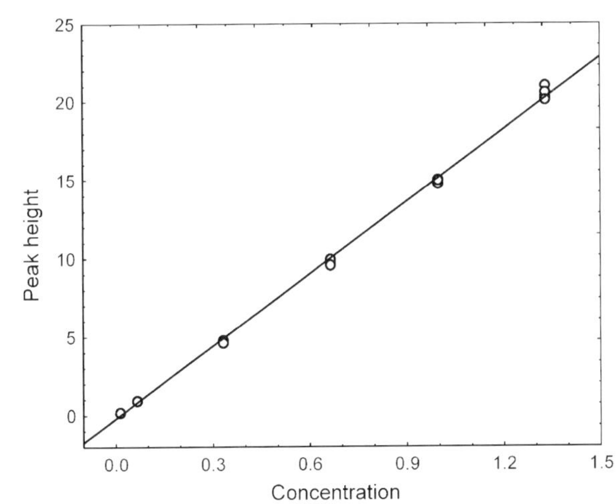

Fig. 5A.3. Chromatographic calibration data for benzatone.

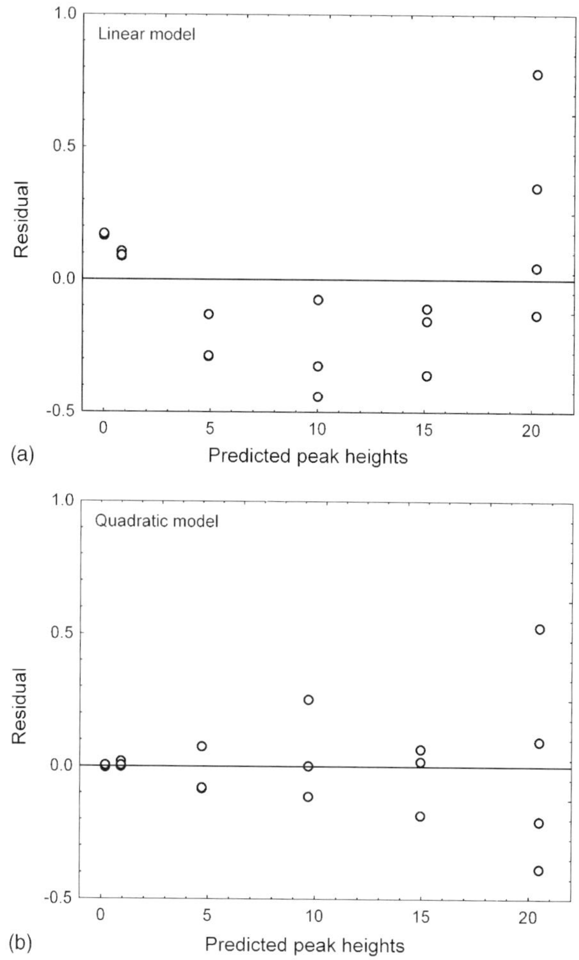

Fig. 5A.4. Residuals of the chromatographic calibration for benzatone.

The residual plots also make it very clear that the variance of the response values increases with the concentration, a common phenomenon in various methods of instrumental analysis. This behavior violates one of the least-squares fit assumptions, and is usually addressed by subjecting the data to a **variance stabilizing transformation** before performing the regression (see, for example, Pimentel and Neto, 1996).

It should be stressed that, depending on the size of the errors we are willing to tolerate in the predictions made from the regression equation, it might be that neither lack of fit nor response **heteroscedasticity** have any practical importance. In any case, it is well to be prepared to treat these problems, should they prove detrimental to the objectives of the experiment.

5A.3. Multivariate calibration

In a study of the simultaneous determination of MnO_4^-, $Cr_2O_7^{2-}$ and Co^{2+}, several mixtures of standard solutions of these three chemical species were prepared and analyzed by UV–visible spectrophotometry, at three different wavelengths: 530, 440 and 410 nm (Scarminio et al., 1993). The data relative to the permanganate ion are given in Table 5A.3. The fitted linear model is

$$\hat{V} = \underset{(\pm 4.404)}{2.642} - \underset{(\pm 10.256)}{3.560}\, A_{530} - \underset{(\pm 6.347)}{37.088} A_{440} + \underset{(\pm 1.845)}{39.636} A_{410}.$$

Since we do not have replicate measurements we cannot rigorously test for lack of fit, but the residual plot (Fig. 5A.5) has a satisfactory appearance, that is, the points do not appear to have a systematic pattern. For this reason we use the value of the residual mean sum of squares, MS_r, to determine the standard errors of the model's coefficients. The model explains 99.74% of the total variation and the value of $MS_R/MS_r =$ 628 is quite high. Note that, as we should expect, the constant term is not statistically significant. In fact, neither is the A_{530} coefficient, indicating that this wavelength is of no help in determining the concentration of the permanganate ion.

Table 5A.3
Multivariate calibration of the MnO_4^- ion by spectrophotometry

Volume of MnO_4^- (mL)	Absorbances		
	A_{530}	A_{440}	A_{410}
7.50	0.084	0.506	0.602
6.60	0.108	0.479	0.561
5.00	0.149	0.426	0.472
3.40	0.182	0.375	0.385
2.50	0.205	0.342	0.334
1.70	0.159	0.409	0.375
5.00	0.149	0.426	0.472
5.00	0.113	0.488	0.523
4.00	0.149	0.432	0.456

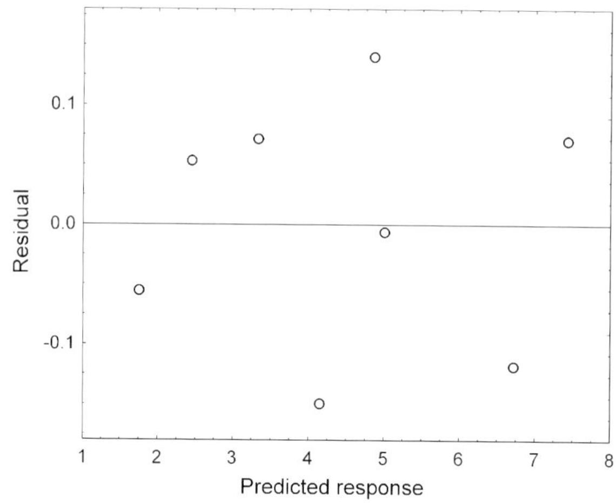

Fig. 5A.5. Residuals of the multivariate calibration for the permanganate ion.

5A.4. Forbidden energy gaps in semiconductors

To determine forbidden energy gaps in a semiconductor, one can investigate how its electronic conductivity (σ) varies with temperature. Semiconductor theory indicates that the logarithm of the conductivity varies linearly with the inverse temperature.[54] The value of the energy gap, E_g, is given by the slope of the ln σ Vs. $1/T$ line. Table 5A.4 contains the values of nine runs performed in duplicate with a germanium intrinsic semiconductor. The data were obtained at the Modern Physics Laboratory of the Physics Department of the Londrina State University, under the supervision of Prof. J. Scarminio.

The fit of a linear model to the data in the table results in the equation

$$\ln \sigma = \underset{(\pm 0.120)}{16.708} - \underset{(\pm 42.2)}{4509.9} \frac{1}{T} \quad \text{with} \quad R^2 = 99.86\%.$$

The two coefficients are highly significant. The constant term represents the value the electronic conductivity would have if there were no energy difference between the highest value of the valence band and the lowest value of the conduction band of the semiconductor. The residual plot (Fig. 5A.6) shows evidence of lack of fit, suggesting that, at least from a statistical point of view, the model could be improved by the inclusion of a quadratic term. This is confirmed by the ANOVA values. However, we must remark that to simplify the experiment the runs were

[54] Two linearizing transformations, isn't it?

Table 5A.4
Conductivity values of a Ge semiconductor at different temperatures

Run	$1/T$, $10^{-3}K^{-1}$	$\ln(\sigma/\Omega^{-1}m^{-1})$	
1	3.19	2.24	2.29
2	3.09	2.74	2.81
3	3.00	3.19	3.22
4	2.91	3.60	3.61
5	2.83	3.95	4.01
6	2.75	4.33	4.33
7	2.68	4.62	4.62
8	2.61	4.92	4.93
9	2.54	5.21	5.21

not performed in random order but following a cooling curve. With this procedure, the pure error mean square could be underestimated.

5A.5. Heat of vaporization determination

Under certain conditions, it can be demonstrated that the vapor pressure of a substance, p_{vap}, is related to the absolute temperature at which it is maintained, T, by the Clausius–Clapeyron equation,

$$\ln p_{\text{vap}} = b_0 - \frac{\Delta H_{\text{vap}}}{R}\left(\frac{1}{T}\right).$$

In this equation, R is the ideal gas constant, b_0 is a constant that varies from one substance to another, and ΔH_{vap} is the heat of vaporization of the substance of interest, assumed to be constant over the temperature range studied. In an experiment to determine the heat of vaporization of carbon tetrachloride, nine runs were performed. Their results are reproduced in Table 5A.5 (Simoni, 1998). If the Clausius–Clapeyron equation holds under these conditions, $\ln p_{\text{vap}}$ can be described by a linear function of $1/T$, from whose slope we can determine the heat of vaporization of carbon tetrachloride.

Fitting the $\ln p_{\text{vap}} = b_0 + b_1(1/T)$ model to the data results in

$$\ln p_{\text{vap}} = \underset{(\pm 0.00689)}{11.20} \; - \underset{(\pm 26.897)}{3901.98}\left(\frac{1}{T}\right),$$

with $MS_R/MS_r = 21,046$ and $R^2 = 99.97\%$.

This corresponds to a heat of vaporization of $32.44 \pm 0.22\,\text{kJ mol}^{-1}$. In spite of these good ANOVA values, however, the residual plot (Fig. 5A.7)

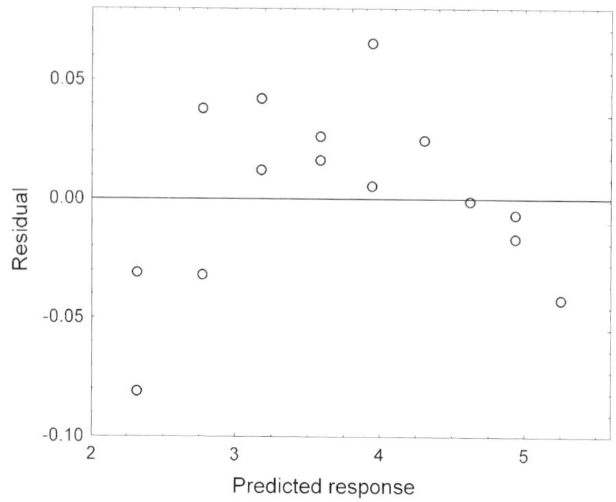

Fig. 5A.6. Residuals of the fit of the $\ln\sigma = b_0 + b_1 \frac{1}{T}$ model to the data of Table 5A.4.

Table 5A.5
Variation of the vapor pressure of CCl_4 with temperature

Run	T (K)	p_{vap} (Torr)
1	273	0.044
2	283	0.075
3	293	0.122
4	303	0.190
5	313	0.288
6	323	0.422
7	333	0.601
8	343	0.829
9	353	1.124

shows a clear systematic pattern, indicating the necessity of augmenting the model with a quadratic term. This in turn produces the new fit

$$\ln p_{vap} = \underset{(\pm 0.14)}{8.00} - \underset{(\pm 86.1)}{1954}\left(\frac{1}{T}\right) - \underset{(\pm 13{,}260)}{300{,}311}\left(\frac{1}{T}\right)^2,$$

with $MS_R/MS_r = 780{,}369$ and $R^2 = 99.9996\%$.

In physical terms, the superiority of the quadratic model indicates that the heat of vaporization cannot be considered constant in this example. Using therefore the quadratic model and assuming that the derivative

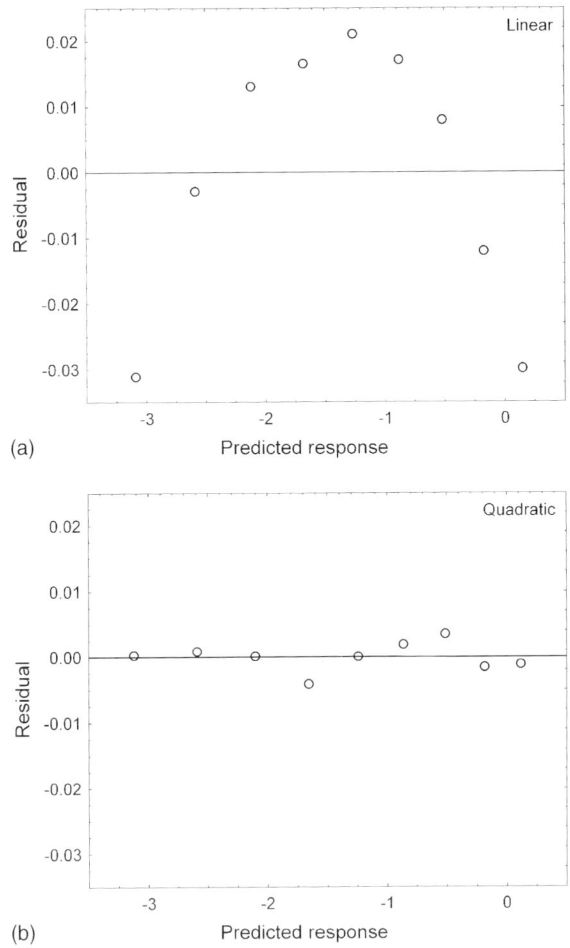

Fig. 5A.7. Residuals of the two fits to the data in Table 5A.5.

$$\frac{d(\ln p_{\text{vap}})}{d(1/T)} = -1954 - 600,622\left(\frac{1}{T}\right)$$

is a more realistic estimate of $-\Delta H_{\text{vap}}/R$, we can conclude that the heat of vaporization of CCl_4 actually varies between 30.39 and 34.54 kJ mol^{-1}, in the temperature range we are considering here.

5A.6. Another calibration

We are leaving this application as an exercise for you, dedicated reader. Table 5A.6 shows the zinc ion concentrations of aqueous solutions and their

Table 5A.6
Atomic absorption calibration data for zinc

$[Zn^{2+}]$	Absorbance		
0.000	0.696	0.696	0.706
0.500	7.632	7.688	7.603
1.000	14.804	14.861	14.731
2.000	28.895	29.156	29.322
3.000	43.993	43.574	44.699

respective absorbances, obtained in triplicate, in an experiment to construct a calibration curve in atomic absorption spectrometry (Lopes, 1999).

(a) Draw a graph of absorbance vs. concentration. What type of model does this graph suggest?
(b) Fit the data with the model suggested in item (a), using absorbance as the dependent variable. Evaluate the fit by an ANOVA. Also make a graph of the absorbances estimated from the model vs. the observed absorbances. Use this model to predict the concentration of a solution whose absorbance is 25.00.
(c) Now look at the graph of the residuals. Can the model be improved? If you think so, fit a better model and compare its results with those of the first model. Compare the two analyses of variances. Which model do you prefer?
(d) With the model you selected, do an inverse calibration, using concentration as the dependent variable. According to this new model, what should be the concentration of a solution with an absorbance value of 25.00? Is there much difference between the predictions of the direct and the inverse calibrations?

CHAPTER 6

Exploring the response surface

Response surface methodology (RSM) is an optimization technique based on factorial designs introduced by G.E.P. Box in the 1950s. Since then, RSM has been used with great success for modeling various industrial processes. In this chapter, we use the concepts introduced in the previous chapters to explain the basic principles of RSM. The interested reader can find more comprehensive treatments in Cornell (1990a), Myers and Montgomery (2002) and in the excellent books and scientific papers of G.E.P. Box and his co-workers (Box and Wilson, 1951; Box, 1954; Box and Youle, 1955; Box and Draper, 1987).

6.1. Response surface methodology

Response surface methodology consists of two distinct stages – **modeling** and **displacement** – that are repeated as many times as necessary, with the objective of discovering an optimum region on the response surface under investigation. Modeling is usually done by fitting simple models (linear or quadratic, as a rule) to response values obtained from standard or augmented factorial designs. The displacement always takes place along the path of steepest ascent of the fitted model, which is the trajectory for which the response varies in the most pronounced way. We will try to clarify these notions with a numerical example.

Suppose a chemist wished to investigate the effect of two factors, concentration of a reagent and stirring speed, on the yield of a certain reaction. She already knows that the process has been operating with these values fixed at 50% and 100 rpm for some time and that the average yields obtained under these conditions have been close to 68%. Now she

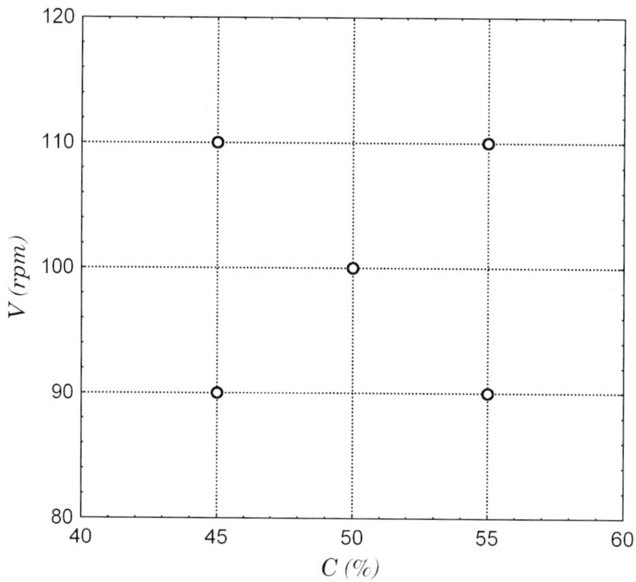

Fig. 6.1. Two-level factorial design augmented with a center point.

would like to know if it is possible to improve the yield, by choosing other levels for the factors.

6.1.1. Initial modeling

The first step is to investigate the response surface in the region around the historical operating conditions, using the factorial design shown in Fig. 6.1. Note that the design has a center point, and therefore contains three levels for each factor, instead of only two. This allows us to test the linear model for lack of fit, a task that would be impossible if we investigated only two levels. Table 6.1 contains the design matrix and the experimentally observed yields for the initial design. Seven runs were carried out, three of these being replicate measurements at the center point.

We start our analysis by assuming that the response surface is a linear function of the factors in the investigated region, and therefore the response can be approximated by

$$\hat{y} = b_0 + b_1 x_1 + b_2 x_2, \tag{6.1}$$

where b_0, b_1 and b_2 are estimators of the parameters of the model and x_1 and x_2 represent the coded factor levels. As we saw in Exercise 5.4, the values of b_0, b_1 and b_2 can be obtained by the least-squares method. In this case, the design matrix **X** will be given by

Table 6.1
Results of the 2^2 design with a center point. x_1 and x_2 are the values of the two factors, coded by the equations $x_1 = (C - 50\%)/5\%$ and $x_2 = (v - 100\ \text{rpm})/10\ \text{rpm}$

Run	C (%)	v (rpm)	X_1	x_2	y (%)
1	45	90	−1	−1	69
2	55	90	1	−1	59
3	45	110	−1	1	78
4	55	110	1	1	67
5	50	100	0	0	68
6	50	100	0	0	66
7	50	100	0	0	69

$$\mathbf{X} = \begin{bmatrix} 1 & -1 & -1 \\ 1 & 1 & -1 \\ 1 & -1 & 1 \\ 1 & 1 & 1 \\ 1 & 0 & 0 \\ 1 & 0 & 0 \\ 1 & 0 & 0 \end{bmatrix}.$$

The first column corresponds to the b_0 term, and the other two contain the coded values of the factor levels. Obviously, we also have for the responses the vector

$$\mathbf{y} = \begin{bmatrix} 69 \\ 59 \\ 78 \\ 67 \\ 68 \\ 66 \\ 69 \end{bmatrix}.$$

Following the usual procedure, we calculate

$$\mathbf{X}^t\mathbf{X} = \begin{bmatrix} 7 & 0 & 0 \\ 0 & 4 & 0 \\ 0 & 0 & 4 \end{bmatrix} \quad \text{and} \quad \mathbf{X}^t\mathbf{y} = \begin{bmatrix} 467 \\ -21 \\ 17 \end{bmatrix}.$$

Then, using Eq. (5.12), we have

$$\mathbf{b} = (\mathbf{X}^t\mathbf{X})^{-1}\mathbf{X}^t\mathbf{y} = \begin{bmatrix} 1/7 & 0 & 0 \\ 0 & 1/4 & 0 \\ 0 & 0 & 1/4 \end{bmatrix} \times \begin{bmatrix} 476 \\ -21 \\ 17 \end{bmatrix} = \begin{bmatrix} 68.00 \\ -5.25 \\ 4.25 \end{bmatrix}. \qquad (6.2)$$

From the triplicate runs at the center point, we calculate $s^2 = 2.33$ as an estimate of the variance of the response values. Substituting this value into Eq. (5.30), we obtain estimates of the variances of the elements of vector **b**:

$$\hat{\mathbf{V}}(\mathbf{b}) = (\mathbf{X}^t\mathbf{X})^{-1}s^2 = \begin{bmatrix} 1/7 & 0 & 0 \\ 0 & 1/4 & 0 \\ 0 & 0 & 1/4 \end{bmatrix} \times 2.33 = \begin{bmatrix} 0.33 & 0 & 0 \\ 0 & 0.58 & 0 \\ 0 & 0 & 0.58 \end{bmatrix}.$$

Taking the square roots of these results we arrive at the standard errors of b_0, b_1 and b_2. With them and the estimates obtained in Eq. (6.2) we can finally write the full equation for the fitted model:

$$\hat{y} = \underset{(\pm 0.58)}{68.00} - \underset{(\pm 0.76)}{5.25x_1} + \underset{(\pm 0.76)}{4.25x_2}. \qquad (6.3)$$

The relatively small sizes of the errors suggest immediately that this model is significant (for a quantitative treatment, see Exercises 6.2 and 6.4). This is confirmed by the analysis of variance results given in Table 6.2. Since the value of the mean square ratio $MS_{\text{lof}}/MS_{\text{pe}}$ is not statistically significant (0.42/2.34 = 0.18), there is no evidence of lack of fit. In the region investigated with this first design, the response surface is satisfactorily described by Eq. (6.3), which defines the plane represented in perspective in Fig. 6.2.

Exercise 6.1. Eliminating the center point from Table 6.1 and Fig. 6.1, we are left with a 2^2 factorial design, which can be analyzed with the methods of Chapter 3.

Table 6.2
Analysis of variance for fitting the $\hat{y} = b_0 + b_1x_1 + b_2x_2$ model to the data of Table 6.1

Source of variation	Sum of squares	Degree of freedom	Mean square
Regression	182.50	2	91.25
Residual	5.50	4	1.38
Lack of fit	0.83	2	0.42
Pure error	4.67	2	2.34
Total	188.00	6	

% explained variation: 97.07
Maximum % explainable variation: 97.52

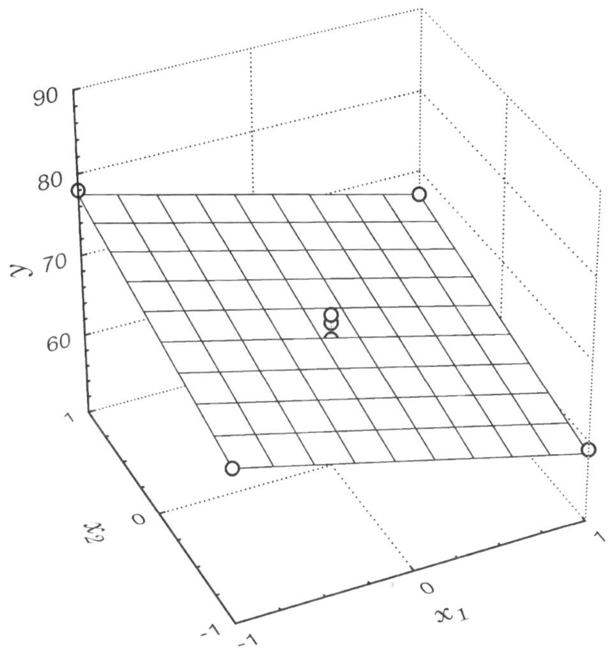

Fig. 6.2. *Plane described by* Eq. (6.3), $\hat{y} = 68.0 - 5.25x_1 + 4.25x_2$.

Calculate the effect values for this factorial and compare them with the values of the coefficients in Eq. (6.3).

Exercise 6.2. Using the estimate $s^2 = \mathbf{2.33}$ obtained from the replicate experiments at the center point of Fig. 6.1, calculate the standard errors of the average of all seven response values and of the effects determined in the preceding exercise. Compare your results with the errors determined for the coefficients of the fitted model (Eq. (6.3)).

Exercise 6.3. Evaluate the statistical significance of Eq. (6.3), using an F-test and the data of Table 6.2. In other words, compare the ratio between the regression mean square and the residual mean square with the appropriate F value.

Exercise 6.4. Use the errors given for the coefficients in Eq. (6.3) to calculate the 95% confidence intervals for β_0, β_1 and β_2. Are all of them significant at this confidence level?

We can obtain a two-dimensional representation of the modeled surface by drawing its response contour lines, that is, lines of constant response. The contour lines of a plane are line segments. For example, if we let $\hat{y} = 70$ in Eq. (6.3) we arrive at the expression

$$x_2 = 1.24x_1 + 0.47,$$

which defines the line where the value of $\hat{y}$ is expected to be equal to 70, according to our model. Doing the same for other values of $\hat{y}$ we obtain

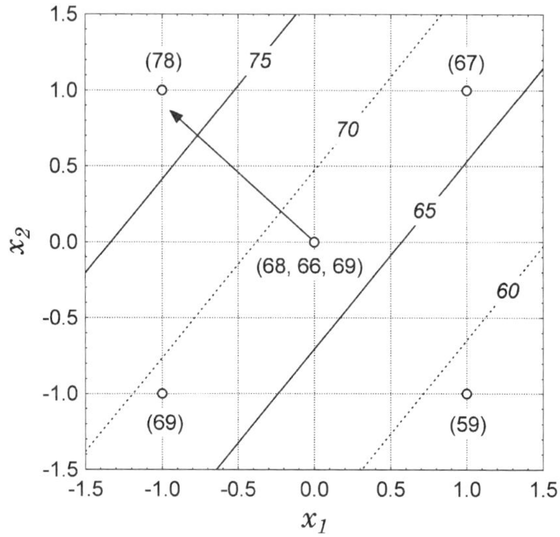

Fig. 6.3. Response contour lines for the plane described by Eq. (6.3). The arrow indicates the steepest ascent path starting at the center of the design. The values in parentheses are the experimentally determined responses.

other contour lines that, taken together, give us an image of the response surface in the region studied (Fig. 6.3). We can clearly see that the model is a plane, obliquely inclined in relation to the coordinate axes, and rising from lower right to upper left. Therefore, if we want to obtain larger yields, we should move the experimental region toward smaller x_1 and larger x_2 values (as, by the way, is already indicated by the signs of the x_1 and x_2 coefficients in Eq. (6.3)). It is clear that progress will be fastest if the displacement is performed along a path perpendicular to the contour lines, that is, if we follow the path of steepest ascent of the fitted surface.

6.1.2. Determining the path of steepest ascent

The arrow in Fig. 6.3 indicates the path of steepest ascent starting at the center point. The direction of this path can be determined algebraically from the model coefficients. For maximum inclination, we must move in steps taken with a b_2/b_1 ratio along the x_2 and x_1 axes.

From Eq. (6.3) we have $b_2/b_1 = 4.25/(-5.25) = -0.81$, which means that for each unit distance receded on the x_1 axis we must advance 0.81 units along x_2. The coordinates of several points along the path of steepest ascent are presented in Table 6.3, in both coded values and in the concentration and stirring speed units actually used in the laboratory.

For the general case of a response surface determined by p factors, the steps along the path of steepest ascent are proportional to the magnitudes

Table 6.3
Points on the steepest ascent path for the models of Figs. 6.2 and 6.3

Step	x_1	x_2	C (%)	v (rpm)	Y (%)
Center	0	0.00	50	100.0	68, 66, 69
Center+Δ	−1	0.81	45	108.1	77
Center+2Δ	−2	1.62	40	116.2	86
Center+3Δ	−3	2.43	35	124.3	88
Center+4Δ	−4	3.24	30	132.4	80
Center+5Δ	−5	4.05	25	140.5	70

and signs of the model coefficients. We can find them easily enough, using the following procedure:

1. First choose one of the factors, say the *i*th one, and change its level by a certain amount, either increasing or decreasing it, as indicated by the sign of its coefficient and the objective of the experiment — maximizing or minimizing the response. Some authors recommend first selecting the factor with the coefficient of largest absolute magnitude in the fitted model. Typically, its initial displacement is one unit (in coded values).
2. Then calculate the displacements of the other factors $j \neq i$, in coded units, by means of

$$\Delta x_j = \frac{b_j}{b_i} \Delta x_i. \tag{6.4}$$

3. Finally, convert the coded displacements back to their laboratory units, and determine the actual new levels of the factors.

Let us consider an example with three factors. In a study to evaluate the influence of some nutrients on chitin[55] production by the fungus *Cunninghamella elegans* a 2^3 factorial design was used, with the levels given in Table 6.4 (Andrade et al., 2000). The linear model satisfactorily fitted the results of this design,

$$\hat{y} = 19.8 + 2.0x_1 + 5.0x_2 + 2.5x_3, \tag{6.5}$$

where y is the amount of chitin produced. Since the model coefficients are all positive and the objective of the study is to maximize chitin production, we should raise the levels of the three factors. Starting with factor x_2 (which has the largest coefficient) we should make the following

[55]A fibrous substance consisting of polysaccharides that is the major constituent of the cell walls of fungi.

Table 6.4
Levels of a 2^3 design with a center point, to study how chitin production by the fungus C. elegans is affected by the concentrations of three nutrients in the culture medium

	Factor	Level		
		−1	0	+1
G (x_1)	D-Glucose ($g\,L^{-1}$)	20	40	60
A (x_2)	L-Asparagine ($g\,L^{-1}$)	1	2	3
T (x_3)	Thiamine ($mg\,L^{-1}$)	0.02	0.05	0.08

displacements to arrive at the first point along the steepest ascent path:

$$\Delta x_1 = \frac{2}{5}(+1) = +0.4 \quad \Delta x_2 = +1 \quad \Delta x_3 = \frac{2.5}{5}(+1) = +0.5.$$

In laboratory units, where the center point is given by $(\mathbf{G}, \mathbf{A}, \mathbf{T}) = (40, 2, 0.05)$, this corresponds to the following experimental conditions:

$$\mathbf{G} = 40 + (0.4 \times 20) = 48 \text{ g L}^{-1} \qquad \mathbf{A} = 2 + (1 \times 1) = 3 \text{ g L}^{-1}$$
$$\mathbf{T} = 0.05 + (0.5 \times 0.03) = 0.065 \text{ mg L}^{-1}$$

Exercise 6.5. Imagine that, in the *C. elegans* example, the researchers had preferred to take the concentration of glucose as the starting factor to determine the steepest ascent path, with an initial displacement of $+25\,g\,L^{-1}$ (note that these are the units actually used in the laboratory). Calculate the coordinates of the third point along the new path, and use Eq. (6.5) to predict chitin yield under these conditions.

We now return to our first example. Having performed the initial modeling and found the direction of steepest ascent, we go on to perform experiments at the conditions specified in Table 6.3. Doing this we obtain the results in the last column of the table, which are also indicated in Fig. 6.4.

At first the yields increase, but they start to decrease after the third run. We can interpret these results by imagining that the response surface is a hill. As the first values show, we are moving up the slope, but after the third run we are already starting to descend the hill on the opposite side.

It is time, therefore, to halt and examine the region where the best response was found in more detail. For that, we set up another design, identical to the first one but centered on the point with the highest result, the third run (35% and about 125 rpm). The new design matrix is presented in Table 6.5, together with the response values obtained from

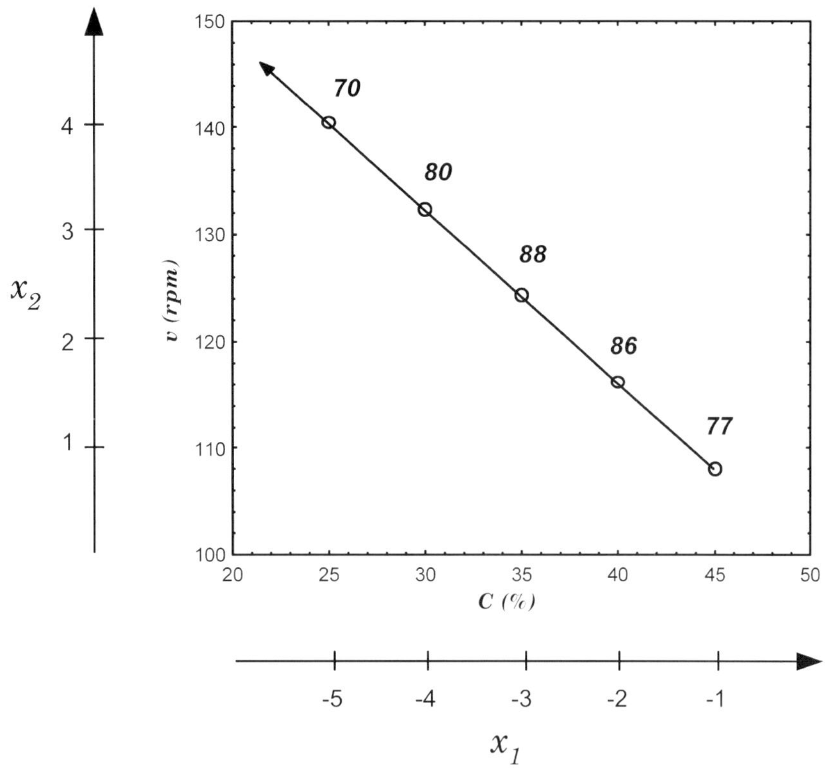

Fig. 6.4. Results obtained in the runs performed along the steepest ascent path of Fig. 6.3.

the new runs. The fit of a linear model to the data in Table 6.5 results in the equation

$$\hat{y} = \underset{(\pm 0.49)}{85.71} + \underset{(\pm 0.65)}{1.25x_1} - \underset{(\pm 0.65)}{2.25x_2}, \tag{6.6}$$

where the standard errors were calculated from a pooled estimate of the variance, combining the replicate runs in the two designs. In relation to the values of the coefficients, the errors are much more important than they were in Eq. (6.3), and the linear dependence of the response on x_1 and x_2 does not seem so assured now.

Exercise 6.6. Use the errors of the coefficients in Eq. (6.6) to calculate 95% confidence intervals for β_0, β_1 and β_2. Are these parameters statistically significant?

The analysis of variance for the new data shows that the fit of the linear model has worsened considerably (Table 6.6). The percent

Table 6.5
Results for the new 2^2 design centered at the third run in Fig. 6.4. x_1 and x_2 now represent the values of the variables coded by the equations $x_1 = (C - 35\%)/5\%$ and $x_2 = (v - 125\ \text{rpm})/10\ \text{rpm}$

Run	C (%)	v (rpm)	x_1	x_2	y (%)
1	30	115	−1	−1	86
2	40	115	1	−1	85
3	30	135	−1	1	78
4	40	135	1	1	84
5	35	125	0	0	90
6	35	125	0	0	88
7	35	125	0	0	89

Table 6.6
Analysis of variance for the fit of the $\hat{y} = b_0 + b_1x_1 + b_2x_2$ model to the data of Table 6.5

Source of variation	Sum of squares	Degree of freedom	Mean square
Regression	26.50	2	13.25
Residual	70.93	4	17.73
Lack of fit	68.93	2	34.46
Pure error	2.00	2	1.00
Total	97.42	6	

% explained variation: 27.20
Maximum % explainable variation: 97.95

explained variation is only 27.20%, while the $MS_{\text{lof}}/MS_{\text{pe}}$ ratio has increased to 34.46, which is larger than $F_{2,2}$ (= 19.0, at the 95% confidence level). This means that, in the region where the steepest ascent path has led us, a linear model does not satisfactorily describe the response surface.

6.1.3. Finding the optimum point

Since the linear model is no longer appropriate, we will substitute it by a quadratic model, whose general expression, for two variables, is

$$\hat{y} = b_0 + b_1x_1 + b_2x_2 + b_{11}x_1^2 + b_{22}x_2^2 + b_{12}x_1x_2. \tag{6.7}$$

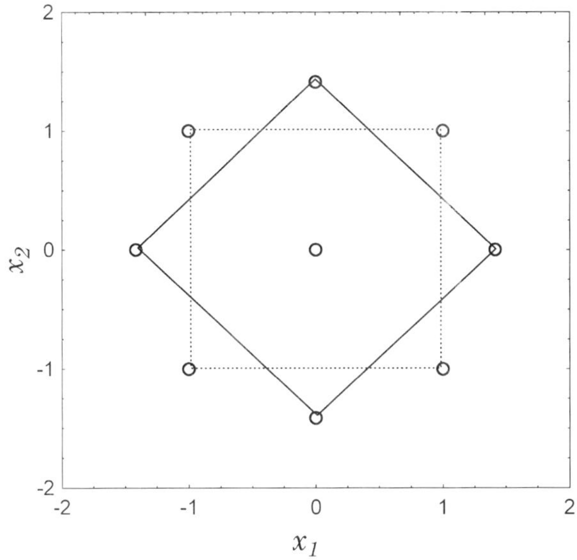

Fig. 6.5. A central composite design for two coded variables (centered factorial combined with star design), corresponding to the data in Table 6.7.

This model has six parameters, and our design has only five "levels", that is, five different combinations of stirring speed and concentration values. Because it is not possible to calculate coefficient estimates when there are more parameters than levels, we must augment our design. This can be done in several ways, the most common being to add a star design to the initial factorial design.

To construct a star design, we rotate the first design by 45°. Combining this new design with the first one, we obtain an octagonal arrangement, as shown in Fig. 6.5. A little geometrical reflection will show that the new points, just like the first ones, are at a distance of $\sqrt{2}$ coded units from the center point. Therefore, all of them lie on a circumference with a $\sqrt{2}$ radius. The coordinates of the star design points are given in the last four lines of Table 6.7. The combination of a factorial with a star design is called a central composite design.

Performing runs at these four new points, our chemist obtained the results shown at the bottom of the last column of Table 6.7, which also contains the values already given in Table 6.5, completing the data for the central composite design.

The $\mathbf{y}$ vector now has 11 values, and the $\mathbf{X}$ matrix has 11×6 dimensions, with its six columns corresponding to the six terms of the quadratic model. By squaring and multiplying the appropriate columns of the design matrix in Table 6.7, we obtain the columns referring to x_1^2, x_2^2 and x_1x_2. Thus, we can write

Table 6.7
Results of the central composite design obtained by augmenting the factorial design in Table 6.5 with a star design. x_1 and x_2 are the values of the variables coded by the expressions in Table 6.5

Run	C (%)	v (rpm)	x_1	x_2	y (%)
1	30	115	−1	−1	86
2	40	115	1	−1	85
3	30	135	−1	1	78
4	40	135	1	1	84
5	35	125	0	0	90
6	35	125	0	0	88
7	35	125	0	0	89
8	28	125	$-\sqrt{2}$	0	81
9	35	139	0	$\sqrt{2}$	80
10	42	125	$\sqrt{2}$	0	86
11	35	111	0	$-\sqrt{2}$	87

$$\mathbf{X} = \begin{bmatrix} 1 & -1 & -1 & 1 & 1 & 1 \\ 1 & 1 & -1 & 1 & 1 & -1 \\ 1 & -1 & 1 & 1 & 1 & -1 \\ 1 & 1 & 1 & 1 & 1 & 1 \\ 1 & 0 & 0 & 0 & 0 & 0 \\ 1 & 0 & 0 & 0 & 0 & 0 \\ 1 & 0 & 0 & 0 & 0 & 0 \\ 1 & -\sqrt{2} & 0 & 2 & 0 & 0 \\ 1 & 0 & \sqrt{2} & 0 & 2 & 0 \\ 1 & \sqrt{2} & 0 & 2 & 0 & 0 \\ 1 & 0 & -\sqrt{2} & 0 & 2 & 0 \end{bmatrix} \quad \text{and} \quad \mathbf{y} = \begin{bmatrix} 86 \\ 85 \\ 78 \\ 84 \\ 90 \\ 88 \\ 89 \\ 81 \\ 80 \\ 86 \\ 87 \end{bmatrix}.$$

Solving Eqs. (5.12) and (5.30) with these matrices, we have

$$\hat{y} = \underset{(\pm 0.75)}{89.00} + \underset{(\pm 0.46)}{1.51x_1} - \underset{(\pm 0.46)}{2.36x_2} - \underset{(\pm 0.54)}{2.81x_1^2} - \underset{(\pm 0.54)}{2.81x_2^2} + \underset{(\pm 0.65)}{1.75x_1x_2}. \qquad (6.8)$$

The standard errors were again calculated from the pooled estimate of the variance obtained from all the replicate runs, including those of Table 6.1. The new analysis of variance is given in Table 6.8. Since the $MS_{\text{lof}}/MS_{\text{pe}}$ value is now only 0.25, there is no evidence of lack of fit for the quadratic model. Consequently, the 0.55 value for the residual mean

Table 6.8
Analysis of variance for fitting the $\hat{y} = b_0 + b_1x_1 + b_2x_2 + b_{11}x_1^2 + b_{22}x_2^2 + b_{12}x_1x_2$ model to the data in Table 6.7

Source of variation	Sum of squares	Degree of freedom	Mean square
Regression	144.15	5	28.83
Residual	2.76	5	0.55
Lack of fit	0.76	3	0.25
Pure error	2.00	2	1.00
Total	146.91	10	

% explained variation: 98.12
Maximum % explainable variation: 98.64

square, MS_r, can also be used as an estimate of error variance, with 5 degrees of freedom.

The response surface and the contour curves corresponding to the fitted model are shown in Figs. 6.6a and b. The region contains a point of maximum response, situated approximately at $x_1 = 0.15$ and $x_2 = -0.37$, that is, at a concentration of 36% and a stirring speed of 121 rpm. At these levels, according to Eq. (6.8), the reaction yield should be close to 89.6%, which represents a 32% improvement on the yield of the starting point, 68%.

Since we have found a maximum, the investigation stops here. If the response values of the second design had been different, the surface fitted to them would also be different, of course. To continue with our topographical analogy, if it were a ridge, for example, instead of a peak, we would have to start another displacement, along a new steepest ascent path. The whole process of modeling → displacement → modeling → ..., would then be repeated, until we eventually found the most favorable region. In practice there should not be many of these steps, because linear models become less efficient as we come close to an extreme (maximum, minimum or saddle point), where surface curvature evidently becomes more important.

Exercise 6.7. Use the data in Table 6.8 to calculate a value showing that Eq. (6.8) is statistically significant.

Exercise 6.8. Though always convenient, a graphical representation is not necessary to locate the maximum point on the response surface. This can be done by taking derivatives of the model equation with respect to all the independent variables and setting their values equal to 0. (a) Use this procedure in Eq. (6.8) to confirm the values presented in the text. (b) What would happen if you tried to do the same with Eq. (6.6)? Why?

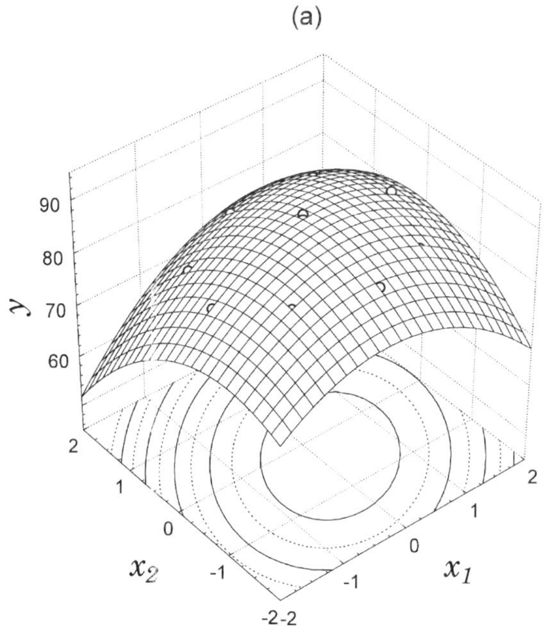

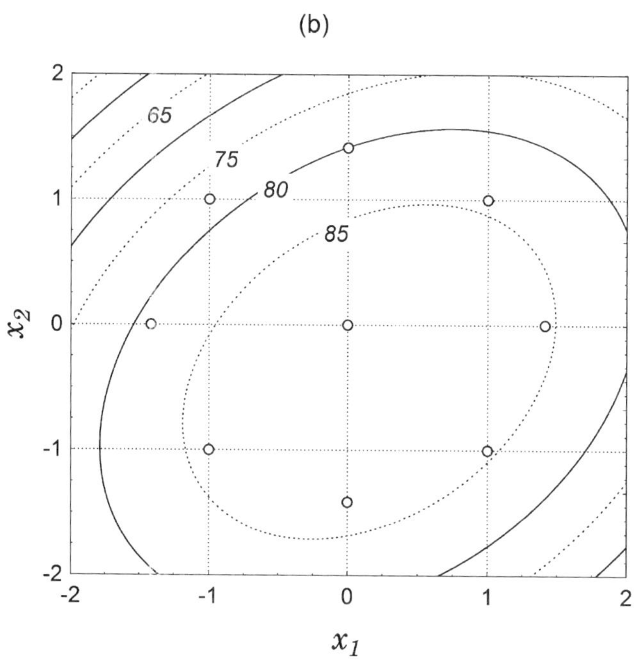

Fig. 6.6. (a) Quadratic response surface described by Eq. (6.8). (b) Its contour curves. The maximum yield (89.6%) occurs at $x_1 = 0.15$ and $x_2 = -0.37$.

6.2. The importance of the initial design

A very important question in RSM is the choice of the initial ranges for varying the factors. This determines the size of the first design and consequently the coding scale and the relative rate at which the subsequent experiments will be displaced over the experimental domain.

Suppose, for example, that in Table 6.1 we had chosen for the second factor — the stirring speed — the limits of 95 and 105 rpm (instead of 90 and 110). Assuming that the response surface is a plane, this decision would have the following consequences:

1. The x_2 coefficient in Eq. (6.3) would be reduced from 4.25 to 2.125, because a unit variation in x_2 would now correspond, in real units, to 5 rpm rather than 10 rpm.
2. With this new coefficient we would have, in Eq. (6.4),

$$\Delta x_2 = \frac{2.125}{-5.25} \Delta x_1 = -0.405 \Delta x_1.$$

3. Consequently, the displacement Δx_2 corresponding to $\Delta x_1 = -1$ would be +0.405, which is now equivalent to $\Delta v = +0.405 \times 5 = 2.03$ rpm. In other words, in terms of the stirring speed, each displacement would correspond to only *one-fourth* of the displacement for the original design. In the last run of Table 6.3, center+5Δ, the value of the stirring speed would be only 110.1 rpm.

Of course, the displacement would be faster if, instead of reducing the scale, we had preferred to expand it. However, we also would be taking risks. Depending on the expansion, we might leave the linear region or even cross the response surface "to the other side of the hill" already on the first displacement, and thus lose the opportunity of discovering the direction of the optimum point.

How should we set about finding the best scale? Unfortunately, the answer is not in this book, or in any other book on statistics, because this choice depends on the specific problem, and usually cannot be known beforehand. The same decision, in fact, must also be made in univariate experiments. All we can say is that researchers should use all the knowledge available to them and try to choose displacements that are not too small to cause detectable effects on the response, nor so large as to make it difficult to fit simple models — linear or quadratic — to the data.

We are the first to admit that this is more easily said than done. Let this point serve as another reminder of how important it is to proceed with the experimental study in a sequential and iterative way, keeping the designs as simple as possible. If the analysis of the first results leads us to modify our original design, the damage will be less than if we perform a lot of experiments right from the start.

6.3. An experiment with three factors and two responses

In RSM there is no restriction on the number of factors and responses studied. RSM can be applied to any number of factors, as well as model several responses at the same time.[56] This is an important characteristic, because many times the product or process of interest must satisfy more than one criterion, such as, for example, present maximum yield with a minimum quantity of impurities, or have minimum production costs while keeping the quality parameters within their specifications. To illustrate the flexibility of RSM, we present a real application in this section, whose objective was the simultaneous maximization of two distinct responses.

R.A. Zoppi, at the Chemistry Institute of Campinas State University, performed a series of experiments on the synthesis of polypyrrole in an EPDM rubber matrix, under the supervision of Professors M.A. de Paoli and M.I. Felisberti. The objective of the study was to obtain a product with electrical properties similar to those of polypyrrole and mechanical properties like those of EPDM rubber. Polypyrrole, a conducting polymer, is very brittle, which prevents its use in applications of practical interest.

The factors chosen for study were reaction time (**t**), oxidizing agent concentration (**C**) and granulation of the oxidizing particles (**P**). The researcher, who lacked formal instruction in experimental design strategy, decided to perform 27 runs in quadruplicate, according to the 3^3 factorial design of Table 6.9. The reaction yield and several mechanical properties, among them Young's modulus, were measured for samples of the final products. The values appearing in the table are averages and standard deviations of four runs[57] performed for each combination of the factor levels, for a total of 106 runs. Note that particle size is not defined in a precise way. The three levels represent size intervals, not specific sizes.

M.R. Vallim and V.F. Juliano analyzed the data of Table 6.9 as an assignment in a chemometrics course, and soon perceived that, since there are 27 different runs, it is possible to fit a function with up to 27 parameters to the data. The linear and quadratic functions for three variables are defined by only 4 and 10 parameters, respectively. Using them to model the data in the table, we will have many degrees of freedom left over to assess lack of fit.

[56]On the other hand, the analysis can become a bit complicated, as we will see later.
[57]In some cases, only three.

Table 6.9
3^3 Factorial design to investigate the effects of reaction time (**t**), oxidant concentration (**C**) and particle size (**P**) on the yield (Y) and Young's modulus (M) of the product of polypyrrole synthesis in a matrix of EPDM rubber

Level	−1	0	+1
T (h)	8	16	24
C (pph)	10	30	50
P (mesh)	>150	150–100	100–60

T	**C**	**P**	Y (%)	M (MPa)
−1	−1	−1	4.55 ± 0.17	0.61 ± 0.07
−1	−1	0	2.77 ± 0.10	0.57 ± 0.03
−1	−1	1	2.01 ± 0.08	0.54 ± 0.02
−1	0	−1	10.75 ± 0.41	0.99 ± 0.10
−1	0	0	7.32 ± 0.28	0.86 ± 0.05
−1	0	1	6.07 ± 0.23	0.74 ± 0.18
−1	1	−1	13.98 ± 0.53	2.13 ± 0.24
−1	1	0	14.59 ± 0.55	2.13 ± 0.18
−1	1	1	12.23 ± 0.46	1.61 ± 0.10
0	−1	−1	4.57 ± 0.17	0.57 ± 0.02
0	−1	0	3.28 ± 0.12	0.50 ± 0.05
0	−1	1	2.37 ± 0.09	0.58 ± 0.05
0	0	−1	11.24 ± 0.43	0.81 ± 0.12
0	0	0	7.37 ± 0.28	0.98 ± 0.09
0	0	1	7.31 ± 0.28	0.79 ± 0.13
0	1	−1	20.02 ± 0.76	2.38 ± 0.48
0	1	0	17.64 ± 0.67	2.07 ± 0.21
0	1	1	16.53 ± 0.63	1.45 ± 0.21
1	−1	−1	5.98 ± 0.23	0.54 ± 0.06
1	−1	0	5.14 ± 0.19	0.45 ± 0.05
1	−1	1	3.27 ± 0.12	0.45 ± 0.12
1	0	−1	13.17 ± 0.50	0.91 ± 0.13
1	0	0	10.78 ± 0.41	0.84 ± 0.06
1	0	1	9.72 ± 0.37	0.77 ± 0.07
1	1	−1	20.34 ± 0.77	2.58 ± 0.18
1	1	0	22.83 ± 0.86	2.06 ± 0.21
1	1	1	18.69 ± 0.71	1.78 ± 0.27

Table 6.10
Analyses of variance for the fits of the linear and quadratic models to the Young's modulus data of Table 6.9. The values for the quadratic model are given in parentheses

Source of variation	Sum of squares	Degree of freedom	Mean square
Regression	37.34 (*43.23*)	3 (*9*)	12.45 (*4.80*)
Residual	8.44 (*2.55*)	102 (*96*)	0.088 (*0.028*)
Lack of fit	6.76 (*0.87*)	23 (*17*)	0.29 (*0.051*)
Pure error	1.68	79	0.023
Total	45.78	105	

% explained variation: 81.56 (94.43)
Maximum % explainable variation: 96.33

The coefficients of the models and their standard errors were calculated as usual, by means of Eqs. (5.12) and (5.30). For Young's modulus, the linear model resulted in the equation

$$\hat{M} = \underset{(\pm 0.03)}{1.13} + \underset{(\pm 0.04)}{0.01\mathbf{t}} + \underset{(\pm 0.04)}{0.74\mathbf{C}} - \underset{(\pm 0.04)}{0.15\mathbf{P}}, \tag{6.9}$$

whereas the quadratic model produced

$$\begin{aligned}\hat{M} = {} & \underset{(\pm 0.09)}{0.86} + \underset{(\pm 0.04)}{0.01\mathbf{t}} + \underset{(\pm 0.04)}{0.74\mathbf{C}} - \underset{(\pm 0.04)}{0.16\mathbf{P}} - \underset{(\pm 0.07)}{0.02\mathbf{t}^2} \\ & + \underset{(\pm 0.07)}{0.44\mathbf{C}^2} - \underset{(\pm 0.07)}{0.05\mathbf{P}^2} + \underset{(\pm 0.05)}{0.07\mathbf{tC}} - \underset{(\pm 0.05)}{0.01\mathbf{tP}} - \underset{(\pm 0.05)}{0.18\mathbf{CP}}\end{aligned} \tag{6.10}$$

The analyses of variance for the two fits are given in Table 6.10. The MS_R/MS_r values are 141.5 and 171.4 for the linear and quadratic models, respectively. Comparing these values with $F_{3,\,102} = 2.71$ and $F_{9,\,96} = 2.00$, at the 95% confidence level, we see that both models are highly significant.

Although there does not appear to be much difference between the two models, a closer examination of Table 6.10 shows that we should prefer the quadratic model. While the MS_{lof}/MS_{pe} ratio is equal to 12.61 for the linear model, a value well above the critical $F_{23,\,79} = 1.67$ value, it is 2.22 for the quadratic model, just a bit higher than the $F_{17,\,79} = 1.75$ 95% confidence value.

The difference between the two models becomes more evident on inspection of the residual graphs (Figs. 6.7a and b). For the linear model, the graph presents a clear curvature. From left to right, the residual values are first positive, then negative and finally positive again. This is not observed for the quadratic model, whose residuals appear to fluctuate randomly about the zero value. In both cases, however, the residual

(a)

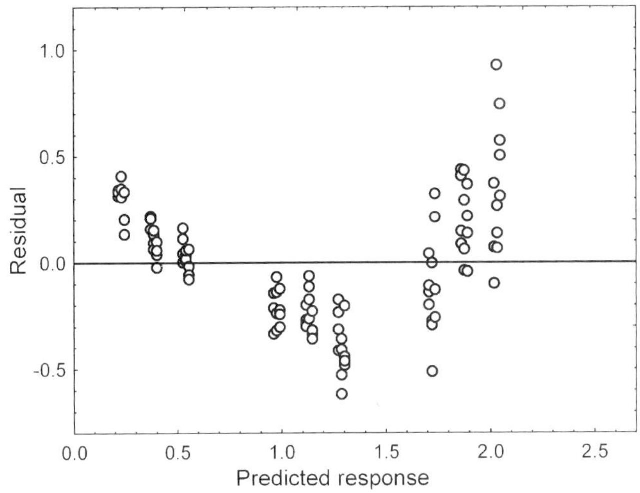

(b)

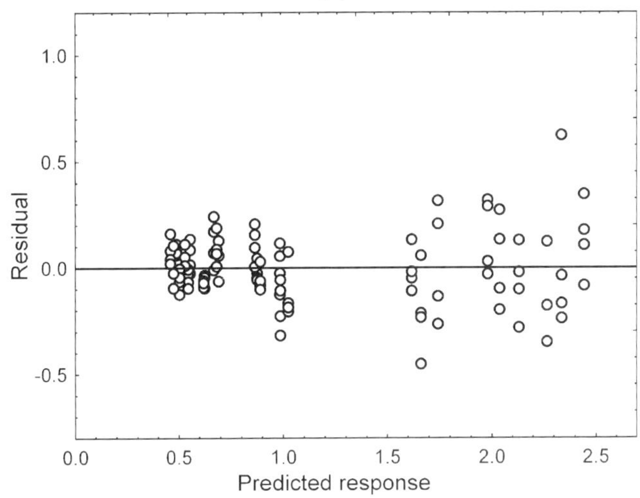

Fig. 6.7. (a) Residuals left by the fit of a linear model to the values of Young's modulus given in Table 6.9. (b) Residuals left by the fit of a quadratic model to the same data.

variance appears to increase with the value of the response, which could indicate that the data are heteroscedastic.

The values of the coefficients of the quadratic terms $\mathbf{C}^2$ and **CP** in Eq. (6.10), 0.44 and –0.18, strengthen the preference for the quadratic model. They are significantly larger than their standard errors (0.07 and 0.05), and therefore both terms should be maintained in the model. Since they are absent from the linear equation, we should not find it strange that the graph of the residuals in Fig. 6.7a shows a curved behavior.

Exercise 6.9. Use the data in Table 6.10 to calculate an experimental error estimate with more than 79 degrees of freedom.

Exercise 6.10. Knowing that the standard error estimate was obtained from the value of MS_{pe} in Table 6.10, find, at the 95% confidence level, the statistically significant coefficients in Eq. (6.10).

Once the model is statistically validated, we can try to interpret Eq. (6.10), to better understand the behavior of Young's modulus (and therefore the mechanical properties) of the samples in question. The results show that Young's modulus depends only on the oxidant concentration and particle size (Exercise 6.10). None of the terms involving the reaction time is statistically significant. As a first approximation, therefore, we can eliminate the terms containing **t**, thus reducing the model to

$$\hat{M} = 0.86 + 0.74\mathbf{C} - 0.16\mathbf{P} + 0.44\mathbf{C}^2 - 0.18\mathbf{CP}. \tag{6.11}$$

The response surface generated by this expression is shown in Fig. 6.8a. There appears to be a kind of valley, situated almost perpendicularly to the concentration axis.

Table 6.11 shows a comparison of the averages of the observed values and the values predicted from Eq. (6.11). The agreement is very good. The absolute value of average error of the predicted values is only 0.06, representing less than 4% of the range of the values in Table 6.9. This confirms that almost all of the variation in the Young's modulus values can be explained by the changes made in the concentration and size of the oxidant particles.

If our objective is to obtain a product with a high value for this parameter, the contour curves of Fig. 6.8b indicate that we should use a concentration of 50 parts in 100 and particles with a granulation larger than 150 mesh. If the model could be extrapolated, we should obtain even higher values by further increasing the concentration and decreasing the size of the particles.[58] In the same way, to obtain smaller values of Young's modulus we should use a low oxidant concentration, about 10 parts in 100, but the size of the particles is not important in this case. All of the experimental results obtained with 10 pph are on the valley bottom, where size variation does not affect the response.

Since the reaction time does not change the value of Young's modulus, we can use any value between 8 and 24 h. If we were interested only in this response, we would not need to control the reaction time. In this

[58]Numerically larger mesh values correspond to smaller particles, that is, to a finer granulation.

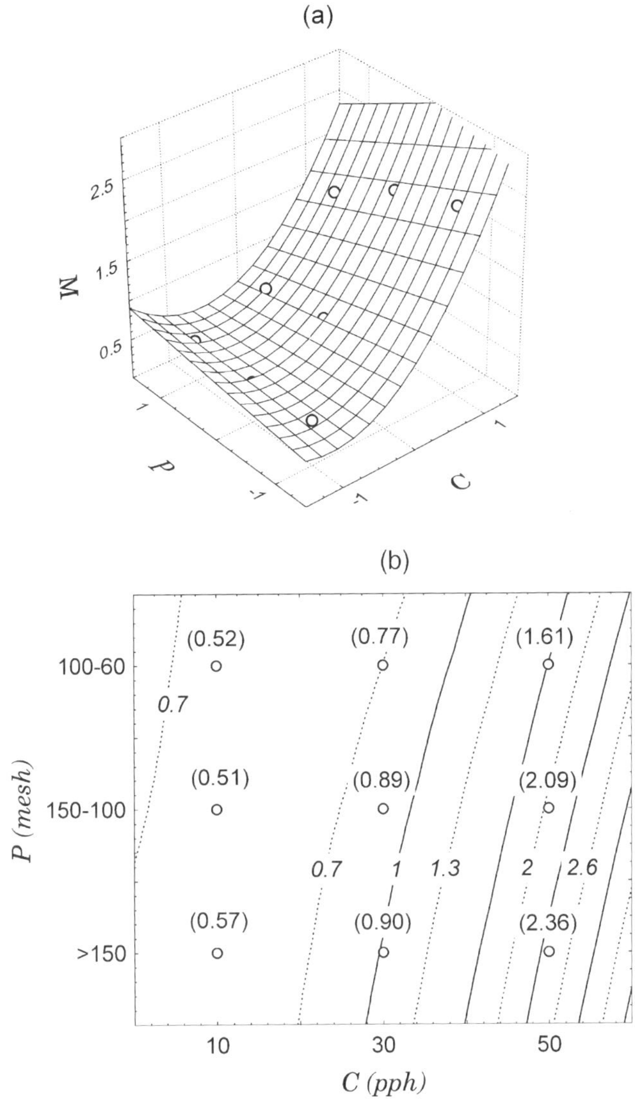

Fig. 6.8. (a) Response surface described by Eq. (6.11), relating Young's modulus to oxidant concentration and granulation. (b) Contour curves for the surface in (a). The values in parentheses are observed average responses.

study, however, the researchers also wanted to increase the reaction yield, and therefore modeled the yield data in the same manner we have described for Young's modulus. This resulted in the equation

$$\hat{Y} = 9.24 + 1.93\mathbf{t} + 6.81\mathbf{C} - 1.47\mathbf{P} + 1.28\mathbf{C}^2 + 1.26\mathbf{tC},$$

Table 6.11
Values predicted by Eq. (6.11) and observed average values for Young's modulus. The absolute average error is given by $\bar{e} = \left(\sum |e|\right)/n = 0.06$ MPa

C	P	$\bar{M}_{obs}$	$\hat{M}$	$e = \bar{M}_{obs} - \hat{M}$
−1	−1	0.57	0.53	0.04
−1	0	0.51	0.55	−0.04
−1	1	0.52	0.57	−0.05
0	−1	0.90	1.03	−0.13
0	0	0.89	0.87	0.02
0	1	0.77	0.71	0.06
1	−1	2.36	2.37	−0.02
1	0	2.09	2.03	0.06
1	1	1.61	1.69	−0.08

where only the statistically significant terms are represented. In this expression time is an important factor. All of the terms involving **t** have positive coefficients, which means that longer times produce larger yields. Setting the time at its maximum value (24 h, or $\mathbf{t} = +1$), we can write

$$\hat{Y} = 11.17 + 8.07\mathbf{C} - 1.47\mathbf{P} + 1.28\mathbf{C}^2. \qquad (6.12)$$

The response surface described by this expression is represented in Fig. 6.9. Comparing it to the one in Fig. 6.8 we can conclude that the region expected to produce higher values of Young's modulus (the lower right-hand corner of the response contour plot) is also expected to produce higher yields. The same parallelism can be observed on the valley bottom: the $\hat{M}$ values of about 0.50 MPa correspond to low yields, about 5%.

Even though the results we have just discussed permitted an adequate description of the response surface, we could have arrived at the same conclusions with a more economical approach. In a first stage, to select a region for a more detailed investigation, we could have probed the factor space with a two-level factorial design. Then, depending on the results, we could

(a) Augment the original design with more runs, adding a star design to transform it into a central composite design, or
(b) Displace the entire design to a more promising region, which could be investigated with a new factorial design.

Alternatively, we could have used a Box–Behnken design, which we will discuss in Section 6.6.

It must be stressed that that these considerations are not meant to belittle the study presented in this section. On the contrary, all the experiments were performed according to a systematic design, permitting

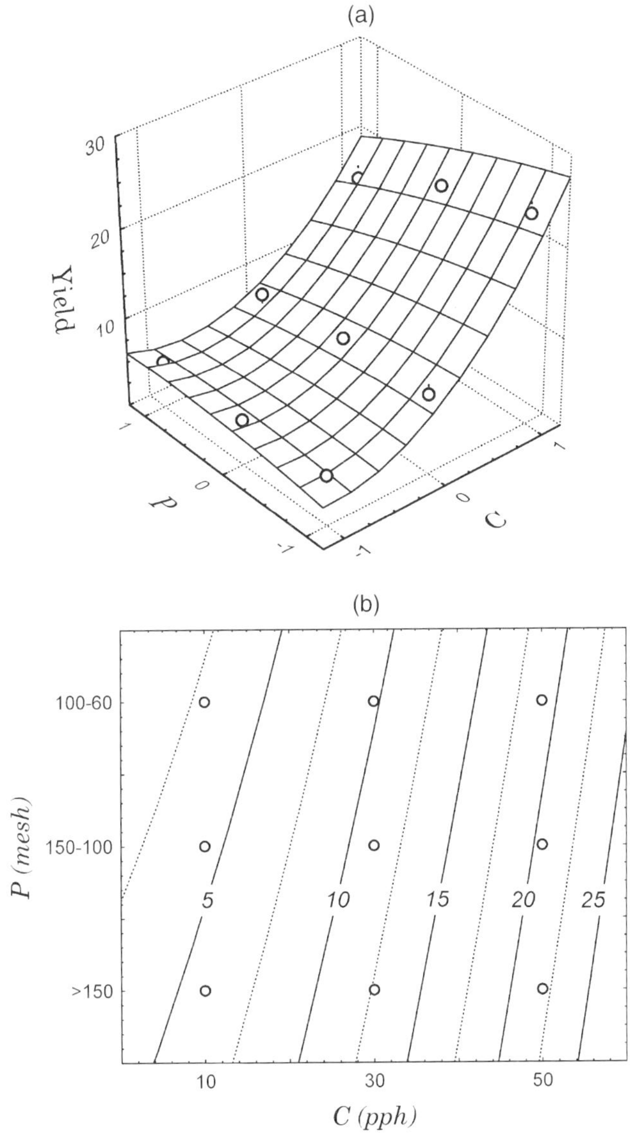

Fig. 6.9. Response surface (a) and contour curves (b) for Eq. (6.12), showing the reaction yield as a function of concentration (C) and oxidant granulation (P) after 24 h of reaction.

a precise characterization of the influence of the investigated factors on the responses of interest. This way of proceeding is undeniably superior to the haphazard one that still seems to prevail in many research laboratories.

6.4. Treating problems with many variables

In our last example, in spite of having three factors to consider, we were able to reduce our analysis to graphs containing only two of them (plus the response). This allowed us to locate the region of interest by straightforward visual inspection of the fitted response surfaces. If all three factors had been found significant, this easy visualization would no longer be possible (it would require four dimensions) and we would need another way to analyze the surfaces. We were also lucky to find that the optimum regions for both responses coincided. If this had not been the case, we would not have been able to optimize both responses simultaneously and would have to settle for a compromise solution. Evidently, as the number of factors increases, the analysis will tend to become more complicated.

Consider a general optimization problem, with several responses $y_1, y_2, \ldots, y_m$, for which we construct models based on a set of coded factors $x_1, x_2, \ldots, x_n$. What can be done to discover the factor levels that will produce the most satisfactory set of responses?

There are several possibilities. If the number of factors, x_i, allows direct visualization of the fitted models, and if the number of responses is not too large, we can overlay the response surfaces — or better yet their contour curves — and locate the best region by visual inspection. That was the procedure followed in the last section.

If, on the other hand, our objective is to maximize or minimize a certain response, while keeping the others subject to certain constraints, we can resort to the linear programming methods — or even non-linear ones — commonly used in engineering.

Finally, if the problem does not fall into either of these two categories, we can attempt to apply the simultaneous optimization methodology proposed by Derringer and Suich (1980), which can be quite useful when used with care.

The Derringer and Suich method is based on the definition of a desirability function for each response, with values restricted to the [0,1] interval. Zero stands for an unacceptable value, while one is assigned to the most desirable value. The nature of the function depends on the objectives of the experiment, as we shall see.

Once the desirability functions have been specified for all the responses, we combine them into an overall desirability, normally calculated as the geometric average of the m individual desirabilities:

$$D = \sqrt[m]{d_1 d_2, \ldots, d_m}. \tag{6.13}$$

With this definition, the simultaneous optimization of the various responses reduces to the problem of maximizing a single value, the overall desirability. Our problem is turned into finding the factor levels that maximize the D value. Another advantage of using Eq. (6.13) is that the geometric average makes the overall desirability always vanish if any response has an unacceptable value, no matter how satisfactory the values of the other responses.

The form of the desirability function for a given response depends on how the problem is formulated. Suppose that the response has an optimum target value, let us say T, somewhere within an acceptance range, whose lower and upper limits are represented by L and U, respectively. In this case, the desirability function for the response is defined by

$$d = \left(\frac{\hat{y} - L}{T - L}\right)^s \quad \text{for } L \le \hat{y} \le T; \tag{6.14a}$$

$$d = \left(\frac{U - \hat{y}}{U - T}\right)^t \quad \text{for } T \le \hat{y} \le U; \tag{6.14b}$$

$$d = 0 \quad \text{for } \hat{y} \text{ outside the } [L, U] \text{ interval.} \tag{6.14c}$$

Note that the value of d is restricted to the [0,1] interval. A desirability equal to 1 will only be obtained if the response exactly coincides with the target value, thus making the numerators of the fractions equal to their respective denominators. As the value of $\hat{y}$ moves away from the target T, the value of the desirability will fall, becoming zero if one of the acceptance limits is reached.

The rate of variation of the desirability with the estimated response is defined by the values of the s and t exponents. We can accelerate or slow down the variation rate by modifying their values, and in this way assign different desirabilities to the different response levels. If we choose high values for the two exponents (e.g., 10), the desirability will fall rapidly, becoming very low unless $\hat{y}$ is very close to the target. Low exponent values, on the other hand, will allow the response to have a much wider variation without diminishing the desirability very much. The choice of the exponents will depend on the priority or relative importance that we attribute to each response. Also, the rate of fall of the desirability does not have to be symmetrical about the target value. The values of s and t can be different. If, for example, it is more acceptable that the response value stay above a certain target rather than below it, we should choose $t \ll s$.[59] Many times we do not have a target value but a one-sided limit, above or below which we want the response to stay. To treat these cases, we should modify one part of the desirability definition, making the target

[59]For example, if the response were the volume of beer in a bottle, the manufacturer certainly would prefer to error in excess rather than risk his reputation by selling bottles containing less beer than stated on the label.

value coincide with one of the extremes and setting $d = 1$ beyond this point. There are two possibilities:

1. Eliminate Eq. (6.14a) and make $d = 1$ for $\hat{y} \leq L$. This means that we are completely satisfied with any response value below the inferior limit, L.
2. If, on the contrary, our objective is to maintain the response above the limit U, we eliminate Eq. (6.14b) and make $d = 1$ for any $\hat{y} \geq U$.

In some optimization problems, it is not possible to clearly specify the acceptance limits for some responses. We cannot, therefore, define the desirability by means of Eq. (6.14). A better alternative in these cases appears to be an exponential function. More details about this subject can be found, for example, in Wu and Hamada (2000).

Even with the interesting methodology of Derringer and Suich, we must be careful when optimizing several responses simultaneously. If we limit ourselves to mechanically applying the search algorithm, we can be led to a set of mathematically "optimized" conditions without practical value, perhaps because some boundary conditions have been relaxed at the beginning of the investigation, or because the exponents have not been properly chosen. Sometimes, we only discover that the problem has not been well formulated when the software points to absurd experimental conditions as the optimum ones.

A good strategy is to feed the optimization algorithm with different choices for the s and t exponents. In this way we will arrive at several sets of optimized conditions, from which we can select the one that best suits our needs. The very variety of these sets will give us a good idea of the robustness of the suggested experimental conditions. If they are relatively insensitive to the variation of the s and t values, this is a good sign.

After discovering the set of conditions that maximize the global desirability D, we should not fail to examine the individual behavior of every response, to make sure that all of them are really in acceptable regions, with all of the constraints satisfied. It is also highly recommended to conduct some confirmatory experiments at the selected conditions and, if possible, in their neighborhood too. Confirmatory experiments are always an excellent idea. When several responses are being investigated, performing them becomes practically mandatory.

To illustrate the application of the Derringer and Suich method, we are going to make a small adaptation of the problem with three factors and two responses presented in Section 6.3. We start with the data in Table 6.9, but initially suppose that we are interested in obtaining a product with a value of Young's modulus as close as possible to 2.0 MPa, and also that we want a yield not much lower than 15%. The algorithm we used (StatSoft, 2004) requires specification of the numerical values of L, U and T, besides s, t and the desirabilities. To arrive at the results we present next, we used the values shown in Table 6.12.

Table 6.12
Parameters used in the simultaneous optimization of the responses of the synthesis of polypyrrole in EPDM rubber. The values in parentheses are the desirabilities of the corresponding response levels

Response	L	T	U	s	t
Young's modulus (MPa)	0.5 (0)	2.0 (1)	2.5 (0)	5	5
Yield (%)	10 (0)	15 (1)	20 (1)	5	1

Note that for Young's modulus we are assuming that any response below 0.5 or above 2.5 MPa is unacceptable. Since we really want to maximize this property, we are modifying the original objective a bit, for illustrative purposes. The high values chosen for the s and t exponents will make the desirability very small if the value of Young's modulus is not close to its target value, 2 MPa. For the yield we are accepting values above 10%, but would prefer it to stay above 15%. Below this value the high value chosen for s (5, again) will make the desirability fall rapidly. Since desirabilities equal to 1 were specified both for 15% and for 20%, any yield in this range will be considered perfectly satisfactory.

Fig. 6.10 shows the results produced by the optimization algorithm. We used a grid of 20 points for each of the three factors, which means that the response values and their corresponding desirabilities were calculated at $20 \times 20 \times 20 = 8000$ different combinations of factor levels. Both Young's modulus and the yield were fitted with full quadratic models. The two graphs in the last column show the profiles of the desirabilities of the two responses, defined according to the values chosen for the limits and exponents in Table 6.12. Note that for Young's modulus the desirability has a very pronounced peak at its target value, a consequence of the high values chosen for its two exponents. For the yield, we have a plateau of desirability equal to 1 above 15%. The dashed vertical lines indicate the conditions of maximum overall desirability, which in this example reached 0.99 and was obtained with $\mathbf{t} = -0.158$, $\mathbf{C} = 0.895$ and $\mathbf{P} = -0.474$, as shown in the last line of graphs. Under these conditions we should have a Young's modulus equal to 2.00 MPa and a yield of 16.5%, as indicated by the values highlighted on their respective axes.

The curves show how the responses vary with each factor, maintaining the levels of the other factors fixed at the specified values. In the second graph of the first column, for example, we see that Young's modulus is practically unaffected by a variation in **t**, as we have already discovered in Section 6.3. The crucial factor for the determination of the optimum point is **C**, the oxidant concentration, which is the factor with the most pronounced slopes. These slopes are all instructive, because they provide an idea of how safe it is to vary the factor levels around the optimized values. The plot of the overall desirability as a function of **t** shows that

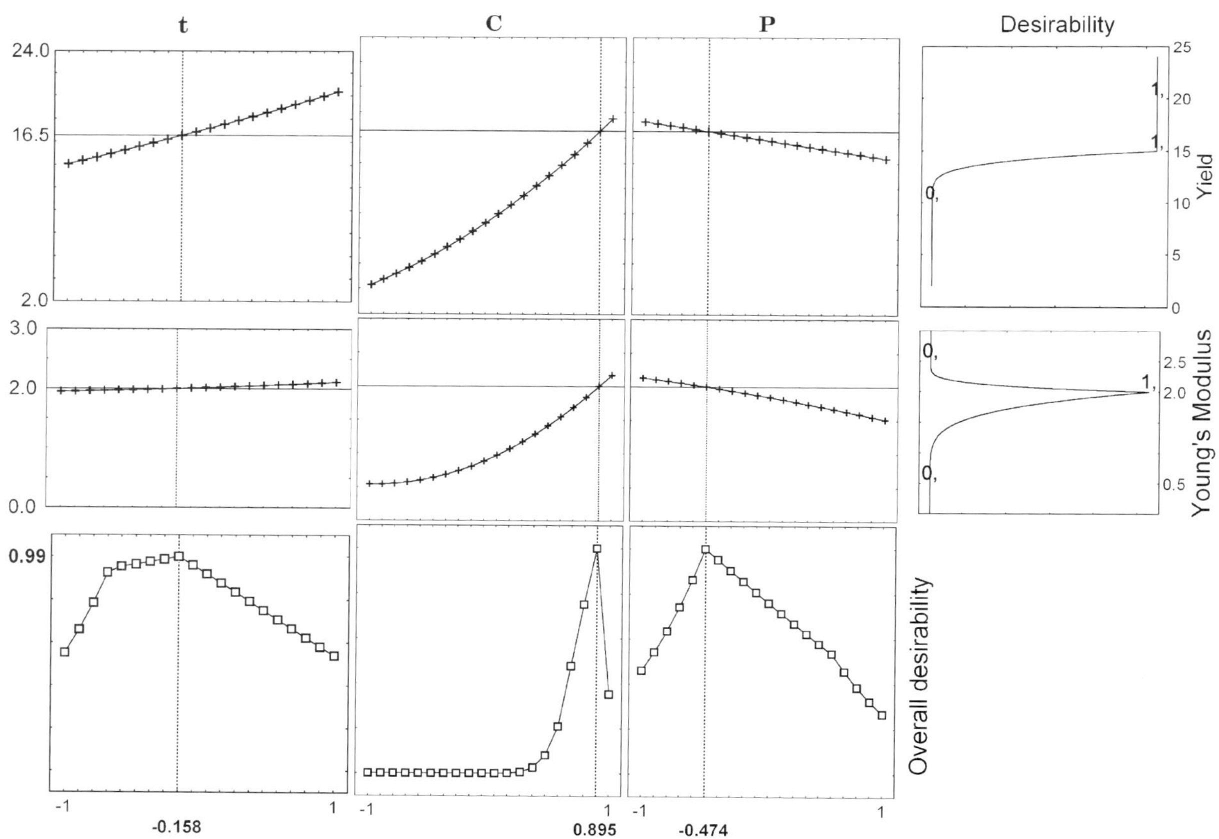

Fig. 6.10. Application of the Derringer and Suich methodology to optimize the properties of the polypyrrole–EPDM rubber system.

this factor can vary in a reasonably wide range without lowering the D value very much. On the other hand, any alteration in the **C** values provokes a sharp drop in the desirability. Therefore, this factor should be kept under strict control.

The graphs involving factor **P** illustrate the problems we were discussing a while ago. They show that an increase in **P** tends to decrease the yield and Young's modulus, more or less in the same proportion. It happens, though, that the granulation **P** is a categorical variable whose only possible values in this experiment are the three levels used in the design: –1, 0 and +1. The optimum value encountered, $\mathbf{P} = -0.474$, is devoid of physical meaning. Of the three possible levels, the less favorable is without doubt the value of +1. To choose between the other two, the researchers should decide if they prefer to stay with responses above or below the "optimum" point encountered ($\mathbf{P} = -1$ or 0, respectively). Remembering now that we really wanted to maximize the two responses, we should choose $\mathbf{P} = -1$, besides preferring $\mathbf{C} = +1$ and $\mathbf{t} = +1$, as is clear from the plots of the responses against the three factors. The algorithm did not make this choice for us because we decided to specify a narrow profile centered on a value of 2 MPa for Young's modulus, to illustrate the method.

6.5. Central composite designs

As we have already seen, the design of Fig. 6.5 is an example of a central composite design for two factors. In general, a central composite design for k factors, coded as $(x_1, \ldots, x_k)$, consists of three parts:

1. A factorial (or cubic) part, containing a total of n_{fact} points with coordinates $x_i = -1$ or +1, for $i = 1, \ldots, k$.
2. An axial (or star) part, formed by $n_{\text{ax}} = 2k$ points with all their coordinates null except for one that is set equal to a certain value α (or $-\alpha$).
3. A total of n_{c} runs performed at the center point, where, of course, $x_1 = \cdots x_k = 0$.

To build a central composite design, we need to specify each of these three parts. We have to decide how many cubic points to use and where they will be, what will be the value of α, and how many replicate runs should we conduct at the center point. For the design in Table 6.7, for example, $k = 2$. The cubic part is defined by the first four runs, the star design by the last four (with $\alpha = \sqrt{2}$) and there are three replicate runs at the center point. The three-factor case is shown in Fig. 6.11, where the origin of the terms chosen to describe the three parts of the design is made clear.

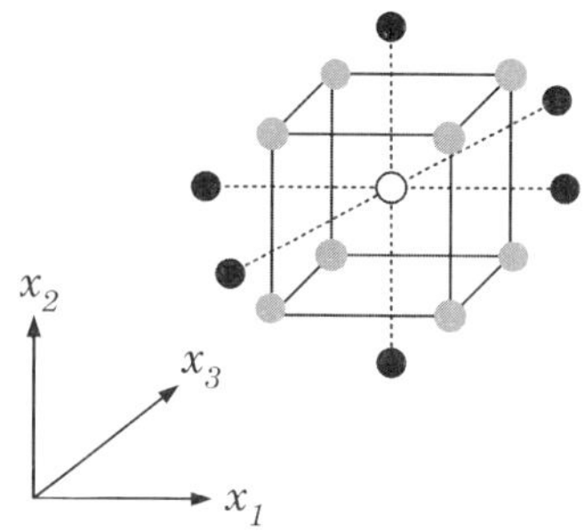

Fig. 6.11. Central composite design for three factors. The gray dots form the cubic part — the runs of a 2^3 factorial. The black dots represent the star part.

The cubic points, as you surely noticed, are the same as those of a factorial design for two factors. In Table 6.7, we have a full factorial design, but this is not strictly necessary. Depending on the number of factors, it might even be unadvisable, because it would require too many experiments.

The total number of distinct levels in a central composite design is $n_{\text{fact}}+2k+1$.[60] The complete quadratic model for k factors contains $(k+1)(k+2)/2$ parameters and is given by

$$y = \beta_0 + \sum_i \beta_i x_i + \sum_i \beta_{ii} x_i^2 + \sum_{i<j} \sum_j \beta_{ij} x_i x_j + \varepsilon. \tag{6.15}$$

With two factors, this model has six parameters. Since the design in Table 6.7 has nine different combinations of levels, we could estimate all of the model's parameters using only two cubic points, corresponding to one of the two 2^{2-1} fractions. In such a simple design, the economy is very small and hardly justifies destroying the symmetry of the cubic part, but this procedure — choosing a fractional design instead of a complete factorial to define the cubic points — becomes more attractive as the number of factors increases. As for the resolution, it is advisable to use a fractional design of resolution V, which allows us to estimate the main effects and the two-factor interactions with relatively low confounding. If we decide to use smaller fractions, however, the choice of the appropriate fraction is not trivial. Just to give an example: when $k = 4$, the 2_{III}^{4-1} fraction generated by **I = 124** is better than the 2_{IV}^{4-1} fraction generated by **I = 1234**, strange as that may seem. An explanation of why this happens is beyond the scope of this book, but let it serve as a warning. A list of the most adequate fractional designs can be found in Wu and Hamada (2000, Chapter 9).

The value of α usually ranges from 1 to $\sqrt{k}$. When $\alpha = \sqrt{k}$, as in Table 6.7, the cubic and axial points are located on the (hyper)surface of a

[60] Can you see why?

(hyper)sphere,[61] and the design is called spherical. In Table 6.7, for example, all of the peripheral points are on the same circumference. At the other extreme, when $\alpha = 1$, the axial points are located at the centers of the faces of the (hyper)cube defined by the cubic part of the design. This type of design is advantageous when the experimental space is cuboidal, which occurs in a natural way when the factors are varied independently of one another. It also has the advantage of only requiring three-factor levels, which can be a big help if one of the factors is qualitative, as we have already seen.

If we choose $\alpha = \sqrt{k}$, the star points will be placed farther from the center point as the number of factors increases. This choice should be made — if it is made at all — with much care, because we run the risk of leaving too much of the intermediate region uninvestigated. With nine factors, for example, α would be equal to 3. The experiments would tell us nothing about the behavior of the response surface in the 1–3 intervals along each axis.

Box and Hunter (1957) proposed the rotatability concept as a criterion for choosing the value of α. A design is called rotatable if the variance of its estimates, $V(\hat{y})$, depends solely on the distance from the center point, that is, if the precision of the predicted response is the same for all points situated on a (hyper)sphere centered at the center of the design. For a design whose cubic portion is a complete factorial or a fractional factorial of resolution V, it can be demonstrated that rotatability is attained when $\alpha = \sqrt[4]{n_{\text{fact}}}$. Even if the resolution is not exactly 5, this expression serves as a guide for choosing an α value, which in any case should also be analyzed for its convenience and feasibility. If, for example, we are interested in investigating the region closer to the faces of the hypercube, then it is better to choose an α value smaller than the rotatable one. On the other hand, it could happen that the chosen α value results in some impracticable runs. In this case, we would have to define new experimental conditions for these runs. Finally, rotatability depends on how the factors are coded. In general, the rotatability of a design will be destroyed if the variables are transformed using different scales. A *near* rotatability seems to be a reasonable criterion (Wu and Hamada, 2000, Chapter 9). Table 6.13 contains α values for constructing rotatable designs for up to six factors.

The replicates at the center point have two main objectives: providing a measure of pure error and stabilizing the variance of the predicted response. To stabilize the variance, a rule of thumb is to make 3–5 repeated runs if α is close to $\sqrt{k}$, and only one or two if it is close to 1. Between these extremes, 2–4 replicates are indicated. To obtain a more

[61]The prefix *hyper* is used here to refer to a geometrical figure in more than three dimensions.

Table 6.13
Values of α for rotatable central composite designs. The total number of runs is $n = n_{\text{fact}} + n_{\text{ax}} + n_C$

K	n_{fact}	N	α
2	4	$8+n_c$	1.414
3	8	$14+n_c$	1.682
4	16	$24+n_c$	2.000
5	32	$42+n_c$	2.378
5	16	$26+n_c$	2.000
6	64	$76+n_c$	2.828
6	32	$44+n_c$	2.378

precise estimate of error, the more replicates the better, as we already know.

Because central composite designs are formed by three distinct parts, we can build them sequentially, according to necessity. If we happen to be in a region of the response surface where curvature is not important, there is no need to fit a quadratic model. The cubic part of the design is sufficient for fitting a linear model, from which we can rapidly move to a more interesting region. Even if we are in doubt about possible curvature, we can test its significance using the runs at the center point. Then, if the curvature is found significant, we can complete the design with the axial points. Actually, we would be performing the runs of the complete design in two blocks — first the cubic and then the axial one.

Suppose that the response values in the axial block contain a systematic error in relation to the response values obtained in the first block. Under certain conditions, this error will not affect the coefficient estimates for the model, that is, the block effect will not confound itself with the effects of the other factors. This will occur if the design blocking is orthogonal, which in turn depends on the α value. Blocking will be orthogonal if

$$\alpha = \sqrt{\frac{n_{\text{fact}}(n_{\text{ax}} + n_{\text{c,ax}})}{2(n_{\text{fact}} + n_{\text{c,fact}})}},$$

where $n_{\text{c,fact}}$ and $n_{\text{c,ax}}$ are the number of runs at the center point in the cubic and axial blocks, respectively (Montgomery, 2004, Chapter 11). In general, when we make a design with orthogonal blocks we are sacrificing its rotatability, but some designs do exist for which the two conditions are approximately satisfied, and others for which both are exactly satisfied. Table 6.14 contains the settings for three blocked designs that also are rotatable.

Response surface methodology is a very important and vast subject, with whole books dedicated to it. If you want to know more about RSM, we

Table 6.14
Three central composite designs that can be sequentially conducted in blocks and still preserve rotatability

Number of factors (k)	2	4	5
Cubic part			
n_{fact}	4	16	16
Number of blocks	1	2	1
$n_{\text{c,fact}}$ (in each block)	3	2	6
Total number of points per block	7	10	22
Axial part (in only one block)			
n_{ax}	4	8	10
$n_{\text{c,ax}}$	3	2	1
α	1.414	2.00	2.00
Total number of points of the design	14	30	33

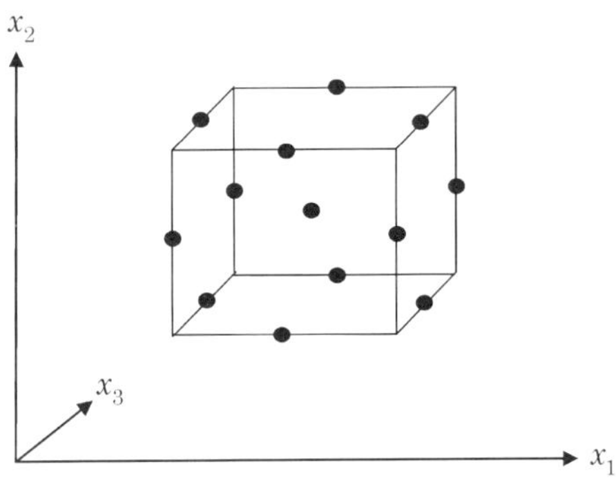

Fig. 6.12. ints representing the experimental runs of a three-factor Box–Behnken design.

recommend the excellent texts by Box and Draper (1987), Myers and Montgomery (2002) and Goupy (1999).

6.6. Box–Behnken designs

Box–Behnken designs constitute an alternative to central composite designs. They are a class of rotatable or nearly rotatable second-order designs based on three-level incomplete factorial designs (Box and Behnken, 1960). Fig. 6.12 shows the Box–Behnken design for three factors, corresponding to the coded values listed in Table 6.15. It is easy to see that this design consist of three parts of four runs. Within each part, two factors are arranged in a full two-level design, while the level of the

Table 6.15
Coded factor levels of a Box–Behnken design for studying polypyrrole synthesis in an EPDM rubber matrix. Values selected from Table 6.9

Run	x_1	X_2	x_3	Yield, Y	Young's modulus, M
1	−1	−1	0	2.77 ± 0.10	0.57 ± 0.03
2	1	−1	0	5.14 ± 0.19	0.45 ± 0.05
3	−1	1	0	14.59 ± 0.55	2.13 ± 0.18
4	1	1	0	22.83 ± 0.6	2.06 ± 0.21
5	−1	0	−1	10.75 ± 0.41	0.99 ± 0.10
6	1	0	−1	13.17 ± 0.50	0.91 ± 0.13
7	−1	0	1	6.07 ± 0.23	0.74 ± 0.18
8	1	0	1	9.72 ± 0.37	0.77 ± 0.07
9	0	−1	−1	4.57 ± 0.17	0.57 ± 0.02
10	0	1	−1	20.02 ± 0.76	2.38 ± 0.48
11	0	−1	1	2.37 ± 0.09	0.58 ± 0.05
12	0	1	1	16.53 ± 0.63	1.45 ± 0.21
13	0	0	0	7.37 ± 0.28	0.98 ± 0.09

third factor is set at 0. The points lie on the surface of a sphere centered at the origin of the coordinate system and tangential to the midpoint of each edge of the cube.

Compared to the central composite design shown in Fig. 6.11, this design has some advantages. It requires only 12 runs plus the replicates at the center point, whereas the three-factor central composite design has 14 non-center points. Also, each factor is studied at only three levels, an important feature in some experimental situations, as we have already mentioned. On the other hand, using $\alpha = 1$ in a central composite design also results in three levels for each factor. In most real applications, these differences are probably not decisive in determining which design to use, at least for this number of factors. However, since Box–Behnken designs do not contain combinations where the factors are all at their higher or lower levels, they may be useful in avoiding experiments under extreme conditions, for which unsatisfactory results might occur. Conversely, they are not indicated for situations in which we would like to know the responses at the extremes, that is, at the vertices of the cube.

Box–Behnken designs for four and five factors can be arranged in orthogonal blocks, as shown in Table 6.16. In this table, each $(\pm 1, \pm 1)$ combination within a row represents a full 2^2 design. Because of the block orthogonality, the second-order model can be augmented to include block effects without affecting the parameter estimates, that is, the effects themselves are orthogonal to the block effects. This orthogonal blocking is a desirable property when the experiment has to be arranged in blocks and the block effects are likely to be large. More information on the

Table 6.16
Box–Behnken designs for four and five factors. The dashed lines separate the orthogonal blocks. The combination (±1, ±1) within each row stands for a full 2^2 design. Use of 3–6 center runs is recommended for these designs

K	N	Design matrix
4	$24+n_C$, $n_C \geq 3$	$\left[\begin{array}{cccc} \pm1 & \pm1 & 0 & 0 \\ 0 & 0 & \pm1 & \pm1 \\ 0 & 0 & 0 & 0 \\ \hdashline \pm1 & 0 & 0 & \pm1 \\ 0 & \pm1 & \pm1 & 0 \\ 0 & 0 & 0 & 0 \\ \hdashline \pm1 & 0 & \pm1 & 0 \\ 0 & \pm1 & 0 & \pm1 \\ 0 & 0 & 0 & 0 \end{array}\right]$
5	$40+n_C$, $n_C \geq 2$	$\left[\begin{array}{ccccc} \pm1 & \pm1 & 0 & 0 & 0 \\ 0 & 0 & \pm1 & \pm1 & 0 \\ 0 & \pm1 & 0 & 0 & \pm1 \\ \pm1 & 0 & \pm1 & 0 & 0 \\ 0 & 0 & 0 & \pm1 & \pm1 \\ 0 & 0 & 0 & 0 & 0 \\ \hdashline 0 & \pm1 & \pm1 & 0 & 0 \\ \pm1 & 0 & 0 & \pm1 & 0 \\ 0 & 0 & \pm1 & 0 & \pm1 \\ \pm1 & 0 & 0 & 0 & \pm1 \\ 0 & \pm1 & 0 & \pm1 & 0 \\ 0 & 0 & 0 & 0 & 0 \end{array}\right]$

properties of Box–Behnken designs can be found in a number of excellent sources, for example Box and Draper (1987) and Myers and Montgomery (2002). Here we will present an application that takes advantage of the fact that Box–Behnken designs require experiments at only three levels of each factor.

Earlier in this chapter we discussed the results of a full 3^3 factorial design, used to study the synthesis of polypyrrole in an EPDM rubber matrix. This design, which requires 27 different level combinations, resulted in the models given by Eqs. (6.10)–(6.12). Now, we shall see that essentially the same information could be obtained from a 13-run Box–Behnken design. Table 6.15 contains the reaction yield and Young's modulus values for the runs of this design, taken from the full design in Table 6.9. All the values in this table are averages and standard deviations of four-fold replicates except one that was done in triplicate. In all, only 51 individual values (in contrast to the 106 of Table 6.9) are used for modeling here. Regression of the individual Young's modulus values on the coded factor values results in the quadratic model

$$\hat{Y} = \underset{(\pm 0.08)}{0.98} - \underset{(\pm 0.03)}{0.03\mathbf{t}} + \underset{(\pm 0.03)}{0.73\mathbf{C}} - \underset{(\pm 0.03)}{0.16\mathbf{P}} - \underset{(\pm 0.05)}{0.04\mathbf{t}^2} + \underset{(\pm 0.05)}{0.36\mathbf{C}^2} - \underset{(\pm 0.05)}{0.09\mathbf{P}^2} + \underset{(\pm 0.04)}{0.01\mathbf{tC}} + \underset{(\pm 0.04)}{0.03\mathbf{tP}} - \underset{(\pm 0.04)}{0.24\mathbf{CP}} \quad (6.16)$$

This model is practically the same as the one given by Eq. (6.10). The differences between the corresponding model coefficients are slight and comparable to their standard errors.

The ANOVA results for the fits of linear and quadratic models to the Young's modulus data of the Box–Behnken design are presented in Table 6.17. The linear model shows significant lack of fit at the 95% confidence level: $MS_{\mathrm{lof}}/MS_{\mathrm{pe}} = 0.35/0.025 = 14.0$, which is larger than $F_{9,\,38} = 2.14$. However, a significant lack of fit at this confidence level was also found with the complete 3^3 data set. When the second-order terms are added to the model, the lack of fit is considerably reduced. The new ratio, $MS_{\mathrm{lof}}/MS_{\mathrm{pe}} = 0.092/0.025 = 3.68$, is a bit larger than the $F_{3,38,95\%}$ value of 2.86, indicating that some lack of fit still remains. The same happened for the quadratic model fitted to all 106 data points. The residual graphs for both models are similar to those shown in Fig. 6.7. The curvature in the residual graph for the linear model is as easy to see as the one in Fig. 6.7a, but the funnel-shaped behavior of the residuals for the quadratic model is not as pronounced as that in Fig. 6.7b.

Retaining only the significant terms at the 95% confidence level, the quadratic models given by the Box–Behnken designs for Young's modulus

Table 6.17
Analyses of variance for the fits of the linear and quadratic models to the Young's modulus data of the Box–Behnken design given in Table 6.15. The figures for the quadratic model are given in parentheses

Source of variation	Sum of squares	Degree of freedom	Mean square
Regression	17.84 (*20.72*)	3 (*9*)	5.95 (*2.30*)
Residual	4.10 (*1.21*)	47 (*41*)	0.088 (*0.030*)
Lack of fit	3.16 (*0.28*)	9 (*3*)	0.35 (*0.092*)
Pure error	0.94	38	0.025
Total	21.94	50	

% explained variation: 81.31 (*94.47*)
Maximum % explainable variation: 95.72

and yield are

$$\hat{M} = 0.98 + 0.73\mathbf{C} - 0.16\mathbf{P} + 0.36\mathbf{C}^2 - 0.24\mathbf{CP}$$

and

$$\hat{Y} = 7.37 + 2.09\mathbf{t} + 7.39\mathbf{C} - 1.73\mathbf{P} + 2.45\mathbf{C}^2 + 1.48\mathbf{tC}.$$

These equations show that the Box–Behnken and the full three-level factorial designs result in essentially the same models for both yield and Young's modulus as functions of time, oxidant concentration and particle size.

6.7. Doehlert designs

The designs we have discussed so far are all symmetrical — that is, they have the same number of levels for each factor under study. Sometimes, however, it is advantageous to use designs where different factors are studied at different numbers of levels. A simple example is a 2×3 factorial design. In Chapter 3, to evaluate the catalyst and temperature effects on reaction yield, two types of catalyst, A and B, were studied at two different temperatures. If we had suspected that the temperature effect was not linear, we would need more levels to study it. Adding one more temperature level, we would have a 2×3 design, which would be the most economical way to test for non-linearities in the temperature. Modern statistical design programs provide a wide variety of mixed k-factorial designs of the type 2^p3^{k-p}, where $p<k$ and p factors are studied at two levels and the other k–p factors at three levels.

Doehlert designs comprise another class of experimental designs, with which different factors can be studied at different numbers of levels (Doehlert, 1970; Doehlert and Klee, 1972; Massart et al., 1997; Ferreira et al., 2004). They are attractive for treating problems where specific information about the system indicates that some factors deserve more attention than others, but they also have other desirable characteristics.

All Doehlert designs are generated from a regular simplex, a geometrical figure containing k+1 points, where k is the number of factors. For two factors, the regular simplex is an equilateral triangle. For Doehlert designs of type D-1, which are the most popular, the coordinates of this triangle are those given in the first three lines of Table 6.18, in coded units. The other runs of the design are obtained by subtracting every run in the triangle from each other, as shown in the table. This corresponds to moving the simplex around the design center.[62]

[62]Simplexes and their displacements are discussed in detail in Chapter 8.

Table 6.18
Coded factor levels for the two-factor Doehlert D-1 design. The runs in bold face are those defining the initial simplex. The other runs are obtained by subtracting every run from each other

Run	x_1	x_2	Subtraction
1	**0.0**	**0.0**	
2	**1.0**	**0.0**	
3	**0.5**	**0.866**	
4	−1.0	0.0	**1−2**
5	−0.5	−0.866	**1−3**
6	−0.5	0.866	**3−2**
7	0.5	−0.866	**2−3**

For any number of factors k, one of the points of the simplex is the origin, and the other k points lie on the surface of a sphere with radius 1.0 centered on the origin, in such a way that the distances between neighboring points are all the same. Each of these points subtracted from the other k points forms k new points, so the design matrix has a total of k^2+k+1 points. Since the points are uniformly distributed on a spherical shell,[63] Doehlert suggested that these designs be called uniform shell designs.

The coordinates of the D-1 designs for up to seven factors can be obtained from Table 6.19. The initial simplexes from which the designs are generated are built sequentially, adding to the previous simplex the row under the appropriate dashed line, and filling the blanks in the last column with zeroes. For example, the basic simplex for $k = 5$ is obtained by adding rows 5, 8 and 12 to the first three rows. This results in the 6×5 matrix

$$\begin{bmatrix} 0 & 0 & 0 & 0 & 0 \\ 1.0 & 0 & 0 & 0 & 0 \\ 0.5 & 0.866 & 0 & 0 & 0 \\ 0.5 & 0.289 & 0.817 & 0 & 0 \\ 0.5 & 0.289 & 0.204 & 0.791 & 0 \\ 0.5 & 0.289 & 0.204 & 0.158 & 0.775 \end{bmatrix},$$

from which the entire D-1 design for five factors can be generated by subtracting the rows from one another.

Despite being spherical, Doehlert designs have none of the classical properties of response surface designs. They are neither orthogonal nor rotatable, and the variance of the predicted values is not uniform over the experimental range. However, they have other interesting features that

[63] Except the origin, of course.

Table 6.19
Doehlert D-1 uniform shell designs. To obtain the coded factor levels for a design in k factors select the upper left matrix of k columns and $\{(k^2+k)/2\}+1$ rows and augment it with the negative of every row except the first, to obtain the full design with k^2+k+1 rows

	K=	2	3	4	5	6	7
1	0	0	0	0	0	0	0
2	1.0	0	0	0	0	0	0
3	0.5	0.866	0	0	0	0	0
4	−0.5	0.866	0	0	0	0	0
5	0.5	0.289	0.817	0	0	0	0
6	−0.5	0.289	0.817	0	0	0	0
7	0	−0.577	0.817	0	0	0	0
8	0.5	0.289	0.204	0.791	0	0	0
9	−0.5	0.289	0.204	0.791	0	0	0
10	0	−0.577	0.204	0.791	0	0	0
11	0	0	−0.612	0.791	0	0	0
12	0.5	0.289	0.204	0.158	0.775	0	0
13	−0.5	0.289	0.204	0.158	0.775	0	0
14	0	−0.577	0.204	0.158	0.775	0	0
15	0	0	−0.612	0.158	0.775	0	0
16	0	0	0	−0.632	0.775	0	0
17	0.5	0.289	0.204	0.158	0.129	0.764	0
18	−0.5	0.289	0.204	0.158	0.129	0.764	0
19	0	−0.577	0.204	0.158	0.129	0.764	0
20	0	0	−0.612	0.158	0.129	0.764	0
21	0	0	0	−0.632	0.129	0.764	0
22	0	0	0	0	−0.646	0.764	0
23	0.5	0.289	0.204	0.158	0.129	0.109	0.756
24	−0.5	0.289	0.204	0.158	0.129	0.109	0.756
25	0	−0.577	0.204	0.158	0.129	0.109	0.756
26	0	0	−0.612	0.158	0.129	0.109	0.756
27	0	0	0	−0.632	0.129	0.109	0.756
28	0	0	0	0	−0.646	0.109	0.756
29	0	0	0	0	0	−0.656	0.756

make them advantageous in some scenarios. Perhaps their most important property is the ability for uniform space-filling, which is unaffected by rotation. This is very convenient when one wishes to cover an experimental range, no matter how irregular, with a uniform grid of points. Doehlert designs then can be displaced to more promising regions while preserving some of the runs already carried out. This property is illustrated in Fig. 6.13. The initial two-factor design is a hexagon, where

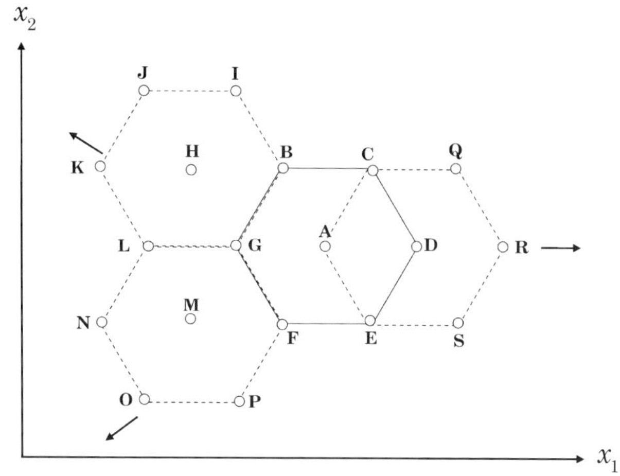

Fig. 6.13. Three possible displacements of a two-factor Doehlert design.

the letters **BCDEFG** denote its vertices. These points, along with the center point **A**, define five levels for the x_1 variable but only three levels for x_2. If the researchers decide to shift the initial design to lower values of x_1 and higher values of x_2, the **BGIJKLH** hexagon can be used, which includes vertices **B** and **G** from the initial configuration. If experimentation is very costly or lengthy, the values already observed for the **B** and **G** vertices can be retained. In case the researchers wish to lower the levels of both x_1 and x_2, displacement to the **FGLMNOP** design is warranted. This time, vertices **F** and **G** belong to the initial hexagon. Finally, if it is decided to shift the levels of only one variable, say x_1, the experimenters might perform new runs only at the **Q**, **R** and **S** vertices to complete the new Doehlert design **ACEDSRQ**, which has two higher levels along x_1 while keeping the same levels for x_2. Note that the space-filling property of Doehlert designs is clearly illustrated in Fig. 6.13. No gaps in the experimental region are left as one design is substituted by another.

Compared to central composite or Box–Behnken designs, Doehlert designs are more economical, especially as the number of factors increase. The basic hexagon in Fig. 6.13 has six points lying on a circumference around the center point, whereas the two-factor central composite design has eight points, also lying on a circumference surrounding its center point. Likewise, the three-factor Doehlert design has 13 points, but the central composite design requires 15. On the other hand, central composite designs are rotatable, a property that Doehlert designs in general do not have. Furthermore, since central composite designs consist of factorial and axial blocks, they provide the basis for an efficient sequential strategy. Linear models can be fitted in a first stage, after which the design can be augmented with complementary points, should quadratic models prove necessary. Finally, using full designs, Doehlert or otherwise, to fit second-order models is hardly justifiable for more than five factors, because it is unlikely that all the coefficients in the quadratic model will turn out to be significant (see, for example, Wu and Hamada, 2000, Chapter 9).

Another very interesting feature of Doehlert designs is the possibility of introducing variation in new factors during the course of an experimental study, without losing the runs already performed. Sometimes we might wish to study first — say — the two factors that seem more promising, analyze the results, and only then introduce variation in a third factor, then in a fourth, and so on. With Doehlert D-1 designs this is possible, provided that all potential factors of interest are introduced in the experiments right from start, set at their average levels (i.e., 0 in coded units).

For example, let us say that there are four factors of potential interest. We can begin with the two-factor design defined by the first four rows in Table 6.19, taking care to keep the levels of factors 3 and 4 fixed at 0 in all runs. Then, when we wish to study the influence of the two factors that have been kept fixed, we only have to add to the initial design the runs corresponding to rows 5–11 in the table (and their negatives, of course).

As can be seen in Table 6.19, a Doehlert design of type D-1 with three or more factors always has one factor at five levels (the first one), one factor at three levels (the last) and the others all at seven levels. Two other Doehlert design types can be generated by different simplexes and result in different level distributions. In the designs of type D-2, the basic simplex is rotated to obtain asymmetrical designs with a minimum number of levels. Table 6.20 contains the coordinates of the D-2 designs for 3–5 factors. Note that the design for $k = 3$ is the same as the corresponding Box–Behnken design.

To construct Doehlert designs of type D-3, one starts with a simplex with a vertex on the origin and totally located in the first quadrant of the coordinate system, having the bisector plane as a plane of symmetry. By the usual subtraction process, this simplex generates symmetrical designs where all factors have seven levels. For k factors, the basic simplex has the general form

$$
\begin{array}{c}
\\ 1 \\ 2 \\ 3 \\ \vdots \\ k \\ k+1
\end{array}
\begin{array}{c}
\begin{array}{ccccc} x_1 & x_2 & \cdots & x_{k-1} & x_k \end{array} \\
\begin{bmatrix}
0 & 0 & \cdots & 0 & 0 \\
a & b & \cdots & b & b \\
b & a & \cdots & b & b \\
\vdots & \vdots & \ddots & \vdots & \vdots \\
b & b & \cdots & a & b \\
b & b & \cdots & b & a
\end{bmatrix}
\end{array}
$$

Table 6.20
Doehlert designs of type D-2, in coded units. To obtain the full designs, the matrices must be augmented with the origin and the negatives of all runs. The integers at the bottom of each design indicate the number of factor levels

$k = 3$

$$\begin{bmatrix} 1.0 & 0.0 & 1.0 \\ 0.0 & 1.0 & 1.0 \\ 1.0 & 1.0 & 0.0 \\ -1.0 & 1.0 & 0.0 \\ 0.0 & 1.0 & -1.0 \\ 1.0 & 0.0 & -1.0 \end{bmatrix}$$
$$\begin{matrix} 3 & 3 & 3 \end{matrix}$$

$k = 4$

$$\begin{bmatrix} -1.118 & -0.5 & -0.5 & -0.5 \\ -1.118 & 0.5 & -0.5 & 0.5 \\ -1.118 & -0.5 & 0.5 & 0.5 \\ -1.118 & 0.5 & 0.5 & -0.5 \\ 0.0 & 1.0 & 0.0 & 1.0 \\ 0.0 & 0.0 & 1.0 & 1.0 \\ 0.0 & 1.0 & 1.0 & 0.0 \\ 0.0 & -1.0 & 1.0 & 0.0 \\ 0.0 & 0.0 & 1.0 & -1.0 \\ 0.0 & 1.0 & 0.0 & -1.0 \end{bmatrix}$$
$$\begin{matrix} 3 & 5 & 5 & 5 \end{matrix}$$

$k = 5$

$$\begin{bmatrix} 0.0 & 1.0 & 0.0 & 1.0 & 0.0 \\ 0.0 & 0.0 & 1.0 & 1.0 & 0.0 \\ 0.0 & 1.0 & 1.0 & 0.0 & 0.0 \\ 0.866 & 0.5 & 0.5 & 0.5 & -0.707 \\ 0.866 & 0.5 & 0.5 & 0.5 & 0.707 \\ 0.0 & -1.0 & 1.0 & 0.0 & 0.0 \\ 0.0 & 0.0 & 1.0 & -1.0 & 0.0 \\ 0.866 & -0.5 & 0.5 & -0.5 & -0.707 \\ 0.866 & -0.5 & 0.5 & -0.5 & 0.707 \\ 0.0 & 1.0 & 0.0 & -1.0 & 0.0 \\ 0.866 & 0.5 & -0.5 & -0.5 & -0.707 \\ 0.866 & 0.5 & -0.5 & -0.5 & 0.707 \\ 0.866 & -0.5 & -0.5 & 0.5 & -0.707 \\ 0.866 & -0.5 & -0.5 & 0.5 & 0.707 \\ 0.0 & 0.0 & 0.0 & 0.0 & 1.414 \end{bmatrix}$$
$$\begin{matrix} 3 & 5 & 5 & 5 & 5 \end{matrix}$$

where $a = \dfrac{k - 1 + \sqrt{k+1}}{k\sqrt{2}}$ and $b = \dfrac{\sqrt{k+1} - 1}{k\sqrt{2}}$.

For two factors, for example, the D-3 design is given by the matrix

$$\begin{array}{c} \\ 1 \\ 2 \\ 3 \\ 4 \\ 5 \\ 6 \\ 7 \end{array} \begin{array}{c} \begin{array}{cc} x_1 & x_2 \end{array} \\ \left[\begin{array}{cc} 0.0 & 0.0 \\ 0.966 & 0.259 \\ 0.259 & 0.966 \\ -0.966 & -0.259 \\ -0.259 & -0.966 \\ -0.707 & 0.707 \\ 0.707 & -0.707 \end{array}\right] \end{array},$$

where the first three lines correspond to the coordinates of the initial simplex.

6.8. Optimal designs

Statistical designs work well because they provide precise estimates of model coefficients. In Chapter 5, we saw that the precision of regression coefficients is calculated by Eq. (5.30),

$$\hat{\mathbf{V}}(\mathbf{b}) = (\mathbf{X}^t\mathbf{X})^{-1}s^2.$$

This means that the variances of the model coefficients are entirely determined by the precision of the responses (measured by the experimental error variance s^2), and the elements of the matrix product $(\mathbf{X}^t\mathbf{X})^{-1}$. For a given model, this matrix, which is called the model's dispersion matrix, solely depends on how the levels of the experimental design are chosen. Consequently, the precision with which the model coefficients are estimated does not depend only on the results obtained for the responses. The particular form chosen for the design matrix partially establishes it, before any experimental run is carried out. Good designs, such as those discussed in this book, have small values for the elements of $(\mathbf{X}^t\mathbf{X})^{-1}$ and hence result in more precise model coefficients.

To obtain narrower confidence intervals for the model's coefficients, designs that minimize some function of the values of the elements of the dispersion matrix have been suggested. These designs are collectively known as alphabetical designs, and differ according to the minimizing criterion on which they are based (Box and Draper, 1987; Aguiar et al., 1995).

The D-optimal designs (D is for "determinant") maximize the determinant of the information matrix, $\mathbf{X}^t\mathbf{X}$, which is the same as minimizing the determinant of its inverse, the dispersion matrix. It can be shown that the eigenvalues $\lambda_1, \lambda_2, \ldots, \lambda_p$ of $(\mathbf{X}^t\mathbf{X})^{-1}$ are proportional to the squared lengths of the p principal axes of the ellipsoid defined by the confidence intervals of the coefficient estimates. Since the determinant of a matrix is the product of its eigenvectors, the square root of the determinant of $(\mathbf{X}^t\mathbf{X})^{-1}$ is proportional to the volume of this confidence ellipsoid. Therefore, minimizing the determinant of $(\mathbf{X}^t\mathbf{X})^{-1}$ corresponds to minimizing the volume of the confidence ellipsoid and consequently the width of the individual confidence intervals for the model coefficients.

Applying the A criterion (A is for "average optimal") to a design corresponds to minimizing the trace of its dispersion matrix, that is, the sum of its eigenvalues. Of course this also minimizes the average of the eigenvalues, and therefore can be interpreted as minimizing the average size of the confidence intervals for the model parameters.

E-optimal designs are those that minimize the size of the largest eigenvalue of $(\mathbf{X}^t\mathbf{X})^{-1}$, which defines the largest confidence interval.[64]

A fourth approach focuses on diminishing the uncertainties of the predicted response values. According to Eq. (5.3), the predicted response at a certain point A is given by

$$\hat{y}_{\mathrm{A}} = \mathbf{x}_{\mathrm{A}}\mathbf{b},$$

where the vectors $\mathbf{x}_{\mathrm{A}}$ and $\mathbf{b}$ contain the coordinates of point A and the coefficient estimates, respectively. It can be shown that the variance of this prediction is given by

$$\hat{V}(\hat{y}_{\mathrm{A}}) = \mathbf{x}_{\mathrm{A}}^t(\mathbf{X}^t\mathbf{X})^{-1}\mathbf{x}_{\mathrm{A}}s^2 = d_{\mathrm{A}}s^2, \tag{6.17}$$

where d_{A} is called the variance function at point A. This shows that the variance of the predicted response only depends on the measure of pure error s^2, the elements of the dispersion matrix and the coordinates of A. That is, the quality of prediction is independent of the coefficient estimates and therefore of the response values themselves. The G-efficiency criterion attempts to minimize the maximum of the variance function over the experimental range.[65]

The use of optimal designs instead of classical designs is normally prompted by two kinds of situations. One happens when experimentation time or resources are limited, and the researcher cannot afford to run the number of experiments required by a classical design. The second occurs when the design space is constrained because some combinations of factor settings are not viable and classical designs cannot be completed. This is

[64]Can you guess what the letter "E" stands for?
[65]G is for "general variance".

usually the case with mixture designs. In almost all practical applications involving mixtures of some complexity, every tested formulation has to include all mixture ingredients. Therefore, the individual ingredient proportions cannot be varied from 0% to 100%. On the contrary, to produce acceptable mixtures they often have to stay within quite short ranges.

Optimal designs are usually generated by computer programs, from a set of candidate points provided by the user. Since the information matrix depends on the design matrix **X**, whose columns are defined by the model used to fit the data, the experimenter must assume a specific model as correct, to find the D- or A-optimal design corresponding to it. If a second, different model turns out to be more adequate, its parameters will not be alphabetically optimal for the design chosen on the basis of the first model.

To illustrate how these optimal designs work, we return to the central composite design results discussed earlier in this chapter, and use the D-optimal and A-optimal designs as examples. The points in Table 6.7 were used as starting points to generate a D-optimal design with only six experimental points. This number of points was chosen because at least six distinct experiments must be performed to fit a quadratic model. The 11 candidate points and a request for a design with six experiments to fit a quadratic model were fed into a commercial computer program (StatSoft, 2004). The Dykstra sequential optimization method was used to generate the D-optimal design given in Table 6.21.

Fitting a quadratic model to the yield data results in the equation

$$\hat{y} = 89.00 + 1.61x_1 - 2.47x_2 - 2.63x_1^2 - 2.75x_2^2 + 1.52x_1x_2.$$

The values of the coefficients of this model are very similar to those in Eq. (6.8), obtained using all 11 yields. This shows that essentially the same model can be obtained from only six experimental results.

On the right-hand side of Table 6.21 we present a design generated by the same computer program and input data except that an A-optimal design was specified. The appropriate yield points were again taken from Table 6.7. The resulting model is

$$\hat{y} = 89.00 + 1.77x_1 - 2.47x_2 - 2.75x_1^2 - 2.75x_2^2 + 1.20x_1x_2,$$

also very similar to the models obtained from the D-optimal design and the central composite design.

In this section, we have only presented a brief introduction to the most common optimal designs for models that can be fitted by least squares. Many other optimal designs have been proposed, based on different criteria, and these include optimal designs for non-linear models. If you wish to know more about this subject, a good place to start is Gauchi (1997).

Table 6.21
D- and *A*-optimal designs generated from the two-factor central composite design given in Table 6.7

Run	D-optimal design			A-optimal design		
	x_1	x_2	Yield[a]	x_1	x_2	Yield[a]
1	$\sqrt{2}$	0	86	0	0	89[b]
2	−1	1	78	1	1	84
3	0	$-\sqrt{2}$	87	$\sqrt{2}$	0	86
4	−1	−1	86	0	$\sqrt{2}$	80
5	0	$\sqrt{2}$	80	0	$-\sqrt{2}$	87
6	0	0	89[b]	$-\sqrt{2}$	0	81

[a] The yield results are from Table 6.7.
[b] The average of the center point yields in Table 6.7 was used.

6A. Applications

6A.1 Mo(VI) catalytic response

The oxidation of the iodide ion by hydrogen peroxide in an acidic medium,

$$3I^- + H_2O_2 + 2H^+ \rightarrow I_3^- + 2H_2O,$$

is catalyzed by several metals in their higher oxidation states. This reaction was used by Andrade et al. (1991) to develop a method for quantifying molybdenum traces using a monosegmented continuous flow system, with the intensity of the iodine absorption at 350 nm as the design response. In an attempt to maximize the sensitivity of the method (i.e., maximize absorption intensity) the researchers used the central composite design whose data are given in Table 6A.1, where x_1 and x_2 represent coded values of the H_2SO_4 and KI concentrations, respectively.

The linear and quadratic models fitted to the data in the table are the following:

$$\hat{y} = \underset{(\pm 0.002)}{0.510} + \underset{(\pm 0.002)}{0.076}\, x_1 + \underset{(\pm 0.002)}{0.058}\, x_2, \quad \text{with } MS_{lof}/MS_{pe} = 90.7;$$

$$\hat{y} = \underset{(\pm 0.003)}{0.542} + \underset{(\pm 0.002)}{0.076}\, x_1 + \underset{(\pm 0.002)}{0.058}\, x_2 - \underset{(\pm 0.002)}{0.055}\, x_1^2 + \underset{(\pm 0.002)}{0.003}\, x_2^2 + \underset{(\pm 0.003)}{0.002}\, x_1 x_2,$$
$$\text{with } MS_{lof}/MS_{pe} = 9.6.$$

The second lack of fit ratio is still larger than $F_{3,4}$ at the 95% confidence level (which is 6.59), but it is clear that the quadratic model provides a much better fit than the linear one, as we can see by inspecting the graphs

Table 6A.1
Central composite design for the optimization study of the Mo(VI) catalytic response

x_1	x_2	Response
−1	−1	0.373
+1	−1	0.497
−1	+1	0.483
+1	+1	0.615
−1.414	0	0.308
+1.414	0	0.555
0	−1.414	0.465
0	+1.414	0.628
0	0	0.538
0	0	0.549
0	0	0.536
0	0	0.549
0	0	0.538

of the predicted responses against the observed values (Figs. 6A.1a, b). The equation coefficients indicate that the improvement is entirely due to the quadratic term in x_1.For x_1 (the H_2SO_4 concentration) both the linear and the quadratic terms are statistically significant. For x_2 (the KI concentration), only the linear term is significant. The interaction between the two concentrations is not significant. The two linear terms have positive coefficients, indicating that the intensity of the absorption signal tends to increase as x_1 and x_2 are increased. In spite of this, since the model also has a *negative* contribution in x_1^2, and its coefficient is similar in size to the coefficient of the linear term, large variations of x_1 in either direction will result in signal reduction. Geometrically, we say that the response surface contains a ridge (Fig. 6A.2). In practical terms, the signal should increase if we increase the KI concentration, at least within the region studied. As for H_2SO_4, there is an optimum region (the ridge), far from which the signal tends to decrease on both sides.

6A.2. Osmotic dehydration of pineapple

One of the principal objectives of research on fruit conservation is the development of products with long shelf-lives and with sensorial and nutritional properties as similar as possible to those of fresh fruit. Evidently, these characteristics increase the probability of product acceptance by consumers.

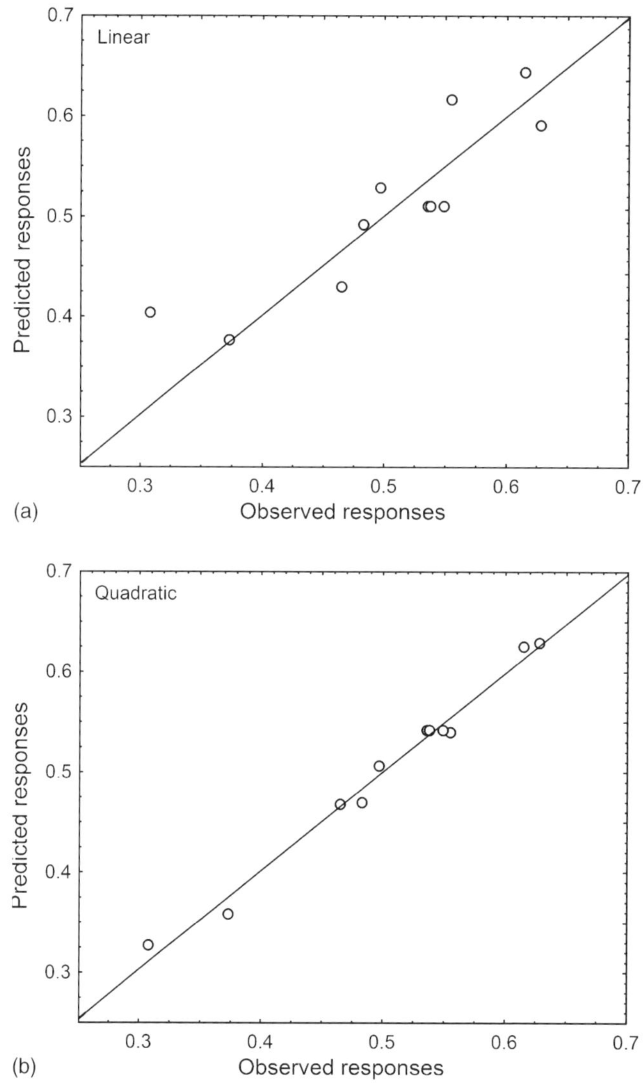

Fig. 6A.1. Comparison of the predictions of the two models with the observed values.

Osmotic dehydration is a technique that permits the reduction of the water content of food and consequently increases the shelf-life of the final product. The process consists of placing the raw material in contact with a very concentrated solution of an osmotic agent that is sensorially compatible with the product one wishes to obtain. This establishes an osmotic gradient that progressively takes water out of the treated fruit.

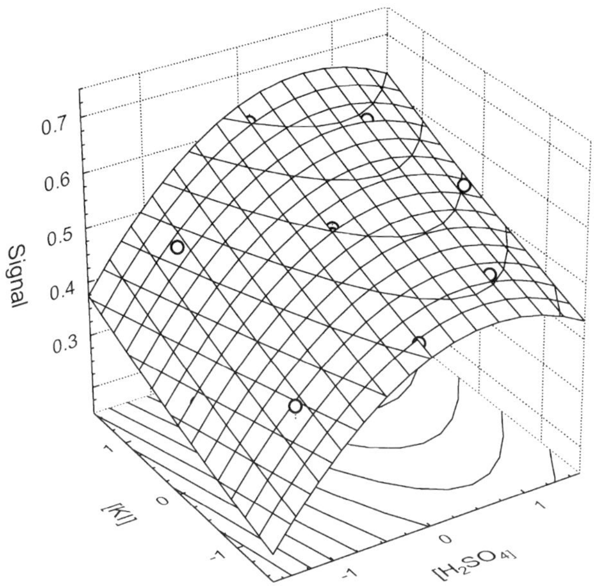

Fig. 6A.2. Response surface for the quadratic model of the Mo(VI) signal.

A central composite design was employed to study how the dehydration of pineapple pieces depends on three factors: contact time (1), process temperature (2) and concentration of osmotic solution (3). The relative loss of weight at the end of each run was taken as a measure of dehydration (Azeredo and Jardine, 2000). The results obtained are presented in Table 6A.2, where x_1, x_2 and x_3 are coded values for the three factors.

Fitting linear and quadratic models to the data in the table, we obtain the following equations:

$$\hat{y} = \underset{(\pm 0.17)}{54.13} + \underset{(\pm 0.20)}{1.72}\, x_1 + \underset{(\pm 0.20)}{2.55}\, x_2 + \underset{(\pm 0.22)}{1.43}\, x_3, \quad \text{with } MS_{\text{lof}}/MS_{\text{pe}} = 8.50;$$

$$\hat{y} = \underset{(\pm 0.34)}{56.32} + \underset{(\pm 0.20)}{1.72}\, x_1 + \underset{(\pm 0.20)}{2.55}\, x_2 + \underset{(\pm 0.20)}{1.43}\, x_3 - \underset{(\pm 0.20)}{0.71}\, x_1^2 - \underset{(\pm 0.20)}{1.47}\, x_2^2$$
$$- \underset{(\pm 0.20)}{0.86}\, x_3^2 - \underset{(\pm 0.27)}{0.85}\, x_1 x_2 + \underset{(\pm 0.27)}{0.21}\, x_1 x_3 + \underset{(\pm 0.27)}{0.42}\, x_2 x_3,$$

with $MS_{\text{lof}}/MS_{\text{pe}} = 2.66$.

It is clear that the quadratic model has a smaller lack of fit, and therefore is better than the linear model. Looking up the F-test table, you can confirm that the quadratic model does not present any evidence of lack of fit at the 95% confidence level. It explains 95.4% of the variation about the average and has $MS_R/MS_r = 20.69$, which is more than six times the value of $F_{9,9}$ (at the same confidence level, of course). This

Table 6A.2
Central composite design for studying the osmotic dehydration of pineapple pieces

Run	x_1	x_2	x_3	Weight loss (%)
1	−1	−1	−1	47.34
2	+1	−1	−1	53.00
3	−1	+1	−1	53.64
4	+1	+1	−1	54.28
5	−1	−1	+1	48.85
6	+1	−1	+1	53.73
7	−1	+1	+1	55.19
8	+1	+1	+1	58.31
9	−1.682	0	0	51.90
10	+1.682	0	0	57.34
11	0	−1.682	0	47.62
12	0	+1.682	0	57.35
13	0	0	−1.682	50.73
14	0	0	+1.682	57.68
15	0	0	0	56.24
16	0	0	0	55.74
17	0	0	0	57.23
18	0	0	0	56.85
19	0	0	0	55.42

means that the regression is significant and also reasonably useful for making predictions, following the Box–Wetz criterion.

The residual plots for the two models (Figs. 6A.3) demonstrate the superiority of the quadratic model. The coefficients of the linear terms are all positive, which means that increasing the levels of all the factors would give us more intense dehydrations. On the other hand, since the coefficients of the quadratic terms are all negative, the same elevation of the factor levels would also tend to diminish the dehydration, in proportion to the square of the variation. All of this suggests that the region of maximum dehydration is not very far away.

6A.3. Reducing cholesterol levels

At the 11th Brazilian Chemical Engineering Congress a paper was presented on the use of quillaja[66] to reduce the cholesterol content of

[66]A commercial saponin preparation obtained from the bark of *Quillaja saponaria molina*.

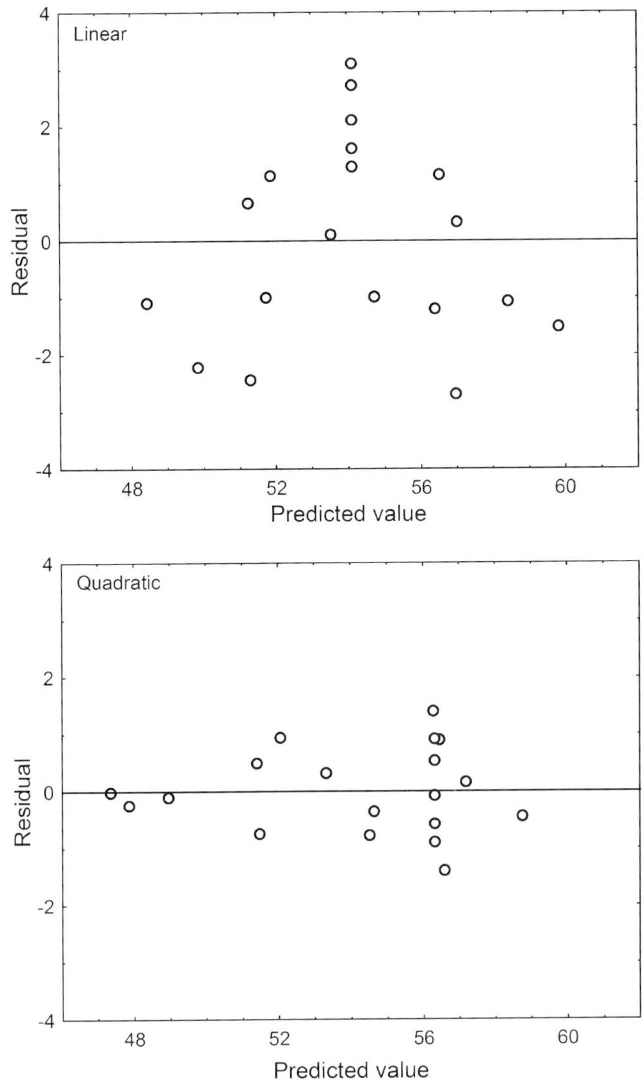

Fig. 6A.3. Residuals left by the two models for the pineapple dehydration study.

butter oil (Brunhara-Salum, 1997). The process consisted of two steps: (a) aggregation of cholesterol with saponin micelles produced in aqueous solution by the quillaja extract and (b) adsorption of these aggregates by diatomaceous earth. Four factors were investigated, with the aim of optimizing the process: concentration of the quillaja solution (**1**), quantity of diatomaceous adsorbent (**2**), contact temperature (**3**) and pH (**4**). The experiment was based on a central composite design whose data are presented in Table 6A.3, in coded form. The residual amounts of cholesterol in the oil were determined chromatographically, after phase separation by decantation.

Proceeding in the usual way, we discover that the quadratic model is superior to the linear one. It shows no evidence of lack of fit at the 95% confidence level and explains 94.3% of the variation about the average. The analysis of variance shows that only the terms involving quillaja concentration and pH are significant, which reduces the model to the equation

$$\hat{y} = 1.861 - 0.189x_1 - 0.083x_1^2 + 0.348x_4 - 0.093x_4^2 + 0.176x_1x_4.$$

The most important coefficients are those of the linear terms, which have opposite algebraic signs — negative for quillaja concentration and positive for pH. Since we want to reduce the cholesterol content, these values tell us that we should increase quillaja concentration and lower the pH value (i.e., use a more acidic medium). However, since the interaction and the quadratic terms are also significant, interpreting these results is more complicated.

Fig. 6A.4 compares the observed responses with the values predicted by the quadratic model. The responses are divided in two groups, with six runs presenting cholesterol contents less than 1.2 mg g^{-1}, while the others are all above 1.6 mg g^{-1}. The six runs with less cholesterol are shown again in Table 6A.4. All of them indeed have $x_1 \geq 0$ and $x_4 \leq 0$, but for the lowest value of all (run 26) three of the factors — one of which is the quillaja concentration — are at their center levels. Considering these facts, don't you think it would be a good idea to make a run with, perhaps, $x_4 = -2$, as in run 26, *but also* with $x_1 = +2$, as in run 21, or even $x_1 = +1$? As Fig. 6A.5 shows, this region was not investigated in this design. The horizontal plane corresponds to the 1.2 mg g^{-1} cholesterol level, and was drawn only to emphasize the separation of the two groups of responses.

6A.4. Laccase production

Laccase, a polyphenol oxidase with several industrial applications, is produced by the fungus *Botryosphaeria* sp. under induction by veratrylic alcohol. Vasconcelos et al. (2000) decided to investigate how the production of this enzyme depends on cultivation time and alcohol concentration. They carried out the central composite design of Table 6A.5, where the response is the enzymatic activity in U mL^{-1}, defined as the number of micromoles of ABTS (a sulfonic acid derivative) oxidized per minute per milliliter of enzyme solution at the standard conditions of the enzymatic run. All the cultures were grown at 28 °C, under constant 180 rpm stirring.

Once more, the quadratic model is found superior to the linear one. It does not present evidence of lack of fit and explains 84.9% of the variation.

Table 6A.3
Central composite design to study cholesterol reduction in butter oil

Run	x_1	x_2	x_3	x_4	Cholesterol (mg g^{-1})
1	−1	−1	−1	−1	1.701
2	+1	−1	−1	−1	1.120
3	−1	+1	−1	−1	1.607
4	+1	+1	−1	−1	0.881
5	−1	−1	+1	−1	1.860
6	+1	−1	+1	−1	0.965
7	−1	+1	+1	−1	1.786
8	+1	+1	+1	−1	0.933
9	−1	−1	−1	+1	2.131
10	+1	−1	−1	+1	2.072
11	−1	+1	−1	+1	2.095
12	+1	+1	−1	+1	2.002
13	−1	−1	+1	+1	2.101
14	+1	−1	+1	+1	2.055
15	−1	+1	+1	+1	2.017
16	+1	+1	+1	+1	1.972
17	0	0	0	0	1.763
18	0	0	0	0	1.840
19	0	0	0	0	1.935
20	−2	0	0	0	1.713
21	+2	0	0	0	1.089
22	0	−2	0	0	1.643
23	0	+2	0	0	1.601
24	0	0	−2	0	1.691
25	0	0	+2	0	1.648
26	0	0	0	−2	0.675
27	0	0	0	+2	2.049
28	0	0	0	0	1.783
29	0	0	0	0	1.983

The maximum explainable variation in this example is 91.2%, because the contribution of pure error to total variation is relatively high. The equation for the fitted quadratic model is

$$\hat{y} = \underset{(\pm 0.35)}{4.93} - \underset{(\pm 0.18)}{1.18}\, x_1 + \underset{(\pm 0.18)}{0.97}\, x_2 - \underset{(\pm 0.23)}{0.70}\, x_1^2 - \underset{(\pm 0.23)}{1.25}\, x_2^2 - \underset{(\pm 0.25)}{0.21}\, x_1 x_2,$$

where all the terms are significant, except the interaction. Since the quadratic terms have negative signs, we must be close to the optimum region. To obtain even better yields, we should use a low level of the veratrylic alcohol concentration (whose linear term is also negative) and a

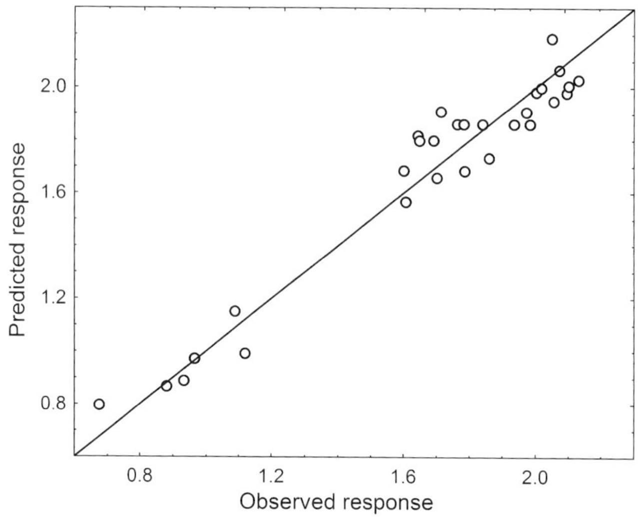

Fig. 6A.4. Predicted responses of the quadratic model.

Table 6A.4
Runs presenting residual cholesterol content less than 1.2 mg g^{-1}

Run	x_1	x_2	x_3	x_4	Cholesterol (mg g^{-1})
2	+1	−1	−1	**−1**	1.120
4	**+1**	+1	−1	**−1**	0.881
6	**+1**	−1	+1	**−1**	0.965
8	**+1**	+1	+1	**−1**	0.933
21	**+2**	0	0	**0**	1.089
26	**0**	0	0	**−2**	0.675

somewhat higher level for the cultivation time. Fig. 6A.6 shows the contour curves for the quadratic model. The point of maximum yield is located in the upper left of the figure, where there appears to be some kind of plateau.

6A.5. Increasing the oxygen in air

In chemical, petrochemical and related industries, separation processes are usually responsible for a major part of the production costs. In the separation of gas mixtures, adsorptive processes like Pressure Swing Adsorption (PSA) are being used by small and medium-sized industries, mainly because they have been found more efficient and economical than traditional separation methods. As part of her research project for a

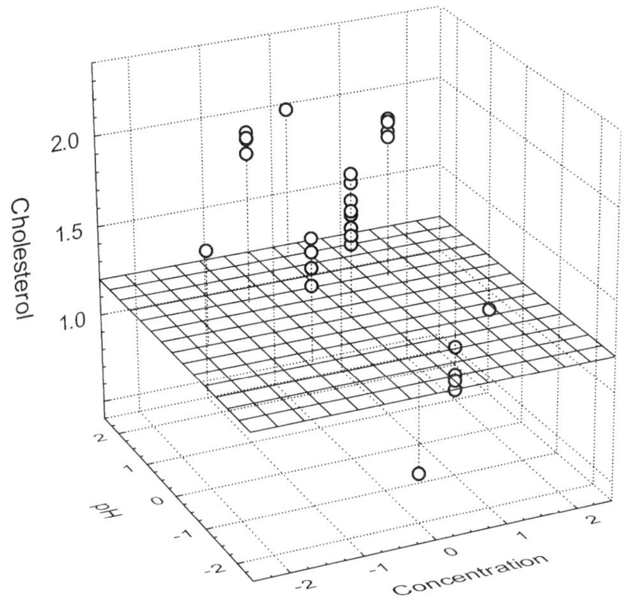

Fig. 6A.5. Observed residual cholesterol content, as a function of pH and quillaja concentration.

doctorate at the Chemical Engineering Department of Campinas State University, C.F.C. Neves developed a PSA gas separation system to produce air enriched with up to 25–50% oxygen, a mixture that is appropriate for some combustion and oxidation processes. Her study was based on a 2^4 factorial design performed in duplicate, which was augmented with 20 runs located inside the hypercube defined by the extreme design levels (Neves, 2000). The response was the oxygen content in the enriched mixture, and the factors were the adsorption and desorption pressures (1 and 2), the feeding rate (3) and the adsorption time (4). Their coded values are given in Table 6A.6, along with the responses obtained.

We start our analysis using the 32 runs of the 2^4 factorial design in duplicate as a training set to fit a regression model. After that, we verify how well this model makes predictions for the rest of the runs, performed at the intermediate levels (the test set).

Fitting a model containing linear, second- and third-order interaction terms to the training set data, we obtain the equation

$$\hat{y} = 36.15 + 2.04x_1 - 2.61x_2 - 5.07x_3 - 3.91x_4 + 0.26x_1x_3 \\ + 0.23x_1x_4 + 0.95x_2x_3 + 0.69x_2x_4 - 0.71x_1x_3x_4,$$

which only shows the significant terms at the 95% confidence level. This equation does not present evidence of lack of fit, and explains 99.57% of

Table 6A.5
Central composite design for studying laccase production

Factors	Levels				
1 Alcohol concentration (mM)	28	30	35	40	42
2 Cultivation time (days)	2.5	3	4	5	5.5
Coded values	−1.41	−1	0	+1	+1.41

Run	x_1	x_2	PPO-I (U mL^{-1})
1	−1	−1	3.50
2	−1	−1	3.20
3	+1	−1	1.17
4	+1	−1	1.70
5	−1	+1	4.10
6	−1	+1	5.40
7	+1	+1	1.90
8	+1	+1	2.10
9	0	0	4.80
10	0	0	5.00
11	0	0	4.70
12	0	0	5.20
13	−1.41	0	5.25
14	−1.41	0	5.41
15	0	+1.41	6.00
16	0	+1.41	3.20
17	+1.41	0	2.30
18	+1.41	0	1.60
19	0	−1.41	0.50
20	0	−1.41	0.50

the total variation about the average, 99.73% being the maximum explainable figure. The cubic graph in Fig. 6A.7 shows the values predicted by the model and demonstrates at once its practical significance: to obtain the maximum oxygen content, we must set the desorption pressure, feeding rate and adsorption time at their low levels.

An efficient model should be capable of making accurate predictions for the response values within the investigated region. The prediction errors, as we already know, should not present any systematic behavior and their size should be comparable to the size of pure error. Fig. 6A.8 shows a comparison of the residuals from model fitting with the prediction errors for responses belonging to the test set. It is clear that something is wrong.

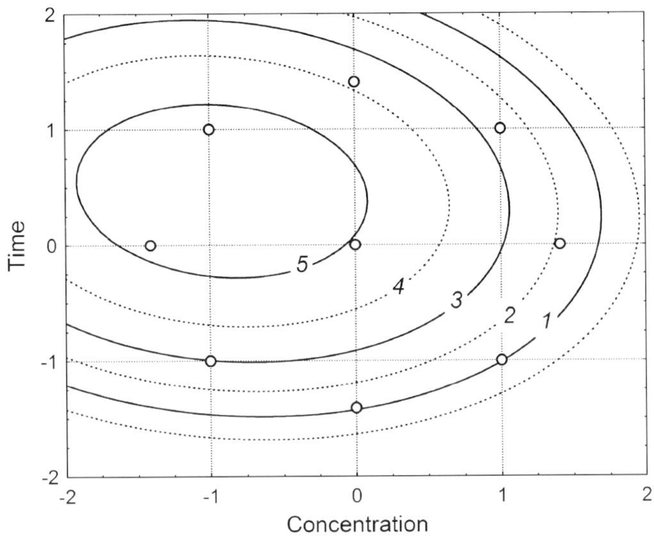

Fig. 6A.6. Contour curves for the production of laccase.

Although our model is well adjusted to the data of the 2^4 factorial runs, it commits systematic errors when predicting values at the intermediate points. All the test set residuals are negative, meaning that the model predictions overestimate the observed response values.[67] The worst prediction is that for run no. 35.

We need a better model, and the next step, naturally, is to include proper quadratic terms, in addition to the cross (interaction) terms already in the model. To have a sufficient number of degrees of freedom, we must include levels other than ± 1 in the training set runs. Using all 52 runs in the table to fit a full quadratic model, we arrive at the equation

$$\hat{y} = 33.01 + 2.00x_1 - 2.56x_2 - 5.09x_3 - 3.96x_4 + 1.67x_2^2 + 1.94x_4^2 - 0.21x_1x_3 + 0.98x_2x_3 + 0.69x_2x_4,$$

which presents a certain amount of lack of fit, at the 95% confidence level, but whose predictions are much better than those of the preceding model, as shown by the new residual graph (Fig. 6A.9). The largest residual again occurs for the result of run 35. The presence of this anomalous point in the training set, as a matter of fact, introduces systematic deviations in the model, causing the negative residuals to be more numerous than the positive ones. Ideally, in view of this result, run

[67]Since the prediction error is, by definition, the observed value minus the predicted one.

Table 6A.6
Experimental design for studying the increase in oxygen content in air

Run	x_1	x_2	x_3	x_4	% oxygen
1	–1	–1	–1	–1	50.1
2	–1	–1	–1	–1	48.1
3	1	–1	–1	–1	50.7
4	1	–1	–1	–1	49.5
5	–1	1	–1	–1	39.4
6	–1	1	–1	–1	39.6
7	1	1	–1	–1	42.1
8	1	1	–1	–1	41.7
9	–1	–1	1	–1	34.5
10	–1	–1	1	–1	33.8
11	1	–1	1	–1	40.3
12	1	–1	1	–1	39.9
13	–1	1	1	–1	30.3
14	–1	1	1	–1	30.2
15	1	1	1	–1	35.6
16	1	1	1	–1	35.2
17	–1	–1	–1	1	37.0
18	–1	–1	–1	1	37.3
19	1	–1	–1	1	43.4
20	1	–1	–1	1	42.1
21	–1	1	–1	1	32.0
22	–1	1	–1	1	32.0
23	1	1	–1	1	37.6
24	1	1	–1	1	36.9
25	–1	–1	1	1	26.6
26	–1	–1	1	1	26.3
27	1	–1	1	1	30.4
28	1	–1	1	1	30.1
29	–1	1	1	1	24.6
30	–1	1	1	1	24.0
31	1	1	1	1	27.9
32	1	1	1	1	27.6
33	–1	0	–1	1	32.8
34	–0.8	0.6	–0.7	–0.09	33.5
35	–0.8	0.87	0.9	–0.91	24.7
36	–0.4	–0.33	–0.4	–0.36	38.0
37	0	–0.07	0	0.55	31.2
38	0	–0.33	–0.2	–0.09	34.6
39	0	0	–1	–1	44.0
40	0	–1	1	1	27.1

Table 6A.6. (*Continued*)

Run	x_1	x_2	x_3	x_4	% oxygen
41	0.2	–0.07	0	0.09	33.3
42	0.4	–0.47	–0.2	0.09	35.8
43	0.4	0.33	0.3	0.27	30.7
44	0.6	–0.73	0.6	–0.73	37.2
45	0.6	0.6	0.6	0.55	29.5
46	0.8	–0.73	–0.8	–0.82	45.3
47	0.8	–0.87	–0.9	0.09	42.8
48	0.8	0.8	0.8	0.82	28.0
49	1	–1	1	0	32.4
50	0	0	0	0	33.6
51	0	0	0	0	33.6
52	0	0	0	0	33.4

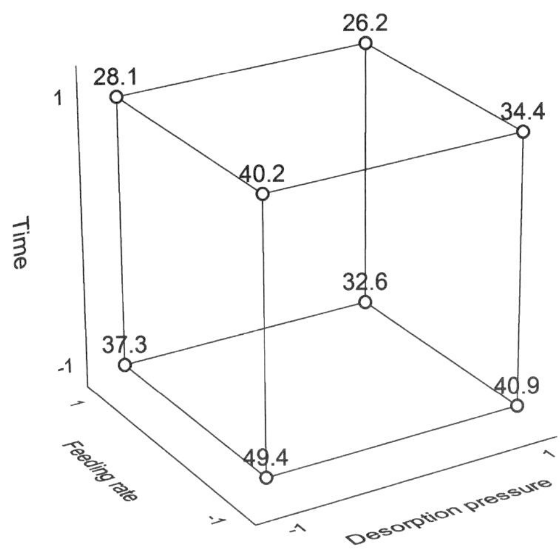

Fig. 6A.7. Cubic plot of the values predicted by the model fitted to the oxygen content.

35 should be repeated, to verify if the value entered in the table is not the result of some error.

The predictions for the first runs (those corresponding to low adsorption times) became worse in the second fit. We conclude, therefore, that the model can still be improved, and that the system under study is quite complex, involving several interactions and non-linear effects. All of this indicates that a more adequate representation will probably require that higher-order terms be included in the model.

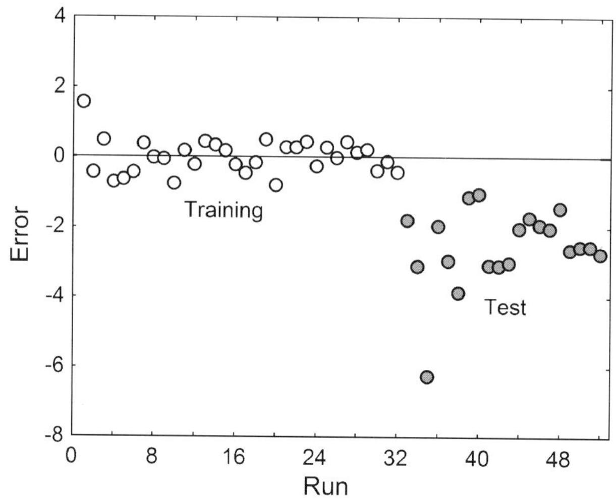

Fig. 6A.8. Prediction errors, given by $e = y - \hat{y}$, for the model fitted to the training set data.

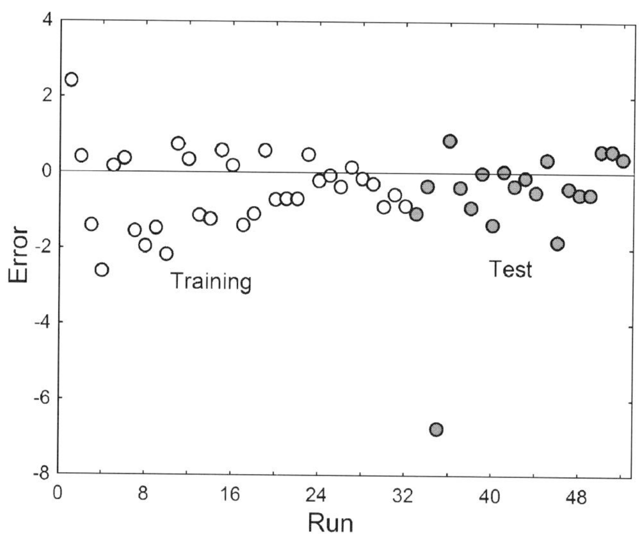

Fig. 6A.9. Residuals left after fitting a quadratic model to all the data.

It is our duty to note that comparing the fits shown in Figs. 6A.8 and 6A.9 is unfair to the first model. While the training set used for fitting in Fig. 6A.8 was restricted to the 32 response values of the two-level factorial, the entire set of 52 values was used to produce the results shown in Fig. 6A.9.

6A.6. Earplug optimization study — concluding phase

In the final phase of the optimization study for earplug production (Sato, 2002) a 3^2 factorial design was carried out to study the dependencies of the three responses, apparent density (*AD*), equilibrium force (*EF*) and retention time (*RT*), on the flow rate of the reticulate solution (**A**) and the quantity of water in the polyol solution (**B**). Only these two factors were found to be important, after the execution of a 2^{7-4} fractional factorial and a blocked 2^4 factorial design (see case studies in Sections 3A.9 and 4A.8). The other factors were fixed at levels chosen to provide the most desirable apparent densities, *EF*s and *RT*s, as well as convenient operational conditions. Table 6A.7 contains the results of the three responses for the three-level factorial experiments, which were performed in duplicate.

The regression results for the linear and quadratic models for the *AD* are

$$AD = \underset{(\pm 0.0010)}{0.1421} - \underset{(\pm 0.0012)}{0.0170}\,x_A - \underset{(\pm 0.0012)}{0.0052}\,x_B$$

and

$$AD = \underset{(\pm 0.0022)}{0.1403} - \underset{(\pm 0.0012)}{0.0170}\,x_A - \underset{(\pm 0.0012)}{0.0052}\,x_B + \underset{(\pm 0.0021)}{0.0028}\,x_A^2 - \underset{(\pm 0.0021)}{0.0002}\,x_B^2 + \underset{(\pm 0.0015)}{0.0020}\,x_A x_B.$$

The ANOVA results for these models are given in Table 6A.8. Neither the linear nor the quadratic model shows significant lack of fit and both are significant at the 95% confidence level. The linear model should be preferred, because it is simpler.

The response surface for the linear model is shown in Fig. 6A.10. Note that the *AD* decreases faster when the reticulate flow rate changes from 15 to 23 g min^{-1} than when 0.1% water is added to the polyol solution. These results are consistent with those of the blocked 2^4 factorial design. The rate of change of the *AD* with the reticulate solution flow rate determined from the 3^2 factorial design, –0.0170 g cm^{-3}, is about half the –0.0372 g cm^{-3} effect value found from the blocked factorial in case study in Section 3A.9. This happens because regression coefficients for coded variables are half-effects of two-level factorials, if the factor range is the same in both cases. Even the reticulate solution flow rate contrast in the case study in Section 4A.8, $l_D = -0.0435$ g cm^{-3}, has a value that seems consistent with this gradient.[68] Note, finally, that the sign of the linear regression coefficient for water in the 3^2 factorial design is opposite to the

[68]Can you explain why this contrast value is different from the –0.0372 g cm–3 effect value?

Table 6A.7
Apparent densities, equilibrium forces and retention times obtained with the duplicate 3^2 factorial design used to study earplug production

Factors	–1	0	+1
A = Reticulate solution flow rate ($g\,min^{-1}$)	15	19	23
B = Quantity of water in polyol solution (%)	0	0.05	0.1

Responses
AD = Apparent density ($g\,cm^{-3}$)
EF = Equilibrium force (lb)[a]
RT = Retention time (s)

Run	A	B	*AD*	*EF*	*RT*
1	–1	–1	0.167	0.190	2.062
2	–1	0	0.169	0.183	2.298
3	–1	1	0.156	0.145	5.292
4	0	–1	0.148	0.100	7.114
5	0	0	0.142	0.097	6.870
6	0	1	0.134	0.053	30.666
7	1	–1	0.127	0.046	31.154
8	1	0	0.126	0.045	41.228
9	1	1	0.124	0.035	95.128
10	–1	–1	0.166	0.179	1.836
11	–1	0	0.154	0.163	3.172
12	–1	1	0.148	0.120	5.080
13	0	–1	0.143	0.099	6.788
14	0	0	0.137	0.082	9.480
15	0	1	0.137	0.065	23.646
16	1	–1	0.132	0.054	23.240
17	1	0	0.125	0.044	39.800
18	1	1	0.122	0.034	102.097

[a] 1 lb = 4.448 N.

one for the blocked factorial effect, because different equations were used for coding the factors.

The analysis of the *EF* data is straightforward. The MS_{lof}/MS_{pe} ratio for the linear model is 5.32, which is larger than $F_{6,9,95} = 3.37$ and so has a significant lack of fit. The same does not happen with the quadratic model,

Table 6A.8
ANOVA table for the linear and quadratic models for the apparent density. The SS (sum of squares) and MS (mean square) values are multiplied by 10,000. df stands for degrees of freedom

Source of variation	Linear model				Quadratic model			
	SS	df	*MS*	*F*	*SS*	df	*MS*	*F*
Regression	37.88	2	18.94	105.22	38.52	5	7.70	45.29
Residual	2.63	15	0.18		1.99	12	0.17	
Lack of fit	0.73	6	0.12	0.57	0.09	3	0.03	0.14
Pure error	1.90	9	0.21		1.90	9	0.21	
Total	40.51	17			40.51	17		

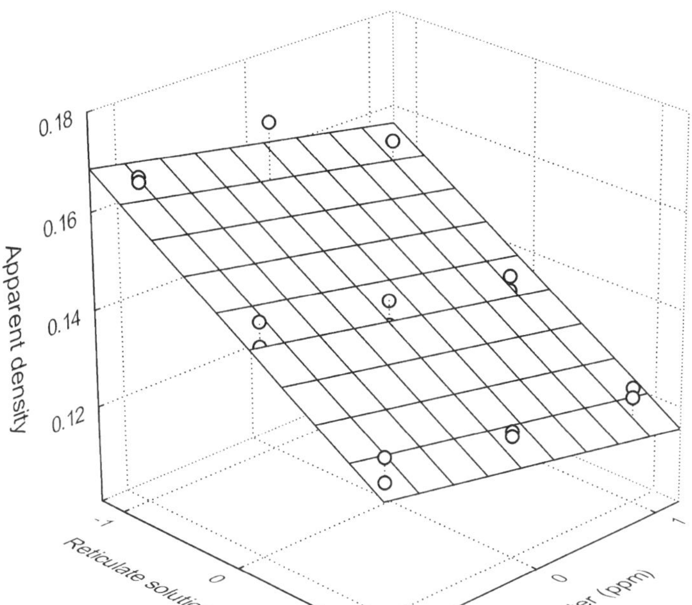

Fig. 6A.10. The apparent density as a function of reticulate solution flow rate and water quantity. The points represent the experimental response values.

$$EF = \underset{(\pm 0.0046)}{0.0887} - \underset{(\pm 0.0025)}{0.0602}\,x_A - \underset{(\pm 0.0025)}{0.0181}\,x_B + \underset{(\pm 0.0044)}{0.0205}\,x_A^2 - \underset{(\pm 0.0044)}{0.0090}\,x_B^2 + \underset{(\pm 0.0031)}{0.0091}\,x_A x_B$$

whose $MS_{\text{lof}}/MS_{\text{pe}}$ ratio is only 0.51. Its $MS_{\text{R}}/MS_{\text{r}}$ ratio, on the other hand, is 130, much larger than $F_{3,\,9,95\%} = 3.86$, which indicates a very significant regression.

The analysis of the RT raw data proved much more complicated. The linear and quadratic models both showed significant lack of fit at the 95% confidence level. Since the RT values spanned two orders of magnitude, the researchers decided to apply a logarithmic transformation to the data. The quality of fitting improves dramatically as a result. The linear model still shows a slight lack of fit, but the quadratic model

$$\log RT = \underset{(\pm 0.0375)}{0.9698} + \underset{(\pm 0.0206)}{0.5993}\, x_A + \underset{(\pm 0.0206)}{0.2630}\, x_B + \underset{(\pm 0.0356)}{0.0180}\, x_A^2 + \underset{(\pm 0.0356)}{0.1348}\, x_B^2 + \underset{(\pm 0.0252)}{0.0344}\, x_A x_B$$

has a $MS_{\text{lof}}/MS_{\text{pe}}$ ratio of 2.01, which is smaller than $F_{3,\,9,95\%} = 3.86$. The $MS_{\text{R}}/MS_{\text{r}}$ ratio of 206 shows that the regression is highly significant.

6A.7. Photodegradability of herbicides

A central composite design on two factors was employed to investigate the degradability, under artificial radiation, of the commercial herbicides tebuthiuron (*Teb*), 2,4-dichlorophenoxyacetic acid (*2,4-d*) and diuron (*Diu*). The degradation of these herbicide molecules is accelerated by a photo-Fenton reaction, an oxidative process involving hydrogen peroxide and iron (introduced as potassium ferrioxalate, in this study). The initial concentrations of hydrogen peroxide and potassium ferrioxalate and the percent degradations of the three herbicides (determined by high resolution liquid chromatography) are presented in Table 6A.9 (Paterlini, 2002).

Linear and quadratic models were fitted to the degradation results of all herbicides. For tebuthiuron, the linear model does not show significant lack of fit but does not have any significant coefficients either. The full quadratic model also shows no significant lack of fit but the quadratic term on the ferrioxalate concentration is statistically significant at the 95% confidence level. These calculations are left to the interested reader. Here, we shall discuss only the results obtained for the other two herbicides.

The linear and quadratic models for the degradation of 2,4-dichlorophenoxyacetic acid are

$$\hat{y} = \underset{(\pm 0.69)}{88.23} - \underset{(\pm 0.84)}{4.53}\, x_1 - \underset{(\pm 0.84)}{0.24}\, x_2$$

and

$$\hat{y} = \underset{(\pm 0.71)}{90.05} - \underset{(\pm 0.51)}{4.53}\, x_1 - \underset{(\pm 0.51)}{0.24}\, x_2 - \underset{(\pm 0.57)}{2.44}\, x_1^2 - \underset{(\pm 0.57)}{0.29}\, x_2^2 - \underset{(\pm 0.71)}{0.35}\, x_1 x_2.$$

Table 6A.9
Central composite design results for the photo-Fenton degradation of tebuthiuron (Teb), 2,4-dichlorophenoxyacetic acid (2,4-d) and diuron (Diu). Hydrogen peroxide and ferrioxalate initial concentrations are in mmol L^{-1}

Run	[Ferriox]	$[H_2O_2]$	x_1	x_2	Teb (%)	2,4-d (%)	Diu (%)
1	0.50	6.0	−1	−1	88.2	91.1	86.1
2	1.00	6.0	1	−1	87.5	84.2	82.1
3	0.50	10.0	−1	1	85.5	91.6	87.8
4	1.00	10.0	1	1	87.7	83.3	83.1
5	0.40	8.0	−1.41	0	81.0	92.4	86.7
6	1.10	8.0	1.41	0	78.1	77.5	84.1
7	0.75	5.2	0	−1.41	90.8	89.8	85.9
8	0.75	10.8	0	1.41	85.7	88.7	85.6
9	0.75	8.0	0	0	87.6	89.3	88.1
10	0.75	8.0	0	0	89.7	91.4	86.9
11	0.75	8.0	0	0	89.6	91.3	89.0
12	0.75	8.0	0	0	86.5	88.2	89.3

The ANOVA results for these models are presented in Table 6A.10. For the linear model, there is no evidence of significant lack of fit at the 95% level: its $MS_{\text{lof}}/MS_{\text{pe}}$ ratio is 2.96, much less than the 95% $F_{6,3}$ critical value, which is 8.94. The $MS_{\text{R}}/MS_{\text{r}}$ ratio for the linear fit is 14.53. Compared with the 95% $F_{2,9}$ value of 4.26 this also indicates a significant — though not very useful[69] — regression.

An analysis of the quadratic model leads to very similar conclusions. There is no significant lack of fit ($MS_{\text{lof}}/MS_{\text{pe}} = 0.66$) and the regression is significant at the 95% confidence level ($MS_{\text{R}}/MS_{\text{r}} = 19.93$ is larger than $F_{5,\,6} = 4.39$). The quadratic model should be preferred, because the 95% confidence interval for the x_1^2 coefficient, which is given by $-2.44 \pm (2.447 \times 0.57) = [-3.83, -1.05]$, does not contain zero and hence this term is significant at this level.[70]

None of the terms involving variable x_2 is significant for the quadratic model. This implies that the degradation of 2,4-dichlorophenoxyacetic acid does not depend on the hydrogen peroxide concentration. Therefore, the percent degradation values for the runs where only the level of this

[69]According to the Box–Wetz criterion, discussed in Section 5.4.

[70]By now it should be clear why the t-point with 6 degrees of freedom (rather than 3) was chosen for calculating this confidence interval. If you are not sure, you may wish to review the example discussed in the last paragraphs of Section 5.6.

Table 6A.10
ANOVA table for the linear and quadratic models for the degradation of 2,4-dichlorophenoxyacetic acid

Source of variation	Linear model				Quadratic model			
	SS	df	*MS*	*F*	*SS*	df	*MS*	*F*
Regression	164.92	2	82.46	14.53	203.72	5	40.74	19.93
Residual	51.06	9	5.67		12.27	6	2.04	
Lack of fit	43.69	6	7.28	2.96	4.90	3	1.63	0.66
Pure error	7.37	3	2.46		7.37	3	2.46	
Total	215.98	11			215.98	11		

factor changes should be the same, within experimental error. This is indeed the case. The degradation values for runs 1 and 3 are 91.1% and 91.6%, respectively. For runs 2 and 4, they are 84.2% and 83.3%. Finally, for runs 7–12, the degradation values 89.8%, 88.7%, 89.3%, 91.4%, 91.3% and 88.2%. In all these sets, the variations are indistinguishable from pure error.

For diuron, the linear and quadratic models fitted to the degradation results are

$$\hat{y} = \underset{(\pm 0.31)}{86.23} - \underset{(\pm 0.38)}{1.55}\, x_1 + \underset{(\pm 0.38)}{0.28}\, x_2$$

and

$$\hat{y} = \underset{(\pm 0.62)}{88.33} - \underset{(\pm 0.44)}{1.55}\, x_1 + \underset{(\pm 0.44)}{0.28}\, x_2 - \underset{(\pm 0.49)}{1.66}\, x_1^2 - \underset{(\pm 0.49)}{1.49}\, x_2^2 - \underset{(\pm 0.62)}{0.175}\, x_1 x_2.$$

The corresponding ANOVA values are presented in Table 6A.11. Although the linear model does not show significant lack of fit at the 95% confidence level ($MS_{\text{lof}}/MS_{\text{pe}} = 4.64$, compared with $F_{6,\,3} = 8.93$), the regression is not significant at this confidence level. The $MS_{\text{R}}/MS_{\text{r}}$ ratio is only 2.49, whereas $F_{2,\,9} = 4.26$. The quadratic model also does not show evidence of lack of fit ($MS_{\text{lof}}/MS_{\text{pe}} = 1.62$, for $F_{3,\,3} = 9.28$) but is slightly significant at the 95% confidence level: $MS_{\text{R}}/MS_{\text{r}} = 6.80$, compared to $F_{5,\,6} = 4.39$.[71] The reader can easily demonstrate that the x_1, x_1^2 and x_2^2 coefficients are significant at the 95% confidence level. The coefficients for x_2 and x_1x_2 are not, since the absolute values of their standard errors are larger than the coefficients themselves.

The three-dimensional representation of the response surface for diuron degradation is presented in Fig. 6A.11 and the corresponding

[71] However, what would the Box–Wetz criterion say about the predictive usefulness of this model?

Table 6A.11
ANOVA table for the linear and quadratic models fitted to the degradation values of diuron

Source of variation	Linear model				Quadratic model			
	SS	df	*MS*	*F*	*SS*	df	*MS*	*F*
Regression	19.81	2	9.90	2.49	46.49	5	9.30	6.08
Residual	35.84	9	3.98		9.16	6	1.53	
Lack of fit	32.36	6	5.39	4.64	5.67	3	1.89	1.63
Pure error	3.49	3	1.16		3.49	3	1.16	
Total	55.65	11			55.65	11		

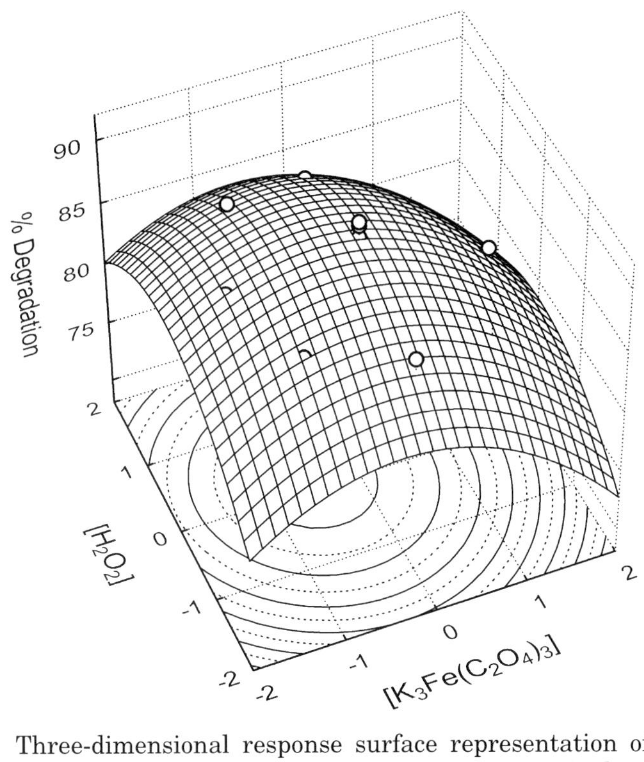

Fig. 6A.11. Three-dimensional response surface representation of the diuron degradation as a function of potassium ferrioxalate and hydrogen peroxide concentrations.

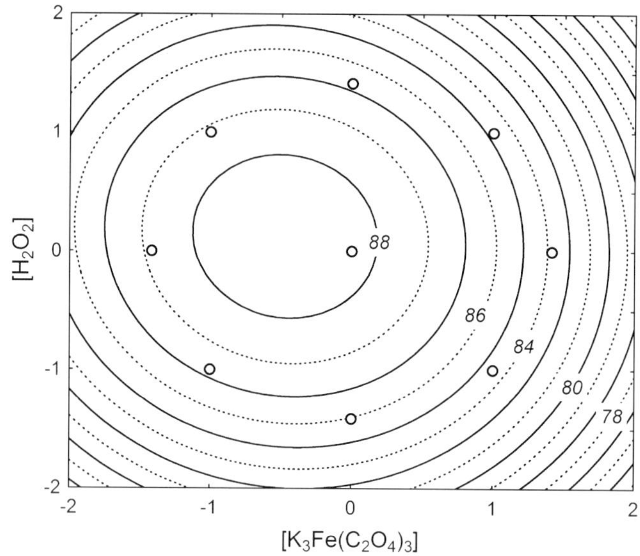

Fig. 6A.12. Percent degradation contour curves for diuron as a function of the potassium ferrioxalate and hydrogen peroxide concentrations.

contour curves in Fig. 6A.12. Maximum degradation is expected to occur at oxalate and peroxide concentrations close to those of the central point, [ferriox] = 0.75 mmol L^{-1} and $[H_2O_2]$ = 8.0 mmol L^{-1}.

CHAPTER 7
Mixture modeling

The experimental designs for mixture studies differ from the response surface designs discussed until now in only one important point. In a factorial design discussed in Chapter 3, we studied the influence of two factors – temperature and concentration – on reaction yield. Imagine that the levels of each of these factors were doubled. As a consequence, we would expect that not only the yield would be affected, but also the properties of the final product, such as, say, viscosity and optical density.

If our system were a mixture, the situation would be different. If we double, for example, the quantities of all the ingredients in a cake mixture, we would expect to obtain a cake only twice as large, but with the same flavor, texture and color, because the properties of a mixture are determined by the *proportions* of their ingredients, not by their absolute values. This explains why cooks can, within reasonable limits, always double or halve recipes to cater for the number of expected dinner guests.

The sum of the proportions of the different components of a mixture is always 100%. For any mixture of q components, we can write

$$\sum_{i=1}^{q} x_i = 1 \quad (i.e., \;\; 100\%), \tag{7.1}$$

where x_i represents the proportion of the ith component. This equation removes 1 degree of freedom from the proportions. To specify the composition of a mixture, we only need to fix the proportions of $q-1$ components, because the proportion of the qth component will always be what is lacking to complete 100%.

If we wish to modify the properties of a mixture by changing its formulation, the new proportions must continue to obey Eq. (7.1). Because of this restriction, the methods that we have discussed until now must be slightly modified to treat the mixture problem. These modified methods have found many applications in science and engineering, particularly in industry (Cornell, 1990b, c; Goupy, 2000).

Exercise 7.1. In several industries the manufacturing process only consists of mixing certain ingredients in adequate proportions to give a final product with desired characteristics. Can you give some examples of industries of this kind, among those operating in the city where you live?

For a binary mixture (i.e., a mixture formed by only two components), Eq. (7.1) reduces to $x_1 + x_2 = 1$. In the coordinate system shown in Fig. 7.1a, this equation is represented by the straight line $x_2 = 1 - x_1$. All possible mixtures of two components correspond to points on this line. If x_1 and x_2 were not proportions but independent factors like temperature and pH, all of the space within the square shown in Fig. 7.1a could be experimentally investigated. In mixture studies, however, the experimental space is restricted to the points lying on the line, that is, the problem becomes univariate. Making this line the abscissa axis, we can use a single graph like the one in Fig. 7.1b to show how several mixture properties vary with its composition.

For systems with three independent continuous factors, we could in principle perform experiments corresponding to any point within the cube of Fig. 7.2a. A study of the variation of a reaction yield with x_1 = time, x_2 = temperature and x_3 = pressure, for example, would be a typical case. If the system is a mixture of three components, however, it would have to obey the constraint $x_1+x_2+x_3 = 1$, which defines an equilateral triangle inscribed in the cube, also illustrated in Fig. 7.2a.

All possible compositions of the ternary mixture are represented by points on the triangle. The vertices correspond to the pure components and the sides to their binary mixtures, whereas the points inside the triangle represent mixtures of three components. The variation of a given property with the composition of the mixture can be described by a response surface drawn above the triangle, as shown in Fig. 7.2b. Representing this surface by its contour curves we obtain a triangular diagram, as illustrated in Fig. 7.2c.[72]

7.1. Two-component mixtures

The investigation of mixture properties follows the same general procedure that we used for systems with independent variables. We start by postulating a model to describe how the properties of interest vary as a function of mixture composition. Then we construct an experimental

[72]If you have studied physical chemistry, you probably have seen similar graphs: phase diagrams for ternary systems.

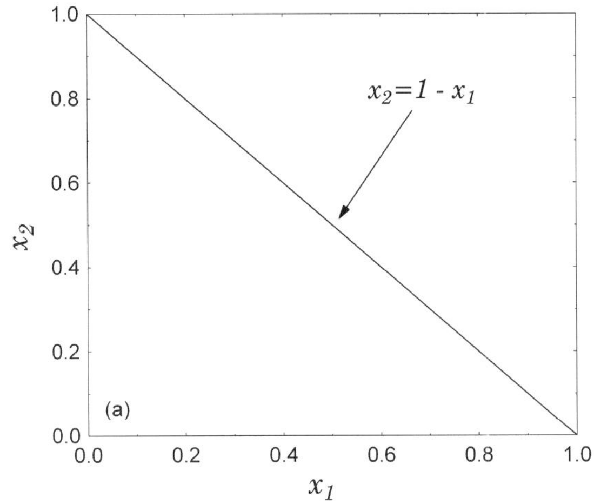

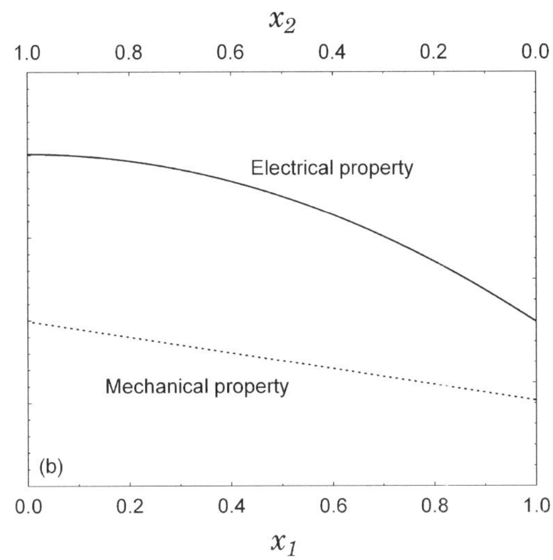

Fig. 7.1. (a) The experimental space for systems with two independent continuous variables includes all the points within the square. The experimental space for mixtures of two components is restricted to the points on the $x_2 = 1 - x_1$ line. (b) Curves representing the variations of two properties of a binary mixture with its composition. Note that the scales of x_1 and x_2 run in opposite directions.

design specifying the mixture compositions to be studied. Finally, we fit the model to the experimental responses, evaluate its quality and compare it to alternative models, when available. The first two stages are closely related. The form of the model initially chosen will determine which compositions are the most adequate to obtain the most precise estimates of its parameters.

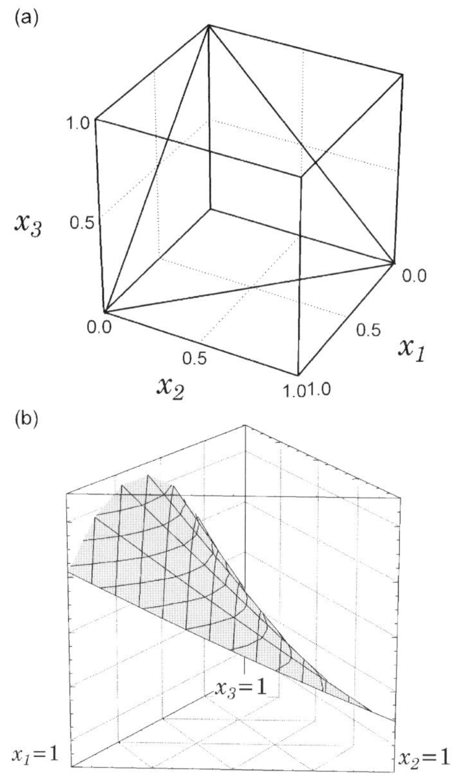

Fig. 7.2. (a) The experimental space for processes with three continuous factors includes all the points within the cube. The experimental space for mixtures of three components is limited to points on the triangle. (b) A response surface for all possible mixtures of components 1–3. (c) Its contour curves.

The simplest model for a two-component mixture is the linear, or additive, model:

$$y_i = \beta_0 + \beta_1 x_1 + \beta_2 x_2 + \varepsilon_i, \tag{7.2}$$

where y_i is an experimental value of the response of interest, β_0, β_1 and β_2 are the parameters of the model and ε_i represents the random error associated with the value of y_i. Fitting the model to the experimental response values leads to the expression

$$\hat{y} = b_0 + b_1 x_1 + b_2 x_2, \tag{7.3}$$

with which we can estimate the average response for a mixture of any composition (x_1, x_2). This equation is identical to Eq. (6.1), but in practice there is a difference. Since it refers to a mixture, factors x_1 and x_2 are not independent variables. As a consequence, the information matrix $\mathbf{X}^t\mathbf{X}$ is singular, and we cannot use Eq. (5.12) to estimate the model coefficients unless we modify the form of the $\mathbf{X}$ matrix.

This could be done by directly substituting the $x_1 + x_2 = 1$ relation into Eq. (7.3) to eliminate x_1 or x_2, but we will use a different strategy, with which we will obtain simpler models. Since x_1+x_2 is always equal to 1, we can introduce it as a coefficient of b_0 in Eq. (7.3) without affecting the equation:

$$\hat{y} = b_0(x_1 + x_2) + b_1x_1 + b_2x_2.$$

With a slight rearrangement, we have

$$\hat{y} = (b_0 + b_1)x_1 + (b_0 + b_2)x_2 = b_1^*x_1 + b_2^*x_2, \tag{7.4}$$

where $b_i^* = b_0 + b_i$. With this simple manipulation, only the two coefficients b_1^* and b_2^* have to be determined, instead of the three parameters in the original equations. To estimate their values, we only need two measurements, made on two mixtures of different compositions.

When $(x_1,\ x_2) = (1{,}0)$, that is, when the "mixture" contains only component 1, Eq. (7.4) reduces to $\hat{y} = b_1^* = y_1$, where y_1 is the response observed for component 1 alone. In the same way, when $(x_1, x_2) = (0{,}1)$ we have $\hat{y} = b_2^* = y_2$. In other words, the two coefficients of the additive model are equal to the response values obtained for the two pure components. If this model is valid, we will be able to predict the properties of any binary mixture without ever having to make an experiment on a binary mixture. This situation is represented geometrically in Fig. 7.3. The response surface, which is one-dimensional in this case, is simply the line joining y_1 to y_2. The response for any mixture of composition (x_1, x_2) will be given by

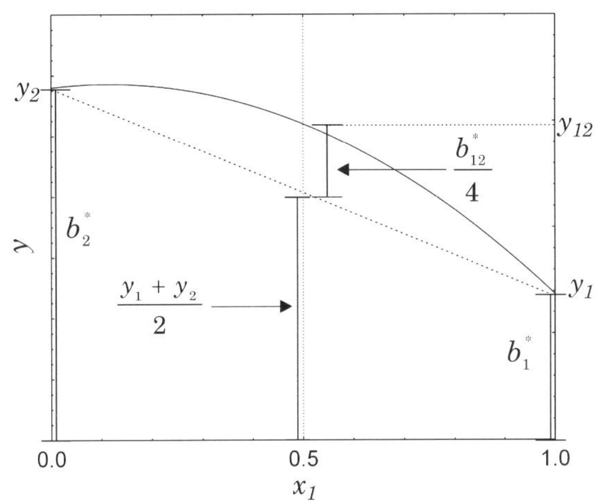

Fig. 7.3. The dashed line represents a linear model for a mixture of the two components, $\hat{y} = b_1^*x_1 + b_2^*x_2$. The b_1^* and b_2^* coefficients are response values for the pure components 1 and 2. The quadratic model is represented by the solid curve. In addition to the terms of the linear model, it contains a term describing the interaction of the two components, $b_{12}^*x_1x_2$.

a weighted average of the responses for the two pure components, with the values of x_1 and x_2 as weights.

We can increase model precision by performing replicates of the runs with the pure components. Then we will have $b_1^* = \bar{y}_1$ and $b_2^* = \bar{y}_2$, where $\bar{y}_1$ and $\bar{y}_2$ are the averages of the replicate responses. The standard errors of the b_1^* and b_2^* values can be obtained directly from the expression we derived for the standard error of an average (Section 2.6):

$$\Delta b_i^* = \frac{s}{\sqrt{n_i}}, \tag{7.5}$$

where s is a pooled estimate of the standard error of a response and n_i the number of observations used to calculate the average $\bar{y}_i$ value.

Even though the results obtained for the pure components completely determine the linear model, we should perform experiments with binary mixtures, to verify if the model is adequate. The effect of the composition on the response can be more complicated, and in that case we will need a more sophisticated model to explain the data.

Exercise 7.2. Two gasolines, A and B, are mixed. When they are pure they yield 14 and 6 km L^{-1}, respectively. (a) Determine the equation of the additive model for the yield of any mixture of these gasolines. (b) Calculate the predicted yield for a 1:1 mixture. (c) Do the same for a mixture containing only 30% of gasoline B.

The simplest augmentation of the linear model is the quadratic one, which is defined by the equation

$$\hat{y} = b_0 + b_1x_1 + b_2x_2 + b_{11}x_1^2 + b_{22}x_2^2 + b_{12}x_1x_2. \tag{7.6}$$

This expression — the same as Eq. (6.7) — contains six parameters. For binary mixtures, however, this number is reduced, because of the constant sum imposed on the proportions of the two components. Substituting $x_1 + x_2 = 1$, $x_1^2 = x_1(1 - x_2)$ and $x_2^2 = x_2\,(1 - x_1)$ into Eq. (7.6), we have

$$\hat{y} = b_0(x_1 + x_2) + b_1x_1 + b_2x_2 + b_{11}x_1(1 - x_2) + b_{22}x_2(1 - x_1) + b_{12}x_1x_2.$$

Grouping the terms in x_1 and x_2 we may write

$$\hat{y} = (b_0 + b_1 + b_{11})x_1 + (b_0 + b_2 + b_{22})x_2 + (b_{12} - b_{11} - b_{22})x_1x_2,$$

and finally

$$\hat{y} = b_1^*x_1 + b_2^*x_2 + b_{12}^*x_1x_2, \tag{7.7}$$

where $b_i^* = b_0 + b_i + b_{ii}$ (for $i = 1,2$) and $b_{12}^* = b_{12} - b_{11} - b_{22}$. In this form, the quadratic model has only one parameter more than the linear model. To obtain a design with the minimum number of runs, we only have to add, to the two runs used to estimate the linear model, a third run made on a binary mixture with any composition. If we have to choose a single mixture, good sense — and also statistics — suggests that the best one is that containing the two components in equal parts (1:1). The observed response for this mixture, which we denote by y_{12}, corresponds to the composition defined by

$$x_1 = x_2 = \frac{1}{2}.$$

Substituting these values into Eq. (7.7) we have

$$y_{12} = b_1^*\left(\frac{1}{2}\right) + b_2^*\left(\frac{1}{2}\right) + b_{12}^*\left(\frac{1}{4}\right). \tag{7.8}$$

You can easily confirm that we still have $b_1^* = y_1$ and $b_2^* = y_2$, that is, these coefficients are the same as those of the linear model. Substituting these equalities into Eq. (7.8), we finally arrive at the expression for the interaction coefficient, b_{12}^*, in terms of the three observed responses:

$$b_{12}^* = 4y_{12} - 2(y_1 + y_2).$$

As always, these calculations can be expressed more concisely in terms of matrices. Equation (7.7), which defines the model, will be given by

$$\hat{y}(x_1, x_2) = \begin{bmatrix} x_1 & x_2 & x_1x_2 \end{bmatrix} \times \begin{bmatrix} b_1^* \\ b_2^* \\ b_{12}^* \end{bmatrix},$$

or simply

$$\hat{\mathbf{y}} = \mathbf{X}\mathbf{b}^*.$$

Substituting for x_1 and x_2 the values corresponding to the pure components and the 1:1 mixture, we can write

$$\begin{bmatrix} y_1 \\ y_2 \\ y_{12} \end{bmatrix} = \begin{bmatrix} 1 & 0 & 0 \\ 0 & 1 & 0 \\ 1/2 & 1/2 & 1/4 \end{bmatrix} \times \begin{bmatrix} b_1^* \\ b_2^* \\ b_{12}^* \end{bmatrix}.$$

Pre-multiplying this equation by $\mathbf{X}^{-1}$, we obtain the vector containing the coefficient estimates

$$\begin{bmatrix} b_1^* \\ b_2^* \\ b_{12}^* \end{bmatrix} = \begin{bmatrix} 1 & 0 & 0 \\ 0 & 1 & 0 \\ -2 & -2 & 4 \end{bmatrix} \times \begin{bmatrix} y_1 \\ y_2 \\ y_{12} \end{bmatrix}.$$

Note that the response measured for the binary mixture affects only the interaction term b_{12}^*. The other two coefficients are entirely determined by the observations made on the respective pure components.

To obtain more precise model coefficients we can perform replicate experiments and use the observed average responses, exactly as we did for the linear model. Then we will have

$$b_i^* = \bar{y}_i \quad \text{and} \quad b_{12}^* = 4\bar{y}_{12} - 2(\bar{y}_1 + \bar{y}_2).$$

A quadratic model is illustrated in Fig. 7.3 by the solid curve situated above the line corresponding to the linear model. The difference between the values predicted by the two models depends on the composition of the mixture. In this example, the value predicted for the y_{12} response by the quadratic model is always larger than the one calculated by the linear model (which is the weighted average of the pure component responses), indicating that the b_{12}^* interaction term in Eq. (7.8) is positive. When this happens, we say that the two components in the mixture present a synergistic effect, that is, they interact synergistically. The response obtained when the two components are mixed is always greater than the sum of their individual responses (properly weighted by their respective proportions). In the opposite case, when $b_{12}^* < 0$, the predictions of the quadratic model are always smaller than those of the additive one. One then says that the interaction between the two components is antagonistic. Common examples of interaction are the warnings written on the instruction sheets of many medicines, about the risk of using them in combination with certain other substances.

More complex models for binary mixtures are easy to construct, if needed. One simply adds the proper higher-order terms, and performs the required additional experiments. In this book, however, we reserve the discussion of more extensive models for mixtures with more components, for which they normally have more practical relevance.

Exercise 7.3. A 1:1 mixture of the two gasolines of Exercise 7.2 yielded 12 km L^{-1}. (a) Determine the coefficients of the quadratic model for mixtures of these two gasolines. Is the interaction between them synergistic or antagonistic? (b) A mixture made up of two parts gasoline A and one part gasoline B has a mileage[73] of 13 km L–1. Is this result in good agreement with the value predicted by the quadratic model?

Exercise 7.4. The table below contains replicate measurements of the viscosity (in 10^5 kg m^{-1} s^{-1}) of two pure melted glasses and also of a mixture of the two in equal parts.

Glass A	1.41	1.47	
Glass B	1.73	1.68	
Glass A–B (50– 50%)	1.38	1.34	1.40

Determine the b_1^*, b_2^* and b_{12}^* values and their standard errors for the quadratic mixture model. Assume that the variance is the same for any mixture, so the variances of the three compositions can be pooled.

[73]For lack of a better word, alas.

7.2. Three-component mixtures

We can obtain models for three-component (or ternary) mixtures by augmenting the binary mixture models. The linear model is given by

$$\hat{y} = b_0 + b_1x_1 + b_2x_2 + b_3x_3, \tag{7.9}$$

subjected to the restriction $x_1 + x_2 + x_3 = 1$.

Substituting the b_0 term by $b_0\,(x_1 + x_2 + x_3)$ and grouping the terms in x_i, we obtain

$$\hat{y} = b_1^*x_1 + b_2^*x_2 + b_3^*x_3, \tag{7.10}$$

with $b_i^* = b_0 + b_i$, for i = 1, 2 and 3.

The interpretation of the b_i^* coefficients is the same as for the two-component case. When $x_i = 1$ (and therefore $x_{j \neq i} = 0$), the response y_i will be equal to the b_i^* coefficient. For example, when $(x_1, x_2, x_3) = (1,\ 0,\ 0)$, we have $y_1 = b_1^*$. We are still able to determine the coefficients of the linear model without making any mixtures. And, as before, we can obtain more precise estimates of the coefficients using average response values from replicate runs.

A response surface for a linear model with three components is shown in Fig. 7.4a. In this example, the surface is an inclined plane with $b_1^* > b_3^* > b_2^*$.

Exercise 7.5. How would you interpret the additive (linear) model of a three-component mixture in which the three coefficients have the same value?

If the linear model does not prove satisfactory, we can try to fit a quadratic model, as we have already seen. For a mixture of three components, the general expression for the quadratic model contains 10 terms:

$$\begin{aligned}\hat{y} = {} & b_0 + b_1x_1 + b_2x_2 + b_3x_3 + b_{11}x_1^2 + b_{22}x_2^2 + b_{33}x_3^2 \\ & + b_{12}x_1x_2 + b_{13}x_1x_3 + b_{23}x_2x_3\end{aligned} \tag{7.11}$$

Substituting the relations $1 \times b_0 = (x_1 + x_2 + x_3) \times b_0$ and $b_{11}x_1^2 = b_{11}x_1(1 - x_2 - x_3)$, in addition to analogous expressions for $b_{22}x_2^2$ and $b_{33}x_3^2$, we have

$$\begin{aligned}\hat{y} = {} & b_0(x_1 + x_2 + x_3) + b_1x_1 + b_2x_2 + b_3x_3 \\ & + b_{11}x_1(1 - x_2 - x_3) + b_{22}x_2(1 - x_1 - x_3) + b_{33}x_3(1 - x_1 - x_2) \\ & + b_{12}x_1x_2 + b_{13}x_1x_3 + b_{23}x_2x_3.\end{aligned}$$

Rearranging, we finally write

$$\hat{y} = b_1^*x_1 + b_2^*x_2 + b_3^*x_3 + b_{12}^*x_1x_2 + b_{13}^*x_1x_3 + b_{23}^*x_2x_3, \tag{7.12}$$

where $b_i^* = b_0 + b_i + b_{ii}$ and $b_{ij}^* = b_{ij} - b_{ii} - b_{jj}$, with $i \neq j$.

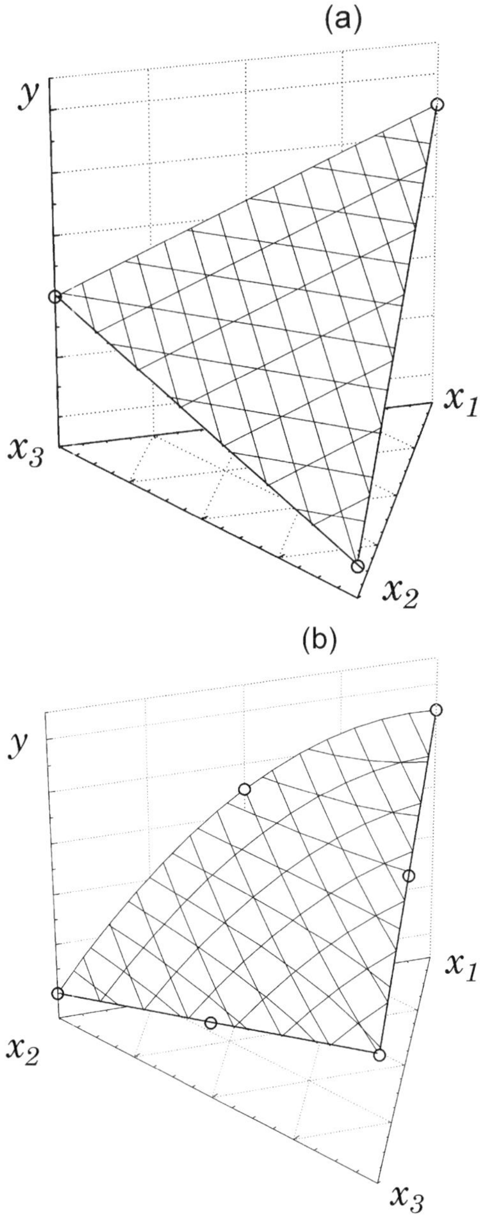

Fig. 7.4. (a) Response surface of a linear model for a mixture of three components, with $b_1^* > b_3^* > b_2^*$. The model can be completely determined using only the response values measured for the three pure components, which are represented at the vertices of the triangular base. (b) Response surface of a quadratic model for a three-component mixture, with $b_1^* > b_3^* > b_2^*$, $b_{12}^* > 0$ and $b_{13}^* = b_{23}^* = 0$. The quadratic model can be determined using only the response values measured for the pure components and for the binary mixtures represented by the midpoints of the three sides of the triangle.

The 10 coefficients of Eq. (7.11) are now reduced to 6. To determine their values we need an experimental design with at least six distinct runs. Performing the runs for the pure components, we obtain the values of the three linear coefficients:

$$y_i = b_i^*, \tag{7.13}$$

for $i = 1$, 2 and 3, as in the linear model.

For the three remaining runs, we can use the three 50–50% binary mixtures, for which

$$y_{ij} = b_i^*\left(\frac{1}{2}\right) + b_j^*\left(\frac{1}{2}\right) + b_{ij}^*\left(\frac{1}{2}\right)\left(\frac{1}{2}\right), \tag{7.14}$$

for $i, j = 1, 2, 3$ and $i \neq j$, an extension of Eq. (7.8) to the case of three components.

The six equations represented by (7.13) and (7.14) can be grouped into a single matrix equation $\mathbf{y} = \mathbf{X}\mathbf{b}^*$. Writing it in detail, we have

$$\begin{bmatrix} y_1 \\ y_2 \\ y_3 \\ y_{12} \\ y_{13} \\ y_{23} \end{bmatrix} = \begin{bmatrix} 1 & 0 & 0 & 0 & 0 & 0 \\ 0 & 1 & 0 & 0 & 0 & 0 \\ 0 & 0 & 1 & 0 & 0 & 0 \\ 1/2 & 1/2 & 0 & 1/4 & 0 & 0 \\ 1/2 & 0 & 1/2 & 0 & 1/4 & 0 \\ 0 & 1/2 & 1/2 & 0 & 0 & 1/4 \end{bmatrix} \times \begin{bmatrix} b_1^* \\ b_2^* \\ b_3^* \\ b_{12}^* \\ b_{13}^* \\ b_{23}^* \end{bmatrix}.$$

Pre-multiplying this equation by $\mathbf{X}^{-1}$, we obtain all six coefficients in a single operation:

$$\mathbf{b}^* = \mathbf{X}^{-1}\mathbf{y},$$

or

$$\begin{bmatrix} b_1^* \\ b_2^* \\ b_3^* \\ b_{12}^* \\ b_{13}^* \\ b_{23}^* \end{bmatrix} = \begin{bmatrix} 1 & 0 & 0 & 0 & 0 & 0 \\ 0 & 1 & 0 & 0 & 0 & 0 \\ 0 & 0 & 1 & 0 & 0 & 0 \\ -2 & -2 & 0 & 4 & 0 & 0 \\ -2 & 0 & -2 & 0 & 4 & 0 \\ 0 & -2 & -2 & 0 & 0 & 4 \end{bmatrix} \times \begin{bmatrix} y_1 \\ y_2 \\ y_3 \\ y_{12} \\ y_{13} \\ y_{23} \end{bmatrix}.$$

Individually, the coefficients are given by the set of equations

$$b_1^* = y_1, \qquad b_{12}^* = 4y_{12} - 2(y_1 + y_2), \tag{7.15a}$$

$$b_2^* = y_2, \qquad b_{13}^* = 4y_{13} - 2(y_1 + y_3), \tag{7.15b}$$

$$b_3^* = y_3, \qquad b_{23}^* = 4y_{23} - 2(y_2 + y_3). \tag{7.15c}$$

These expressions, once more, are similar to the equations derived for two-component mixtures. The b_i^* coefficients are the average response values for the pure components, and the b_{ij}^* values are obtained from the runs made with the pure i and j components and with the 50–50% binary mixtures of i and j. The model is completely determined without any data from ternary mixtures.

If we want to obtain more precise coefficient values, we proceed as before, performing replicate runs for each composition. With these results, we solve Eqs. (7.15a–c) using average values instead of individual ones.

A response surface for a quadratic model with $b_1^* > b_3^* > b_2^*$, $b_{12}^* > 0$ and $b_{13}^* = b_{23}^* = 0$ is illustrated in Fig. 7.4b. The boundaries of the response surface situated above the sides of the triangle (and therefore representing binary mixtures) are analogous to the response curves of Fig. 7.3. Since in this example $b_{13}^* = b_{23}^* = 0$, the curves for the binary mixtures of components 1 and 3 (and 2 and 3) are straight lines, showing that the responses for these mixtures vary linearly with the proportions of their components. The curve located above the side representing the binary mixtures of components 1 and 2 is similar to the solid curve in Fig. 7.3, because $b_{12}^* > 0$ (i.e., the components interact synergistically). For ternary mixtures, corresponding to points in the interior of the triangle, the interpretation is more complicated. The response becomes a sum of six terms: the three linear contributions owing to the pure components plus the three binary contributions.

The experimental design represented by the six points in Fig. 7.4b is used often enough to deserve its own name. In the mixture design literature it is commonly called a simplex lattice design.

7.3. An example of a three-component mixture

Before discussing more complicated models and systems with more than three components, we present a mixture design problem with real data, performed in the laboratory of Professor G. Oliveira Neto (Chemistry Department, Campinas State University).

Some biological substrates can be analyzed with selective electrodes. A critical part of these electrodes is a membrane, whose properties have a large influence on the analytical sensitivity of the electrode. The objective of this study was to determine the composition of the membrane that produces the largest possible analytical signal. The researchers used a simplex lattice design, and measured the responses given in Table 7.1. The compositions of the mixtures are represented in the triangle of Fig. 7.5a, together with the response values obtained with them.

Pure components 1 and 3 produce average signals of 3.10 and 0.35 cm, respectively. For a 1:1 mixture of these two components, the average

Table 7.1
Study of a membrane for making a selective electrode. Mixture compositions and observed analytical signals are presented

i	x_1	x_2	x_3	Signal			$\bar{y}_i$	s_i^2
1	1	0	0	3.2	3.0	—	3.10	0.020
2	0	1	0	0.5	0.4	—	0.45	0.005
3	0	0	1	0.4	0.3	—	0.35	0.005
4	1/2	1/2	0	1.9	1.2	2.0	1.70	0.190
5	1/2	0	1/2	3.9	4.4	4.1	4.13	0.063
6	0	1/2	1/2	0.3	0.3	0.2	0.27	0.003

Note: The analytical signal is the peak height, in centimeters.

signal observed was 4.13 cm. Because this value is much larger than the average of the response values obtained for the pure components, we conclude that the additive model is not appropriate, and proceed right away to fit the data with a quadratic model.

Substituting the average response value observed for each mixture in Eqs. (7.15a–c), we arrive at the following values for the coefficients of the quadratic model:

$$\begin{aligned} b_1^* &= 3.10, \quad b_{12}^* = -0.30, \\ b_2^* &= 0.45, \quad b_{13}^* = 9.62, \\ b_3^* &= 0.35, \quad b_{23}^* = -0.52. \end{aligned}$$

The large value of the b_{13}^* coefficient immediately suggests a strong synergistic interaction between components 1 and 3. However, good statistical practice requires that before trying to interpret these results we obtain estimates of their errors. Since the runs were replicated, we can use the variances observed for the response values of each distinct run (last column of Table 7.1) to calculate a pooled estimate of the variance of an individual response. From there, by means of Eq. (5.30), we arrive at estimates of the standard errors of the coefficients. With these values we can finally write the complete equation for the fitted model:

$$\hat{y} = \underset{(\pm 0.17)}{3.10x_1} + \underset{(\pm 0.17)}{0.45x_2} + \underset{(\pm 0.17)}{0.35x_3} - \underset{(\pm 0.75)}{0.30x_1x_2} + \underset{(\pm 0.75)}{9.62x_1x_3} - \underset{(\pm 0.75)}{0.52x_2x_3}.$$

Only the b_1^* and b_{13}^* coefficients have values that are much larger than their respective standard errors. Therefore, we can adopt a simplified model to represent the data:

$$\hat{y} = 3.10x_1 + 9.62x_1x_3. \tag{7.16}$$

This model tells us that the presence of component 1 in the mixture leads to more intense analytical signals. Component 2 does not appear in

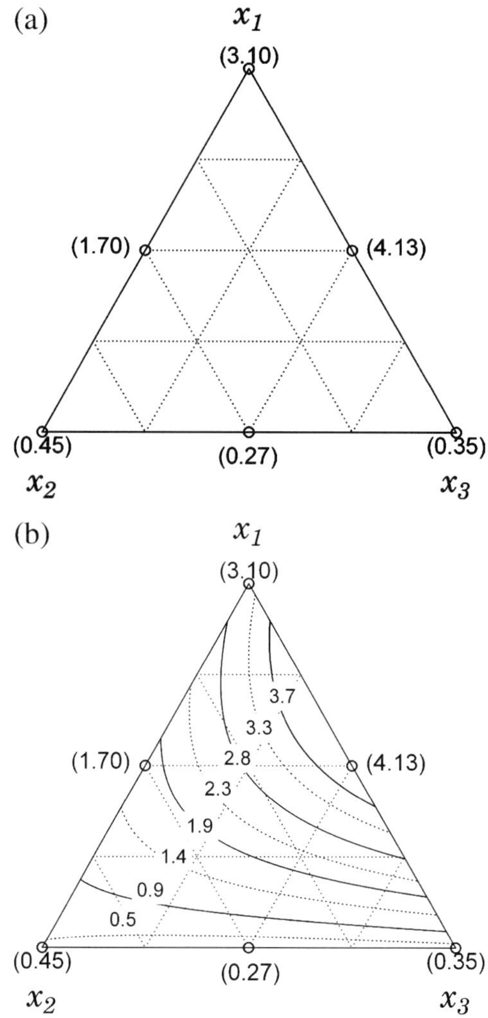

Fig. 7.5. (a) Simplex lattice design and the average analytical signals observed for the mixtures represented by the points. (b) Response curves for the quadratic model of the analytical signal, Eq. (7.16).

the model, and for this reason does not contribute to increasing or decreasing the analytical signal, although it may be important for other properties of the membrane. As we have already noted, component 3 has a synergistic interaction with component 1. This means that the simultaneous presence of the two components in the mixture will produce stronger signals than those predicted by the additive model.

According to Eq. (7.16), the maximum value of the analytical signal for this kind of membrane should be 4.2 cm, obtained with a membrane containing only components 1 and 3, in the 66–34% proportion.

The response curves corresponding to Eq. (7.16) are shown in Fig. 7.5b. Each curve connects mixture compositions that, theoretically, are expected to give an analytical signal of the same height, specified by the corresponding numerical value. The interpretation of these curves is analogous to the interpretation of phase diagrams of ternary mixtures, studied in physical chemistry. To obtain analytical signals approximately 4 cm high, we need to prepare membranes containing two parts of component 1 for each part of component 3.

Exercise 7.6. Do a t-test to check if the coefficients of the full quadratic model for the analytical signals of the membrane study are significant.

Exercise 7.7. Use the simplified equation of the quadratic model, Eq. (7.16), to find the composition of the mixture that should produce the maximum value of the analytical signal and what this value should be.

Exercise 7.8. The average results listed below were obtained in a project performed at the Pirelli research center in Santo André, SP (Brazil), with the objective of improving electrical cable insulation (Costa et al., 1991).

Mixture	Al_2O_3	Fe_2O_3	Co_3O_4	Mass loss	Tracking
1	1	0	0	2.84	94.26
2	0	1	0	5.24	8.95
3	0	0	1	3.80	11.52
4	1/2	1/2	0	1.18	125.00
5	1/2	0	1/2	2.18	103.00
6	0	1/2	1/2	3.38	10.55

(a) A high tracking value is desirable, because it means that the cable can resist electrical tension at a higher level for more time, without leaking current. Which of the three components would you use in higher proportions to obtain an electrical insulation with high tracking values?
(b) Determine the equations of the quadratic models for both responses. Do they confirm your answer to part (a)?

7.4. Cubic models for three-component mixtures

Besides the terms of the additive model, the quadratic model of Eq. (7.12) contains cross terms that describe interactions between two components, and for this reason it is usually able to reproduce the response values at the vertices and on the sides of the concentration triangle, since they represent pure and binary mixtures, respectively. It should not surprise us, however, that non-additive effects involving the simultaneous presence of three components will be important to describe response values for ternary mixtures (the points in the interior of the triangle). If this is the case, the quadratic model will prove unsatisfactory, and we will need to add cubic terms to the model.

The full cubic model for a mixture of three components is given by the equation

$$\hat{y} = b_0 + \sum_{i=1}^{3} b_i x_i + \sum_{i\leq j}^{3} \sum_{j}^{3} b_{ij} x_i x_j + \sum_{i\leq j}^{3} \sum_{j\leq k}^{3} \sum_{k}^{3} b_{ijk} x_i x_j x_k. \qquad (7.17)$$

Introducing the identity $1 = x_1 + x_2 + x_3$ and making the appropriate substitutions, with a little handiwork we can arrive at the expression

$$\begin{aligned} \hat{y} &= b_1^* x_1 + b_2^* x_2 + b_3^* x_3 + b_{12}^* x_1 x_2 + b_{13}^* x_1 x_3 + b_{23}^* x_2 x_3 \\ &\quad + d_{12}^* x_1 x_2 (x_1 - x_2) + d_{13}^* x_1 x_3 (x_1 - x_3) + d_{23}^* x_2 x_3 (x_2 - x_3) \\ &\quad + b_{123}^* x_1 x_2 x_3. \end{aligned} \qquad (7.18)$$

Since this equation has 10 terms, we will have to perform at least 10 distinct runs to determine the values of all the coefficients. Often, however, adding a single cubic term is sufficient to transform the model into a satisfactory description of the experimental region. Eliminating the d_{ij}^* terms in Eq. (7.18), we arrive at the expression for the special cubic model, which contains only one more term than the quadratic model, and therefore requires only one more run:

$$\hat{y} = b_1^* x_1 + b_2^* x_2 + b_3^* x_3 + b_{12}^* x_1 x_2 + b_{13}^* x_1 x_3 + b_{23}^* x_2 x_3 + b_{123}^* x_1 x_2 x_3. \qquad (7.19)$$

The experimental design normally used to determine the values of the coefficients of the special cubic model is called the simplex centroid, which we obtain by simply adding a center point, corresponding to a 1:1:1 ternary mixture, $(x_1, x_2, x_3) = \left(\frac{1}{3}, \frac{1}{3}, \frac{1}{3}\right)$ to the simplex lattice design. The coefficient of the cubic term is given by

$$b_{123}^* = 27y_{123} - 12(y_{12} + y_{13} + y_{23}) + 3(y_1 + y_2 + y_3),$$

where y_{123} is the response observed for the (1:1:1) ternary mixture. The other coefficients have the same expressions as for the quadratic model.

In a duplicate experiment using the (1/3, 1/3, 1/3) mixture for the membrane of the selective electrode, analytical signals of 3.40 and 3.60 cm were obtained. Combining this information with the responses for the pure components and the binary mixtures already presented, we obtain a value of 33.00 for the b_{123}^* coefficient, which is highly significant.

Exercise 7.9. Suppose that, in the membrane experiments, the observed response for the (1:1:1) ternary mixture had been 2.50 cm (average of two runs), instead of 3.50 cm. (a) Calculate the value of the b_{123}^* coefficient. (b) Using 0.056 cm^2 as an estimate of the variance of the analytical signal, calculate the standard error of the new value of b_{123}^*. Is it significant?

Eliminating the terms with non-significant coefficients, we reduce the special cubic model to the equation

$$\hat{y} = 3.10x_1 + 9.62x_1x_3 + 33.00x_1x_2x_3. \tag{7.20}$$

Note that the values of the non-cubic terms are the same as those for the quadratic model, as we have already remarked. The contour curves corresponding to this expression are shown in Fig. 7.6, together with the experimental results of the simplex centroid design. Close to the sides of the triangle the predictions of the cubic model are very similar to those of the quadratic model, because one of the three proportions is necessarily close to zero, reducing the impact of the cubic term on the response value. In the central region of the triangle the predictions of the two models differ quite a bit, because there the cubic term makes contributions of the same order of magnitude as the other two.

The cubic model predicts a maximum analytical signal of 4.2 cm, the same value predicted by the quadratic model. To produce this signal the mixture should have 62%, 4% and 34% of components 1, 2 and 3, respectively. This is practically the same composition of the mixture indicated by the quadratic model (66%, 0% and 34%). Actually, as shown in Fig. 7.6, the optimal region is a kind of plateau, where we can vary the concentrations without affecting the response very much. For example, substituting $(x_1, x_2, x_3) = (0.70, 0.05, 0.25)$ in Eq. (7.20) we obtain $\hat{y} = 4.14$,

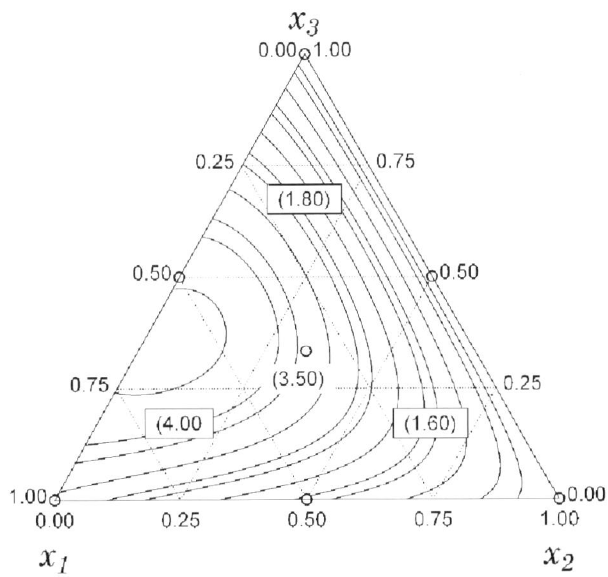

Fig. 7.6. Simplex centroid design, with average responses for the mixtures represented by the points and contour curves for the special cubic model, Eq. (7.20). The three boxed response values close to each vertex were used to test the quality of the fit.

which is, for all practical purposes, the same as the mathematical optimum. This is an interesting situation, because it permits us to change the composition of the mixture to improve other criteria, without sacrificing the analytical signal sensibility. For example, if component 3 is more expensive than component 1, it is more advantageous to use the latter mixture rather than the mixture corresponding to the maximum response predicted by the model.

Exercise 7.10. What is the value predicted by the quadratic model for the average response of the membrane with 1:1:1 composition? How does it compare with the prediction of the special cubic model?

Exercise 7.11. The cubic model can also be determined from a matrix equation. Write the matrix equation that we should solve to obtain the values of the seven coefficients of the special cubic model for the membrane study.

7.5. Model evaluation

We are sure you will not be very surprised when we say that fitting mixture models is nothing more than a special case of least-squares fitting. What we did in the last section, actually, was the same as solving Eq. (5.12) for a set of 17 observations: the 15 appearing in Table 7.1, plus the two observations made at the center point of the triangle. We used two models for this: the quadratic one, with six parameters, and the special cubic one, with seven. The statistical significance of these models can be evaluated by an analysis of variance.

The cubic model cannot be tested for lack of fit, since it has as many parameters as there are distinct runs. It is indeed better at describing the interior of the triangle than the quadratic model, but we do not have a way to decide whether the cubic model is the most adequate to represent the data over the entire experimental region. Consequently, we can split the total variance about the average into only two contributions, one for the regression and the other for the residuals. In other words, the ANOVA in this case will be similar to the one of Table 5.2. The results are in Table 7.2.

The values of MS_R/MS_r are 31.63 and 111.33 for the quadratic and special cubic models, respectively. As we know, they must be compared with the values of $F_{5,11}$ and $F_{6,10}$. At the 99% confidence level these values are only 5.32 and 5.39, showing that both models are highly significant. The larger value for the special cubic model indicates that it explains a greater percentage of the total variation, but we must also consider that it has one more parameter, and that a model with more parameters will always explain more variation.

When we add another term to the model, we transfer 1 degree of freedom from the residual mean square to the regression mean square. To

Table 7.2
Analysis of variance for the fits of the quadratic and special cubic models to the data in Table 7.1, augmented by the results observed for the mixture with $x_1 = x_2 = x_3 = 1/3$ (duplicate response values with an average of 3.50 cm), which increases the number of observations to 17

Source of variation	Sum of squares	Deg. of freedom	Mean square
Regression	37.96 (*40.06*)	5 (*6*)	7.59 (*6.68*)
Residual	2.66 (*0.56*)	11 (*10*)	0.24 (*0.06*)
Total	40.62	16	

Note: Values within parentheses refer to the special cubic model.

decide if this is worthwhile, we can apply a new F-test, to compare the reduction in the residuals caused by the presence of the new term with the residual mean square of the extended model. If the test is not significant, then introducing the additional parameter was not worthwhile. This procedure is not restricted to mixtures. It can be applied to any least-squares model.

We consider the general case of any two models, I and II, where II has d more parameters than I. Model I leaves a residual sum of squares $SS_{r,I}$, which is reduced to $SS_{r,II}$ when the d terms are added. The F-test relation is

$$F = \frac{(SS_{r,I} - SS_{r,II})/d}{MS_{r,II}}. \tag{7.21}$$

In our example, I is the quadratic model, II the special cubic model and $d = 1$. We simply have

$$F = \frac{SS_{r,\text{quad}} - SS_{r,\text{cub}}}{MS_{r,\text{cub}}},$$

where the quad and cub subscripts refer to the quadratic and special cubic models. Using the values of Table 7.2 we have

$$F = \frac{2.66 - 0.56}{0.06} = 35.0.$$

Comparing this result with $F_{1,10} = 10.0$ (99% confidence level), we can conclude that the cubic model is indeed better than the quadratic one, for the data in our example.

A better way of comparing the quality of the two models is to obtain new response values for mixtures that were not used in modeling, and compare the observed values with the values predicted by each model. In the membrane study, three more runs were carried out, as indicated by

Table 7.3
Study of a membrane for a selective electrode. Comparison between the average observed analytical signals ($\bar{y}_{obs}$) and the values predicted by the quadratic ($\hat{y}_{quad}$) and special cubic models ($\hat{y}_{cub}$)

Mixture	x_1	x_2	x_3	$\hat{y}_{quad}$	$\hat{y}_{cub}$	$\bar{y}_{obs}$
1	1	0	0	3.10 (*0*)	3.10 (*0*)	3.10
2	0	1	0	0 (*0.35*)	0 (*0.35*)	0.35
3	0	0	1	0 (*0.35*)	0 (*0.35*)	0.35
4	1/2	1/2	0	1.55 (*0.15*)	1.55 (*0.15*)	1.70
5	1/2	0	1/2	3.96 (*0.17*)	3.96 (*0.17*)	4.13
6	0	1/2	1/2	0 (*0.28*)	0 (*0.28*)	0.28
7[a]	1/3	1/3	1/3	2.10 (*1.40*)	3.33 (*0.17*)	3.50
8[b]	2/3	1/6	1/6	3.14 (*0.86*)	3.75 (*0.25*)	4.00
9[b]	1/6	2/3	1/6	0.79 (*0.81*)	1.40 (*0.20*)	1.60
10[b]	1/6	1/6	2/3	1.59 (*0.21*)	2.20 (*–0.40*)	1.80

Note: Values within parentheses are the residuals left by the models, calculated as $e = \bar{y}_{obs} - \hat{y}$.
[a] Duplicate with individual results of 3.60 and 3.40 cm.
[b] Single measurements.

the boxed values in Fig. 7.6. The new mixtures have component proportions of $(x_1, x_2, x_3) = (\frac{2}{3}, \frac{1}{6}, \frac{1}{6})$ $(\frac{1}{6}, \frac{2}{3}, \frac{1}{6})$ and $(\frac{1}{6}, \frac{1}{6}, \frac{2}{3})$.

The predictions of the two models are compared with the observed responses in Table 7.3. Since the two models are identical except for the $x_1x_2x_3$ term, their predictions only differ for the ternary mixtures. As we might have expected, the cubic model is better. For the three test mixtures, the average absolute residual left by the special cubic model is 0.28. For the quadratic model this value is 0.63 — more than two times larger.

A third way to compare the efficiencies of the two models is to perform an analysis of variance including the lack of fit contribution, now that we have more distinct runs than parameters for all the models of interest. The results of this analysis are presented in Table 7.4. Their interpretation, which we leave as an exercise, confirms everything we have just said concerning the two models.

7.6. Pseudocomponents

In practice, mixture optimization problems normally require the presence of all the components, to obtain a satisfactory product. For example, to produce the membrane for the selective electrode, we need to

Table 7.4
Analyses of variance for the quadratic and special cubic models. The number of distinct mixtures is 10, which permits us to test for lack of fit of the two models

Source of variation	Sum of squares	Deg. of freedom	Mean square
Regression	41.81 (*43.91*)	5 (*6*)	8.36 (*7.32*)
Residual	3.05 (*0.95*)	14 (*13*)	0.22 (*0.073*)
Lack of fit	2.49 (*0.39*)	4 (*3*)	0.62 (*0.13*)
Pure error	0.56	10	0.056
Total	44.86	19	

% Explained variation: 93.20 (97.88)
Maximum % explainable variation: 98.75

Note: Values within parentheses are those of the special cubic model.

mix solutions of pyrrole, KCl and $K_4Fe(CN)_6$. A membrane only forms if all three components are present. In spite of this, in our previous discussion we used values corresponding to pure components and binary mixtures. How could that be, if no membrane is formed with these compositions?

We misled a bit, for didactic purposes. In fact, the "components" 1, 2 and 3 whose proportions appear in Table 7.1 are pseudocomponents, that is, mixtures of the real chemical components.

Experience shows that for a membrane to form we have to use at least 10% of each of the three components (pyrrole, KCl and $K_4Fe(CN)_6$ solutions). Also, these lower limits lead to upper bounds on the component proportions. The maximum proportion that any component can have in a mixture is 80%, because each one of the other components must be present at a level of at least 10%. In this example, letting c_i represent the proportion of the ith component in the mixture, we see that $0.10 \le c_i \le 0.80$.

This situation can be generalized for any mixture in which the proportions of the pure components have to obey non-zero lower limits, which we shall generically call a_i. Obviously, the sum of all these limits must be less than one, otherwise, the mixture would be impossible to prepare. Considering the general case of a mixture containing q components, we can write

$$0 \le a_i \le c_i \quad \text{and} \quad \sum_{i=1}^{q} a_i < 1, \quad i = 1, 2, \ \ldots \ , q. \tag{7.22}$$

In the membrane example, since all the lower limits are equal to 0.10, the value of the sum is 0.30.

The levels of the mixture components in terms of pseudocomponents, denoted by x_i, are given by the expression

$$x_i = \frac{c_i - a_i}{1 - \sum_{i=1}^{q} a_i}, \tag{7.23}$$

which is a kind of variable coding. For the membrane example we have

$$x_i = \frac{c_i - 0.1}{0.7}, \quad i = 1, 2, 3. \tag{7.24}$$

Thus, for example, the pseudocomponent composition (1, 0, 0) in Table 7.1 is really a mixture with proportions of 0.8, 0.1 and 0.1 of pyrrole, KCl and $K_4Fe(CN)_6$ solutions, respectively.

Exercise 7.12. What are the compositions, in terms of the real components, of the mixtures corresponding to the last four lines in Table 7.3?

Substituting Eq. (7.24) into the equation for the special cubic model in terms of pseudocomponents, Eq. (7.20), we can express the response as an explicit function of the proportions of pyrrole, KCl and $K_4Fe(CN)_6$ solutions in the mixture:

$$\begin{aligned}\hat{y} = &-0.34 + 3.43c_{\text{pyrrole}} + 0.97c_{\text{KCl}} - 1.00c_{\text{K}_4\text{Fe(CN)}_6} - 9.67c_{\text{pyrrole}}c_{\text{KCl}} \\ &+ 9.98c_{\text{pyrrole}}c_{\text{K}_4\text{Fe(CN)}_6} - 9.67c_{\text{KCl}}c_{\text{K}_4\text{Fe(CN)}_6} + 96.74c_{\text{pyrrole}}c_{\text{KCl}}c_{\text{K}_4\text{Fe(CN)}_6}.\end{aligned} \tag{7.25}$$

This expression not only has more terms than Eq. (7.20), but interpreting its coefficients is much more complicated. The corresponding contour curves are shown in Fig. 7.7. Note that in terms of concentrations of the pure components instead of the pseudocomponents, the experimental region is limited to the internal triangle. Using Eq. (7.25) to predict response values outside this region would be extrapolating, and should be viewed with a good deal of skepticism, as always. In our specific example, these points correspond to mixtures that are not expected to form acceptable membranes. Note, finally, that the response surfaces in Figs. 7.7 and 7.6 are identical, their only difference being the way of describing the composition. What is interpreted as a pure component in Fig. 7.6 is actually a ternary mixture. For example, $x_1 = 1$ corresponds to the mixture of composition $c_{\text{pyrrole}} = 0.8$, $c_{\text{KCl}} = c_{\text{K}_4\text{Fe(CN)}_6} = 0.1$.

7.7. Other mixture designs

When the component proportions must obey lower limits, the experimental region becomes constrained and the design results are much easier to analyze in terms of pseudocomponents, as we have just seen. In

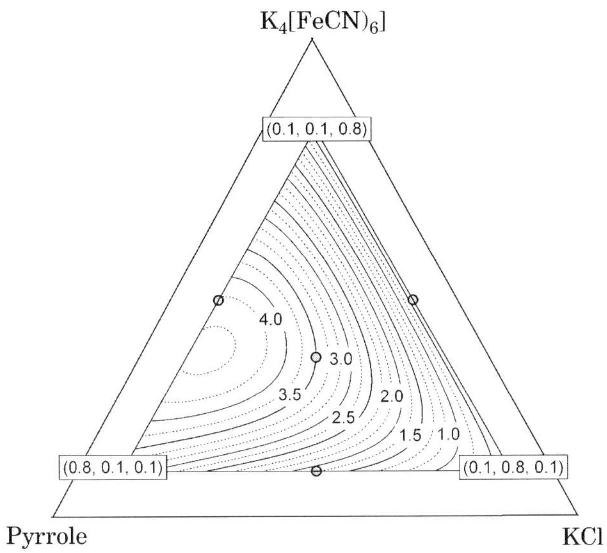

Fig. 7.7. Response contour curves for the special cubic model as a function of the proportions of pyrrole, KCl and $K_4Fe(CN)_6$ solutions.

many formulations, the composition of the mixture has to satisfy not only lower limits but also upper ones. This further restricts the region to be studied and complicates the selection of the mixture points to be used in the design. In this section we present an example of this type of problem, also studied in the Chemistry Department of Campinas State University, in the laboratory of Professor I. Joekes (Rubo, 1991). We will just try to describe a bit of the appropriate methodology. Fuller treatments can be found in more specialized works, such as Cornell (1990a, b) and Goupy (2000).

The goal of the investigation was to study the elongation until rupture and swelling, in dioxane, of polymeric films made from polyisobutene (**PIB**), polyethylene (**PE**) and paraffin wax (**PW**). For technical reasons, the proportions of these components in the mixtures were restricted to the following intervals:

$$0.50 \le c_{\text{PIB}} \le 0.65;$$
$$0.15 \le c_{\text{PE}} \le 0.35;$$
$$0.10 \le c_{\text{PW}} \le 0.25.$$

Because all the intervals are different, these inequalities define an irregular hexagon inside the concentration triangle, shown in Fig. 7.8a. The points belonging to this hexagon represent the mixtures that in principle can be studied. With these specifications, the pseudocomponents are defined by the expressions:

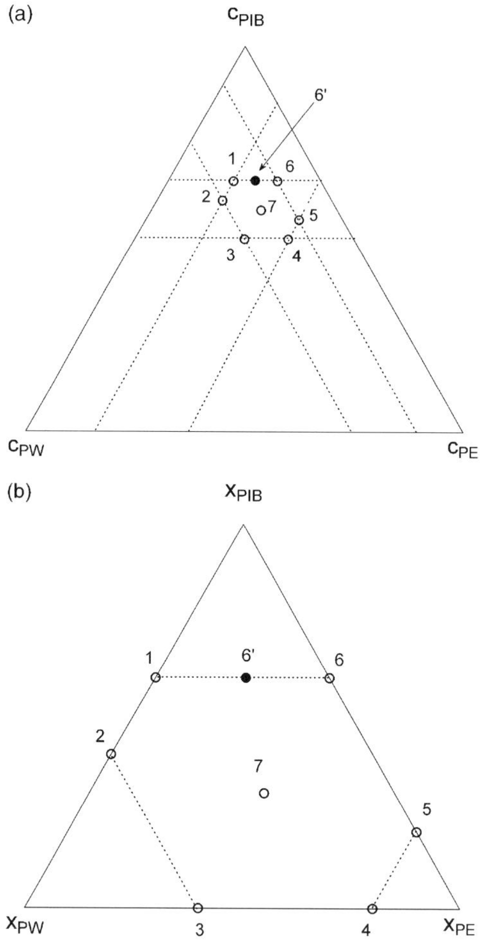

Fig. 7.8. (a) Concentration triangle for polyisobutene (PIB), polyethylene (PE) and paraffin wax (PW). The region studied is determined by the lower and upper concentration limits of these components and results in an irregular hexagon whose vertices are numbered from 1 to 6. (b) The same region, in terms of pseudocomponents.

$$x_{\mathrm{PIB}} = \frac{c_{\mathrm{PIB}} - 0.50}{0.25};$$

$$x_{\mathrm{PE}} = \frac{c_{\mathrm{PE}} - 0.15}{0.25};$$

$$x_{\mathrm{PW}} = \frac{c_{\mathrm{PW}} - 0.10}{0.25}.$$

To define the design, we need to consider what models could be appropriate to describe the behavior of the two responses — film elongation and swelling. Unless we already know a good deal about the system under study, this usually cannot be done before the experiments are made. Also, different responses may require different models. Since it is possible that a description of the results will require a special cubic model, we had better choose a design with at least seven different mixtures.

As in any least-squares fit, propagating experimental error to the estimates of the model coefficients is done by means of Eq. (5.30), which contains the design matrix, **X**. Depending on the design, these estimates will be more or less affected by experimental errors. In general, errors in the coefficient estimates are smaller when the design points are uniformly spread over the experimental region to be studied.[74] That is why we prefer to use vertices, midpoints of sides and centroid points. In this example two designs were initially considered:

a. The six vertices of the hexagon plus the central point
b. The midpoints of the sides and the central point

Examining the $(\mathbf{X}^t\mathbf{X})^{-1}$ matrix for these two designs, one can conclude that the first design (points 1–7 in Fig. 7.8a) produces coefficient estimates 10–70% more precise than those of the second design, and for this reason it was chosen. When the experiments were performed, mixture number 6 did not produce a viable film and had to be substituted by mixture 6′, with composition defined by the midpoint of the 1–6 side.

Perhaps, you are wondering how we would choose the best levels for a mixture study with many components, for which we are not able to visualize the mixture simplex surface. Fortunately, this choice need not be of concern to the researcher. Various commercial computer programs are available to do this for us, using certain statistical criteria.

The design matrix, whose elements are the proportions used to prepare the various mixtures, is presented in Table 7.5, in terms of components and of pseudocomponents. Fig. 7.8b shows the geometric representation of the design in terms of pseudocomponents. The values of the two responses were determined in duplicate for every mixture. These values are also shown in Table 7.5 and were used to obtain a pooled estimate of the experimental variance, from which we can calculate the standard errors of the parameter estimates.

Equation (5.12) was used to fit linear, quadratic and special cubic models to the data from each response. The analysis of the results led to the following conclusions:

[74]For more on optimal designs, you may wish to review Section 6.8.

Table 7.5
Film compositions of PIB–PE–PW mixtures, in the real values for the chemical components (c_i) and in pseudocomponents (x_i), and the measured values for elongation until rupture and for swelling in dioxane. The responses were determined in duplicate

Film	c_{PIB}	c_{PE}	c_{PW}	x_{PIB}	x_{PE}	x_{PW}	Elongation (%)		Swelling (m3 kg^{-1})	
1	0.65	0.15	0.20	0.60	0.0	0.40	214	232	176	177
2	0.60	0.15	0.25	0.40	0.0	0.60	130	236[a]	172	182
3	0.50	0.25	0.25	0.00	0.40	0.60	114	137	161	153
4	0.50	0.35	0.15	0.00	0.80	0.20	111	91	139	134
5	0.55	0.35	0.10	0.20	0.80	0.00	227	189	165	165
6′	0.65	0.20	0.15	0.60	0.20	0.20	408	394	177	184
7	0.575	0.25	0.175	0.30	0.40	0.30	303	265	165	165

[a] This value was excluded from the model calculations, because the film formed with much difficulty and the elongation value was not considered reliable.

- For elongation the linear model

$$\hat{y}_{\text{elong}} = \underset{(\pm 18)}{479}\, x_{\text{PIB}} + \underset{(\pm 12)}{176}\, x_{\text{PE}} + \underset{(\pm 20)}{20}\, x_{\text{PW}}$$

  proved the most adequate. The response surface corresponding to this equation is an inclined plane, whose contour lines are shown in Fig. 7.9a.
- The quadratic model was the best for swelling:

$$\hat{y}_{\text{swell}} = \underset{(\pm 15)}{202}\, x_{\text{PIB}} + \underset{(\pm 5)}{140}\, x_{\text{PE}} + \underset{(\pm 14)}{212}\, x_{\text{PW}} + \underset{(\pm 32)}{76}\, x_{\text{PIB}}x_{\text{PE}} - \underset{(\pm 54)}{126}\, x_{\text{PIB}}x_{\text{PW}} - \underset{(\pm 34)}{110}\, x_{\text{PE}}x_{\text{PW}}.$$

  Its contour curves are shown in Fig. 7.9b.

The goal of the researchers was to obtain a film that simultaneously had a high elongation value and a low swelling value. According to Fig. 7.9a, we can obtain the elongation optimum value for mixtures with compositions similar to those of the 6′ point. On the other hand, to obtain low swelling values we should prepare mixtures close to point 4, which is directly opposite to point 6′. The conclusion is clear, though somewhat frustrating. With these chemical components, producing a film with acceptable properties will require a compromise between the desired responses, high elongation and low swelling. The method of Derringer and Suich, discussed in Section 6.4, could help us arrive at such a compromise.

7.8. Mixtures with more than three components

Models for mixtures containing more than three components are simple extensions of three-component mixture models. For the general case of q

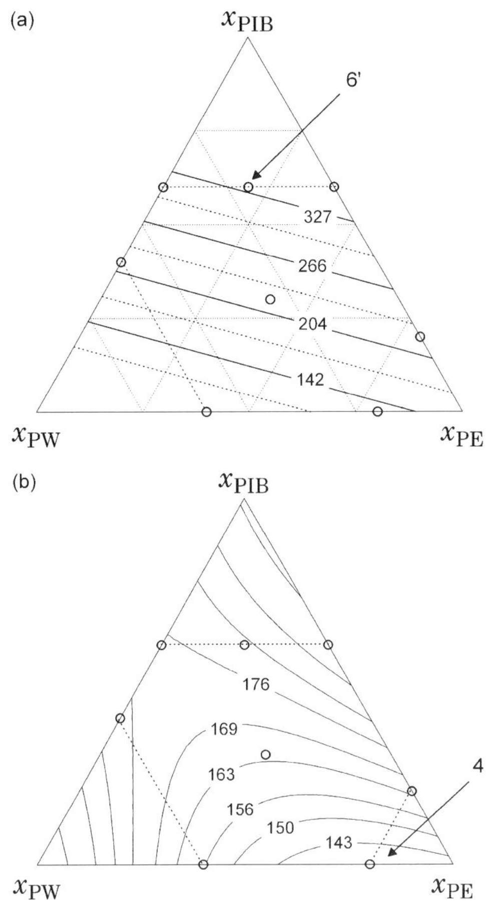

Fig. 7.9. Contour curves for properties of PIB–PE–PW films, in terms of pseudocomponents. (a) Linear model for elongation. (b) Quadratic model for swelling. The desired characteristics are obtained for compositions similar to those of the 6′ mixture (high elongation) and mixture 4 (low swelling).

components, the linear, quadratic and special cubic models are, respectively, given by

$$\hat{y} = \sum_{i=1}^{q} b_i^* x_i,$$

$$\hat{y} = \sum_{i=1}^{q} b_i^* x_i + \sum_{i<j}^{q} \sum_{j}^{q} b_{ij}^* x_i x_j,$$

and

$$\hat{y} = \sum_{i=1}^{q} b_i^* x_i + \sum_{i<j}^{q} \sum_{j}^{q} b_{ij}^* x_i x_j + \sum_{i<j}^{q} \sum_{j<k}^{q} \sum_{k}^{q} b_{ijk}^* x_i x_j x_k.$$

The calculations needed to determine the coefficients of each model are extremely fast for those having access to a microcomputer. In a nutshell, everything comes around to solving Eqs. (5.12) and (5.30) with the appropriate matrices. The main difficulty lies in the graphical representation of the results. In some cases, the equations can be simplified if it is found that one component is not important. In others, we can visualize contour curves of cross sections taken at appropriate angles, but in general graphical representation is problematic, and becomes still more complicated when several responses are in question. In algebraic terms, however, the models are interpreted in the same way as those involving fewer components.

For a four-component mixture, the quadratic model has 10 terms, whose coefficients can be estimated using a {4, 2} simplex lattice design, illustrated by the tetrahedron in Fig. 7.10a. Each face of the tetrahedron has an array of points that is equal to the one used for fitting the quadratic model to three-component mixtures. The total number of points in the {4, 2} lattice is equal to the number of terms in the quadratic model. In general, for q-component mixtures, the number of runs in a $\{q, 2\}$ simplex lattice design is equal to the number of terms contained in the expression for the quadratic model. To fit the quadratic model, therefore, this design is the most economical.

The special cubic model for four-component mixtures has 14 terms, and its coefficients can be estimated from the design shown in Fig. 7.10b. The points on each face now reproduce the arrangement corresponding to the simplex centroid design, which we used to fit the special cubic model for three-component mixtures.

If we wish to add a fourth-order term to the model, $b_{1234}x_1x_2x_3x_4$, we need to perform at least one more run, with a mixture of composition specified by the central point of the tetrahedron, $x_1 = x_2 = x_3 = x_4 = \frac{1}{4}$. A black star indicates this point in Fig. 7.10b. In general, models of this type for q components are defined by the expression

$$\hat{y} = \sum_{i=1}^{q} b_i^* x_i + \sum_{i<j}^{q} \sum_{j}^{q} b_{ij}^* x_i x_j + \sum_{i<j}^{q} \sum_{j<k}^{q} \sum_{k}^{q} b_{ijk}^* x_i x_j x_k + \cdots + b_{12\ldots q}^* x_1 x_2 \ldots x_q$$

and have a total of $2^q - 1$ terms, whose coefficients can be determined using the $2^q - 1$ points of a simplex centroid design. The geometrical representation of such a design, however, would require a $q - 1$ dimensional space.

7A. Applications

7A.1. Solvent influence on Fe(III) ion complexation

In aqueous solution, the Fe(III) ion has a behavior that varies a great deal with the conditions of the dissolving medium, owing to its ability to

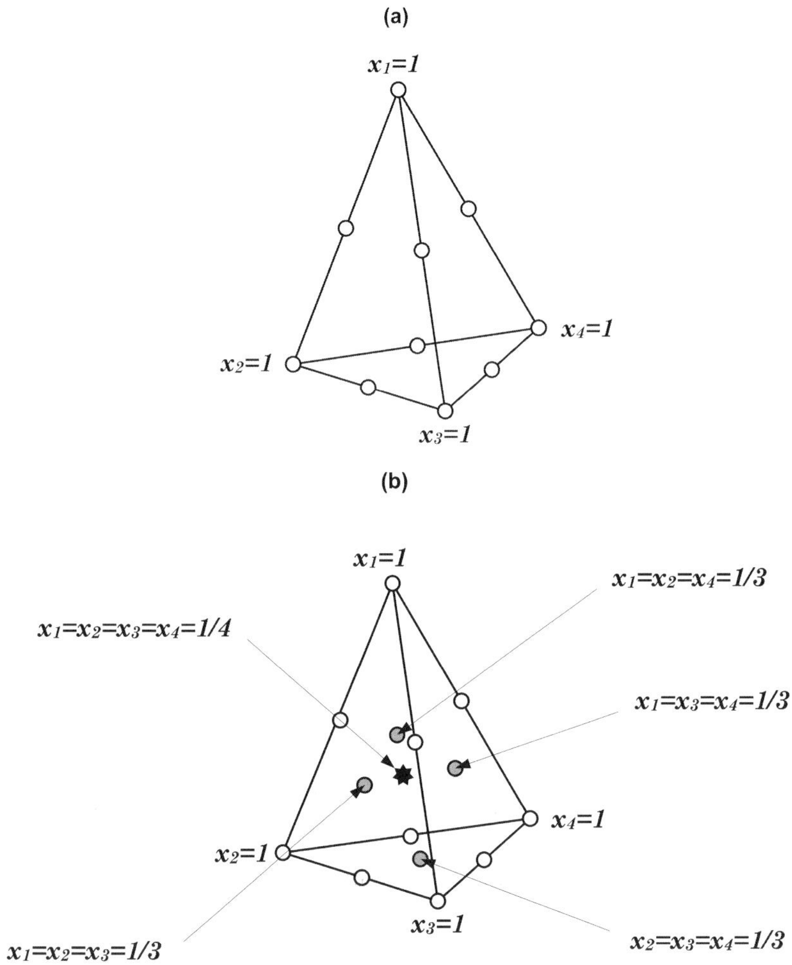

Fig. 7.10. Experimental designs for studying four-component mixtures. (a) Simplex lattice design. (b) Simplex centroid design.

form different complexes and its tendency to suffer hydrolysis, even in acidic solutions. In the presence of thiocyanate ions, the iron (III) ion produces a red-colored solution containing several Fe(III) complexes with the SCN^{-1} ion, produced by this general chemical reaction:

$$[Fe(H_2O)_6]^{3+} + nSCN^- \rightarrow [Fe(SCN)_n(H_2O)_{6-n}]^{3-n} + nH_2O.$$

To study solvent effects on this complexation reaction, Bruns et al. (1996) prepared 16 ternary mixtures of water, ethanol and acetone, to which fixed quantities of Fe(III) and SCN^- ions were added. The mixture compositions were specified by the approximately hexagonal design of Table 7A.1, where the concentrations are expressed as proportions of

Table 7A.1
The approximately hexagonal design for studying the complexation of Fe(III) with thiocyanate ions. x_1, x_2 and x_3 are the proportions, in pseudocomponents, of water, ethanol and acetone, respectively

Run	x_1	x_2	x_3	Absorbance	
1	0.667	0.000	0.333	0.411	0.412
2	0.333	0.667	0.000	0.612	0.607
3	0.000	0.333	0.667	0.614	0.605
4	0.667	0.333	0.000	0.451	0.450
5	0.000	0.667	0.333	0.693	0.688
6	0.333	0.000	0.667	0.456	0.464
7	0.555	0.222	0.222	0.461	0.455
8	0.222	0.555	0.222	0.608	0.605
9	0.222	0.222	0.555	0.521	0.531
10	0.111	0.444	0.444	0.607	0.615
11	0.444	0.111	0.444	0.468	0.457
12	0.444	0.444	0.111	0.520	0.524
13	0.333	0.333	0.333	0.533	0.534
14	0.778	0.111	0.111	0.412	0.403
15	0.111	0.111	0.778	0.528	0.519
16	0.111	0.778	0.111	0.682	0.699

pseudocomponents. The concentration of the complexes was monitored by recording the absorbance of the solutions at 623 nm. All runs were performed in duplicate.

Fitting the linear, quadratic and special cubic models to the data in the table, we arrive at the following equations:

$$\hat{y} = \underset{(\pm 0.006)}{0.322}\, x_1 + \underset{(\pm 0.006)}{0.754}\, x_2 + \underset{(\pm 0.006)}{0.531}\, x_3;$$

$$\hat{y} = \underset{(\pm 0.008)}{0.357}\, x_1 + \underset{(\pm 0.008)}{0.791}\, x_2 + \underset{(\pm 0.006)}{0.518}\, x_3 - \underset{(\pm 0.029)}{0.205}\, x_1x_2 - \underset{(\pm 0.029)}{0.012}\, x_1x_3$$
$$- \underset{(\pm 0.029)}{0.017}\, x_2x_3;$$

$$\hat{y} = \underset{(\pm 0.011)}{0.355}\, x_1 + \underset{(\pm 0.011)}{0.789}\, x_2 + \underset{(\pm 0.011)}{0.516}\, x_3 - \underset{(\pm 0.047)}{0.194}\, x_1x_2 - \underset{(\pm 0.047)}{0.001}\, x_1x_3$$
$$- \underset{(\pm 0.047)}{0.006}\, x_2x_3 - \underset{(\pm 0.168)}{0.051}\, x_1x_2x_3.$$

The analyses of variance results for the three models are given in Table 7A.2. The linear model shows lack of fit. The other two do not. This is due to the inclusion of the interaction term between the first and second components (water and alcohol), since the other binary interactions and

Table 7A.2
ANOVA for the models fitted to the data in Table 7A.1

Model	MS_{lof}/MS_{pe}	ν_{lof}	ν_{pe}	$F_{lof,pe}$ (95%)	R^2
Linear	7.45	13	16	2.40	98.6%
Quadratic	1.70	10	16	2.49	99.6%
Special cubic	1.87	9	16	2.54	99.6%

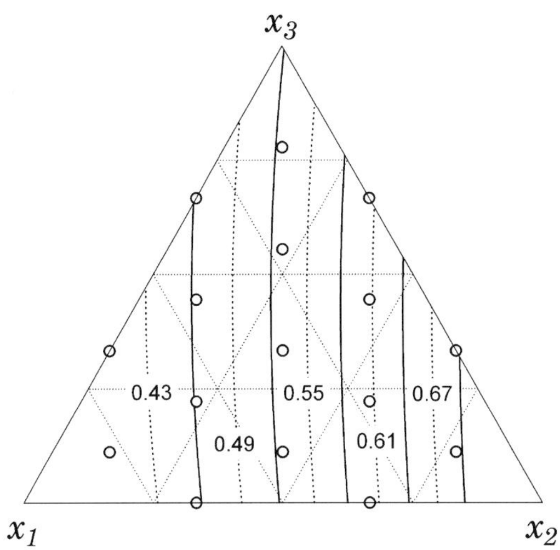

Fig. 7A.1. Response contour curves for the quadratic model fitted to the Fe(III) complexation data.

the cubic term are not statistically significant. When we add the $x_1x_2x_3$ term to the quadratic model, we take a degree of freedom from the lack of fit mean square, and actually worsen the fit, as shown by the values in the second column of the table. Since the design has 16 different combinations of levels, we could fit models with even more terms, such as the full cubic model. Doing this, however, we would discover that none of the new terms is significant. This should come as no surprise, since the quadratic model already provides a satisfactory fit.

In short, we should stay with the quadratic model, whose contour curves are shown in Fig. 7A.1. The most intense absorbance values are obtained in the vicinity of the lower right vertex, which corresponds to ethanol-rich mixtures. The absence of significant interactions involving the third component (acetone) is reflected in the almost vertical contour curves. The graph of the predicted responses against the observed ones (Fig. 7A.2) visually confirms the quality of the fit. Note that the responses are stratified in five groups, corresponding to the five columns of points in Fig. 7A.1.

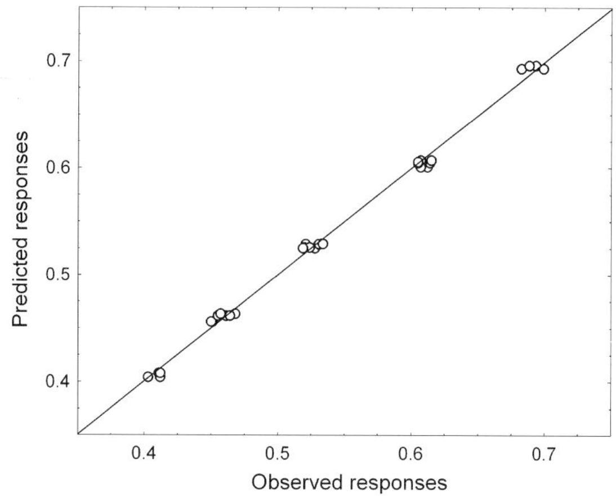

Fig. 7A.2. Responses predicted by the quadratic model plotted against the observed responses.

7A.2. Tensile strength of polymeric materials

Preparing blends — that is, mixtures — is one of the ways in which researchers try to optimize the properties of polymeric products. Among the principal properties described in patents of such materials are resistance to impact, ease of processing and tensile strength. In an attempt to maximize the tensile strength of blends of poly(vinylidene fluoride), poly(methyl metacrylate) and polystyrene — PVDF, PMMA and PS, respectively — Siqueira et al. (1993) obtained the results in Table 7A.3. The first 16 runs in the table correspond to a simplex centroid design with 9 replicate measurements, whereas in runs 17–20 the PVDF:PS proportion is maintained equal to 1, and the quantity of PMMA is varied.

Proceeding as in the last example, we arrive at the following fitted models:

$$\hat{y} = \underset{(\pm 5.99)}{44.25}\,x_1 + \underset{(\pm 4.66)}{29.68}\,x_2 + \underset{(\pm 5.06)}{19.65}\,x_3;$$

$$\hat{y} = \underset{(\pm 5.10)}{49.72}\,x_1 + \underset{(\pm 2.96)}{20.79}\,x_2 + \underset{(\pm 3.62)}{18.66}\,x_3 + \underset{(\pm 20.22)}{61.77}\,x_1x_2$$

$$- \underset{(\pm 14.39)}{36.89}\,x_1x_3 + \underset{(\pm 14.38)}{57.89}\,x_2x_3;$$

$$\hat{y} = \underset{(\pm 4.44)}{51.09}\,x_1 + \underset{(\pm 2.56)}{21.32}\,x_2 + \underset{(\pm 3.14)}{19.34}\,x_3 + \underset{(\pm 20.45)}{36.21}\,x_1x_2 - \underset{(\pm 13.29)}{48.18}\,x_1x_3$$

$$+ \underset{(\pm 13.06)}{48.22}\,x_2x_3 + \underset{(\pm 97.87)}{234.99}\,x_1x_2x_3.$$

Table 7A.3
Design for studying polymeric mixtures of PVDF, PMMA and PS (see text)

Run	x_{PVDF}	x_{PMMA}	x_{PS}	Tensile strength (MPa)
1	1.0	0.0	0.0	51.2
2	0.0	1.0	0.0	20.0
3	0.0	0.0	1.0	20.2
4	0.5	0.5	0.0	44.8
5	0.5	0.0	0.5	23.5
6	0.0	0.5	0.5	35.7
7	0.333	0.333	0.333	45.4
8	0.0	1.0	0.0	20.8
9	0.0	1.0	0.0	23.0
10	0.0	0.0	1.0	18.6
11	0.0	0.5	0.5	23.0
12	0.0	0.5	0.5	38.0
13	0.5	0.0	0.5	28.3
14	0.5	0.0	0.5	22.6
15	0.5	0.0	0.5	25.4
16	0.333	0.333	0.333	46.5
17	0.490	0.020	0.490	21.5
18	0.475	0.050	0.475	23.5
19	0.450	0.100	0.450	35.1
20	0.400	0.200	0.400	33.6

Table 7A.4
ANOVA results for the models fitted to the data of Table 7A.3

Model	MS_{lof}/MS_{pe}	ν_{lof}	ν_{pe}	$F_{lof,pe}$ (95%)	R^2(%)
Linear	9.50	8	9	3.23	28.3
Quadratic	2.46	5	9	3.48	82.0
Special cubic	1.44	4	9	3.63	87.6

The analysis of variance shows that the linear model is unsatisfactory, and that the other two models do not show lack of fit (Table 7A.4). This time, however, the special cubic model seems superior. The explained variation is larger, the MS_{lof}/MS_{pe} ratio is smaller, and the cubic term is statistically significant. The contour curves for the cubic model are presented in Fig. 7A.3. The largest tensile strength values are obtained close to the base of the triangle, toward the left vertex, which corresponds to a blend that is predominantly PVDF, with little or no polystyrene.

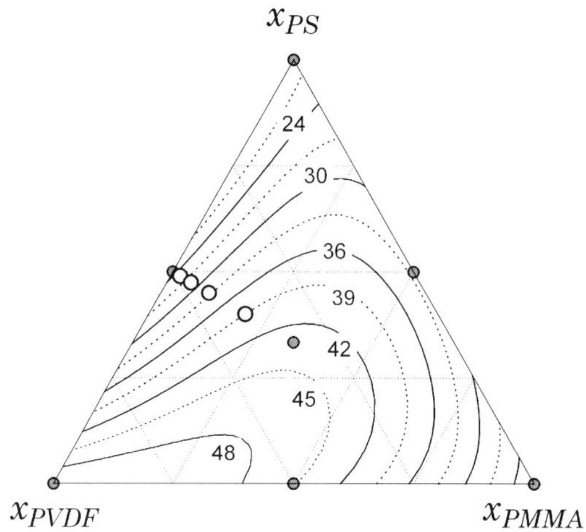

Fig. 7A.3. Response contour curves for the special cubic model fitted to the tensile strength data.

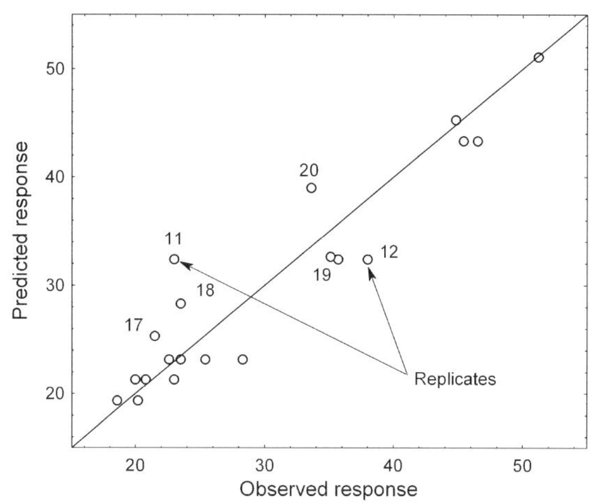

Fig. 7A.4. Observed against predicted responses for the special cubic model.

Unfortunately, this fit is still far from being satisfactory, as we can see from Fig. 7A.4, the plot of the predicted against the observed responses. In the first place, the predictions for three of the four runs that did not belong to the simplex design (17, 18 and 20) are significantly overestimated, indicating that the model is not accurate in this region. But the most worrying aspect is that points 11 and 12 are *replicate* runs, and the difference between their responses is half the difference between the extreme values of the whole table. Since the measure of pure error is calculated from replicate runs, this huge variation inflates MS_{pe} and

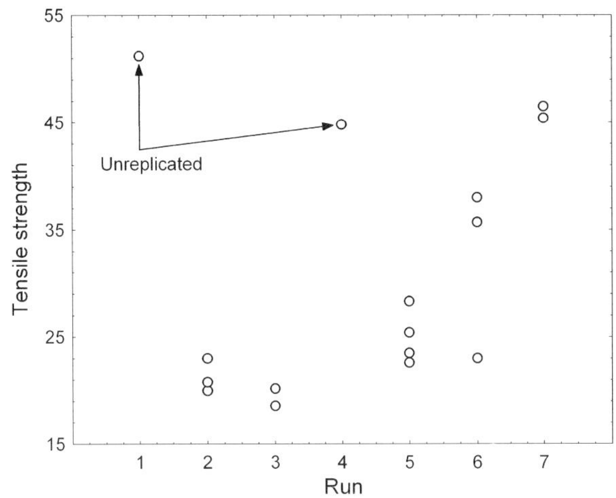

Fig. 7A.5. Responses of the seven runs of the centroid simplex model fitted to the tensile strength data. Is something amiss with any replicate value?

makes the models look as though they do not suffer from lack of fit when in fact they do.

Fig. 7A.5 shows all the response values of the simplex design as a function of run number. Only runs 1 and 4 have not been repeated. Can you detect something strange in the responses of the other runs? Does any value seem to be anomalous?

Surely you noticed that the response for one of the three replicates of run no. 6 is much lower than the other two. If this value is removed from the data as a suspected outlier, the pure error mean square drops from 17.4 to only 3.6. The downside of this decision is that in comparison with the smaller error value even the special cubic model shows lack of fit. This suggests that a more extensive model is needed to adequately represent these data.

This example illustrates once again the importance of performing careful graphical analyses of any fitted model. If we had based our judgment solely on the ANOVA values, it would be very difficult to suspect that something was wrong with the modeling process in this case. Despite these statistical problems, however, the researchers were able to interpret the data, using chemical information relevant to this complex polymeric system.

7A.3. Cr(VI) catalytic determination

In a full factorial design, the runs at all possible combinations of the selected factor levels should be performed in random order, as a preventive measure against bias in the response values. Sometimes this is not convenient, because the levels of some factors are harder to change than the levels of others. An alternative, in such cases, is to perform a split-plot design instead of a full factorial. In a split-plot design, the levels of some factors are varied while the levels of the others — those more difficult to change — are kept fixed at a certain combination.

In this example, the researchers wanted to develop an analytical method to determine the concentration of the Cr(VI) ion, based on the catalytic reaction of o-dianisidine with H_2O_2, in an acidic medium. In the original study, a system controlled by the concentrations of the three reagents (HCl, o-dianisidine and H_2O_2) and by the solvent medium (a mixture of water, acetone and N,N-dimethylformamide) was optimized using a split-plot design (Reis et al., 1998). In Table 7A.5, we present only the results for the variations of the proportions of the three solvents, for a certain fixed combination of the reagent concentrations. x_1, x_2 and x_3 are the proportions, in pseudocomponents, corresponding to water, acetone and *N*, *N*-dimethylformamide, respectively. The response is the absorbance, whose value we wish to maximize. All the runs were conducted in duplicate.

The linear model shows lack of fit. The quadratic and special cubic models are represented by the expressions

$$\hat{y} = \underset{(\pm 0.091)}{\mathbf{1.036}}\, x_1 + \underset{(\pm 0.026)}{\mathbf{0.265}}\, x_2 + \underset{(\pm 0.026)}{0.048}\, x_3 + \underset{(\pm 0.211)}{0.106}\, x_1x_2 - \underset{(\pm 0.211)}{\mathbf{0.555}}\, x_1x_3 - \underset{(\pm 0.116)}{0.202}\, x_2x_3 \quad \textit{and}$$

$$\hat{y} = \underset{(\pm 0.095)}{\mathbf{1.106}}\, x_1 + \underset{(\pm 0.025)}{\mathbf{0.275}}\, x_2 + \underset{(\pm 0.025)}{\mathbf{0.056}}\, x_3 - \underset{(\pm 0.235)}{0.110}\, x_1x_2 - \underset{(\pm 0.235)}{\mathbf{0.771}}\, x_1x_3 - \underset{(\pm 0.124)}{\mathbf{0.304}}\, x_2x_3 + \underset{(\pm 0.782)}{1.330}\, x_1x_2x_3,$$

Table 7A.5
Design for the catalytic determination of Cr(VI). x_1, x_2 and x_3 are the pseudocomponent proportions of water, acetone and N,N-dimethylformamide

Run	x_1	x_2	x_3	Absorbance	
1	0.625	0.375	0.000	0.810	0.777
2	0.000	1.000	0.000	0.279	0.288
3	0.000	0.000	1.000	0.054	0.046
4	0.625	0.000	0.375	0.507	0.498
5	0.313	0.687	0.000	0.474	0.486
6	0.000	0.500	0.500	0.115	0.077
7	0.313	0.000	0.687	0.277	0.232
8	0.313	0.344	0.344	0.409	0.370
9	0.468	0.266	0.266	0.548	0.492
10	0.156	0.211	0.633	0.192	0.109

where the significant coefficients are shown in bold face. The analysis of variance results are very similar for the two models:

$$\text{Quadratic model} \quad MS_{\text{lof}}/MS_{\text{pe}} = 3.92 \quad (F_{4,10} = 3.48) \quad R^2 = 97.8\%,$$
$$\text{Cubic model} \quad MS_{\text{lof}}/MS_{\text{pe}} = 3.66 \quad (F_{3,10} = 3.71) \quad R^2 = 98.2\%.$$

Although the coefficient of the cubic term is not significant at the 95% confidence level, the special cubic model seems to be slightly superior to the quadratic one, judging from the residual plots shown in Figs. 7A.6a and b. On the other hand, it might be argued that this is too little to show for the additional degree of freedom it takes. The contour curves of the cubic model — which anyway are almost identical to those for the quadratic model — are shown in Fig. 7A.7. The larger absorbance values are found close to the left vertex, that is, where the solvent mixture is rich in water.

7A.4. Polymer blend conductivity

Electrolytic polymeric films used for manufacturing miniature electrochemical devices can be made by incorporating a relatively high salt concentration into a flexible polymer. To reduce the risk of crystallization, and also to increase segmented mobility (which can bring about higher conductivity), one usually adds a plasticizing agent. Almost all the optimization studies reported in the literature for these systems are based on univariate strategies, where the proportion of one ingredient — normally the salt — is maintained constant. Recently, a multivariate optimization was reported for a mixture of lithium perchlorate and poly(ethylene oxide) (POE), with ethylene carbonate as a plasticizing agent (Silva et al., 2000). The results are presented in Table 7A.6, where c_1, c_2 and c_3 are the real proportions of polymer, salt and plasticizer, respectively. Two responses were measured: the ionic conductivity of the blend (here the logarithmic scale is used) and the glass transition temperature, T^*. Figs. 7A.8a and b show the design pattern in terms of the true component values and in terms of the pseudocomponent values.

Trying to fit the usual models to the data in the table, we find that this system is more complex than the others we have seen in this chapter. All of the simpler models presented lack of fit, and only the *full* cubic model proved satisfactory for both responses. In terms of the original quantities, these models are given by the equations below, where only the significant terms (at the 95% confidence level) are retained.

$$\text{Log(cond)} = -9c_1 + 327c_2 - 601c_1c_2 - 920c_2c_3 + 1090c_1c_2c_3$$
$$+334c_1c_2(c_1 - c_2)$$
$$T^* = -9080c_3 + 15732c_1c_3 - 7589c_1c_3(c_1 - c_3) - 10846c_2c_3(c_2 - c_3).$$

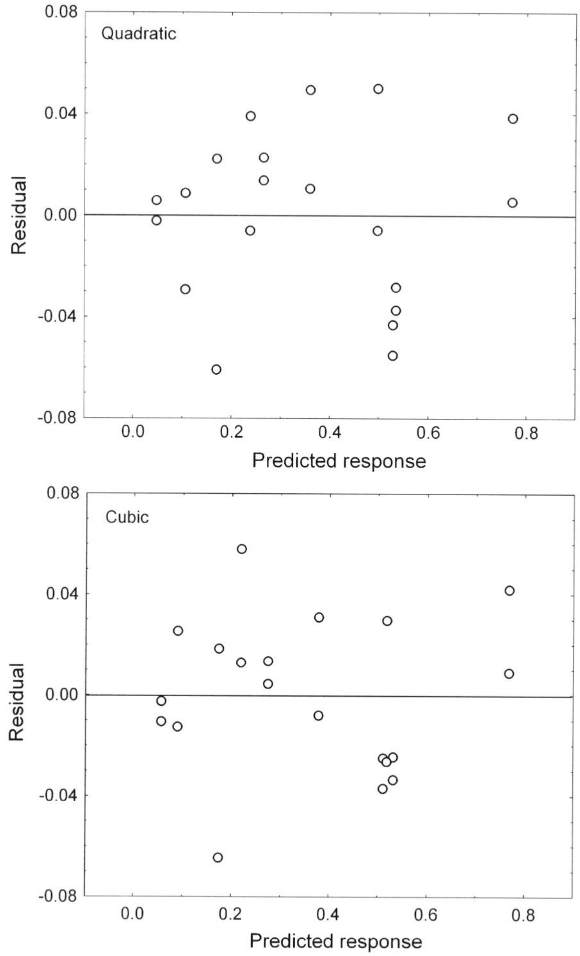

Fig. 7A.6. Residuals left by the two models fitted to the Cr(VI) data.

The explained variations for these models are 99.27% and 93.99%, respectively. Their contour curves are shown in Figs. 7A.9a and b, where the vertices represent pseudocomponents. If we would like to obtain higher conductivities (i.e., less negative logarithms) we should move toward to the right side of the graph in Fig. 7A.9a, which corresponds to increasing the proportion of salt in the blend. To raise the transition temperature, however, the most favorable region is on the opposite side. This means that the simultaneous maximization of both responses is not feasible. We must be content with a compromise solution, or favor the optimization of one response at the cost of tolerating undesirable values from the other.

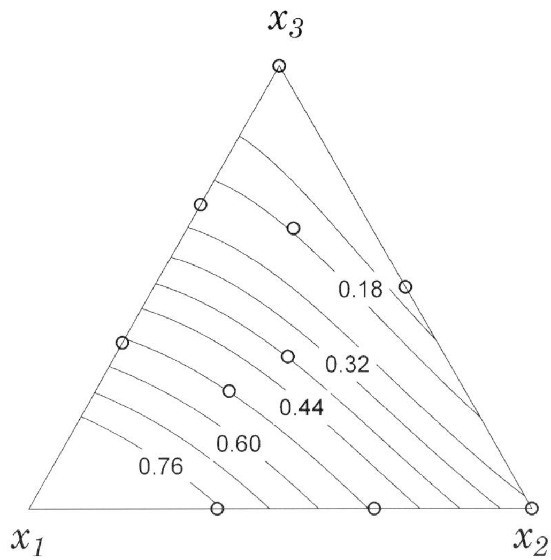

Fig. 7A.7. Contour curves for the special cubic model.

Table 7A.6
Ternary mixture design for the polymer blend conductivity study. c_1, c_2 and c_3 are the real proportions of polymer, salt and plasticizer, respectively

Run	c_1	c_2	c_3	Log (cond)	T[a] (°C)
1	0.86	0.07	0.07	−6.000	−47
2	0.75	0.18	0.07	−4.699	−48
3	0.64	0.29	0.07	−4.523	−67
4	0.50	0.29	0.21	−3.523	−67
5	0.50	0.21	0.29	−3.398	−68
6	0.64	0.07	0.29	−4.398	−64
7	0.75	0.07	0.18	−6.000	−45
8	0.66	0.17	0.17	−4.155	−56
9	0.66	0.17	0.17	−4.097	−63
10	0.76	0.12	0.12	−5.000	−51
11	0.76	0.12	0.12	−4.699	−50
12	0.60	0.25	0.15	−4.155	−71
13	0.60	0.25	0.15	−4.000	−69
14	0.60	0.15	0.25	−3.699	−50
15	0.60	0.15	0.25	−3.699	−49

[a] Transition temperatures were corrected for materials with crystallinity above 20%.

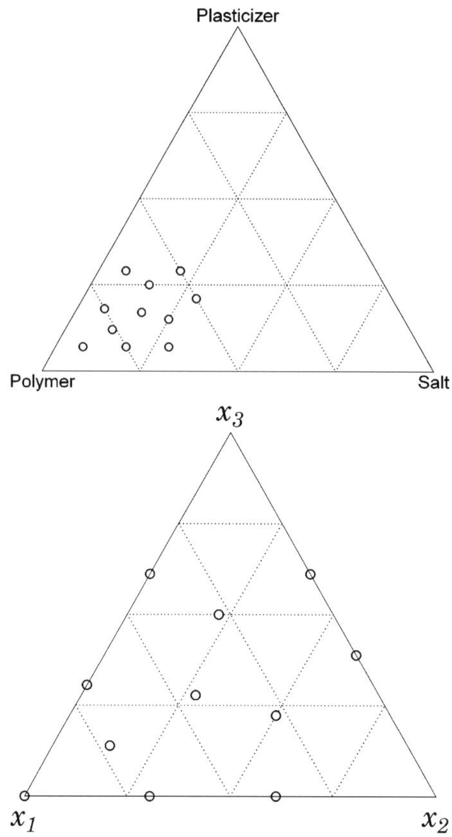

Fig. 7A.8. The mixture design of Table 7A.6, as real composition values and as pseudocomponents.

7A.5. The proof of the pudding is not in the eating

Industrial food products are more or less complex mixtures, whose properties should be recognized and appreciated by consumers. In the development of a new product, engineers and technical staff must take into consideration its chemical, compositional, structural and textural properties, which are determining factors for the product's marketability. In addition, as in any industry, economic aspects must be considered, to decide if the developed product is to be commercialized or not.

Sugar, starch and powdered milk are components normally present in large proportions in the preparation of puddings. The proportion of starch is highly influential on the textural and structural properties of these confections. Braga Neto (1998) investigated how five pudding characteristics are affected by variations in the proportions of the three main

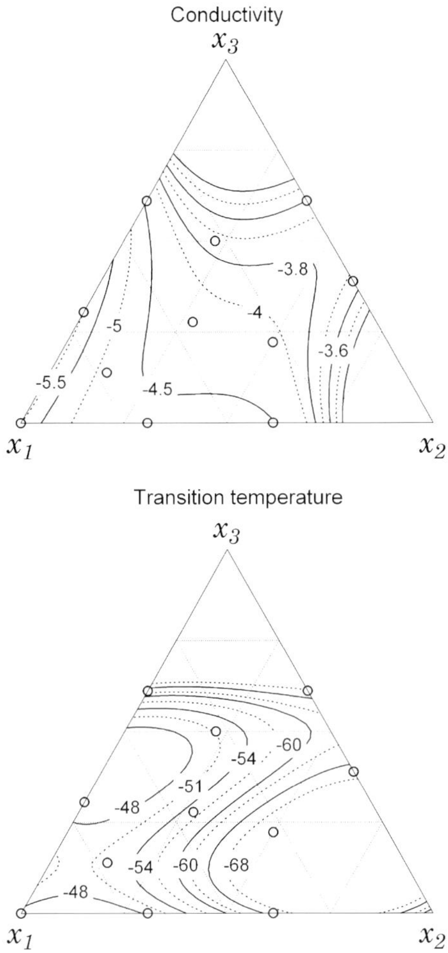

Fig. 7A.9. Contour curves for the two full cubic models fitted to the data of Table 7A.6.

ingredients and obtained the results of Table 7A.7, where the units are omitted to save space.

The coefficients of the statistically significant terms of the fitted models for the five responses are given in Table 7A.8. The special cubic model is the one that best fits all properties except cohesivity, for which the quadratic model is sufficient. Of all fitted models, only that for firmness shows some lack of fit, but the number of level combinations in the design is not sufficient for fitting a full cubic model. The coefficient values clearly indicate that starch (x_1) is the most important component, but its effect depends on the proportions of the other two ingredients, sugar and powdered milk.

Table 7A.7
Data from the pudding formulation study

Order	Starch (x_1)	Powdered milk (x_2)	Sugar (x_3)
	Pseudocomponent		
4	0.176	0.000	0.824
7	0.000	0.000	1.000
9	0.000	0.412	0.588
6	0.000	0.824	0.176
2	0.176	0.412	0.412
3	0.088	0.824	0.088
5	0.088	0.000	0.912
1	0.088	0.412	0.500
8	0.176	0.824	0.000

	Firmness			Fracturability			Adhesivity			Cohesivity			Elasticity		
4	42.0	41.0	45.0	87.5	82.0	92.0	1.50	1.80	1.60	0.37	0.38	0.37	2.00	1.85	2.00
7	11.0	11.0	10.0	11.7	11.5	10.5	0.53	0.55	0.65	0.70	0.82	0.97	2.28	2.38	2.25
9	10.0	10.5	11.0	7.0	8.5	9.5	0.10	0.15	0.10	0.94	0.85	0.89	1.65	1.78	1.65
6	9.0	9.5	10.0	6.0	7.0	7.0	0.10	0.10	0.10	0.87	1.00	0.96	1.50	1.75	2.00
2	34.0	31.0	32.5	44.0	45.0	47.5	2.40	2.80	2.20	0.49	0.57	0.50	2.30	2.50	2.20
3	24.0	24.5	26.0	19.0	23.0	22.0	1.17	1.30	1.25	0.71	0.55	0.69	2.32	2.27	2.29
5	28.0	25.5	28.5	39.0	45.0	42.5	1.00	0.90	0.90	0.42	0.44	0.47	2.28	2.05	1.98
1	17.0	18.5	20.0	20.0	22.5	21.5	1.00	1.30	0.95	0.59	0.60	0.63	2.45	2.48	2.43
8	48.0	49.5	50.0	45.0	51.0	50.0	2.20	2.70	2.30	0.60	0.52	0.51	2.05	2.18	2.20

Table 7A.8
Coefficients of significant terms in the fitted models. Pseudocomponents as in Table 7A.7

	x_1	x_2	x_3	x_1x_2	x_1x_3	x_2x_3	$x_1x_2x_3$
Firmness	273.8	8.5	11.1	—	—	—	−478.2
Fracturability	876.6	7.8	10.4	−768.8	−530.1	—	−693.1
Adhesivity	28.1	—	0.6	−19.1	−26.1	−1.5	22.6
Cohesivity	10.8	0.9	0.8	−14.5	−15.1	—	—
Elasticity	−19.3	1.9	2.2	27.2	23.6	−1.6	22.4

The quality of the fit for the five properties is illustrated in the plots of the predicted against the observed responses (Figs. 7A.10a–e). The firmness and fracturability values are predicted quite well, the adhesivity values show a bit more scatter, but the large spreads of the replicate values for cohesivity and for elasticity make the fits for these two properties appear less impressive.

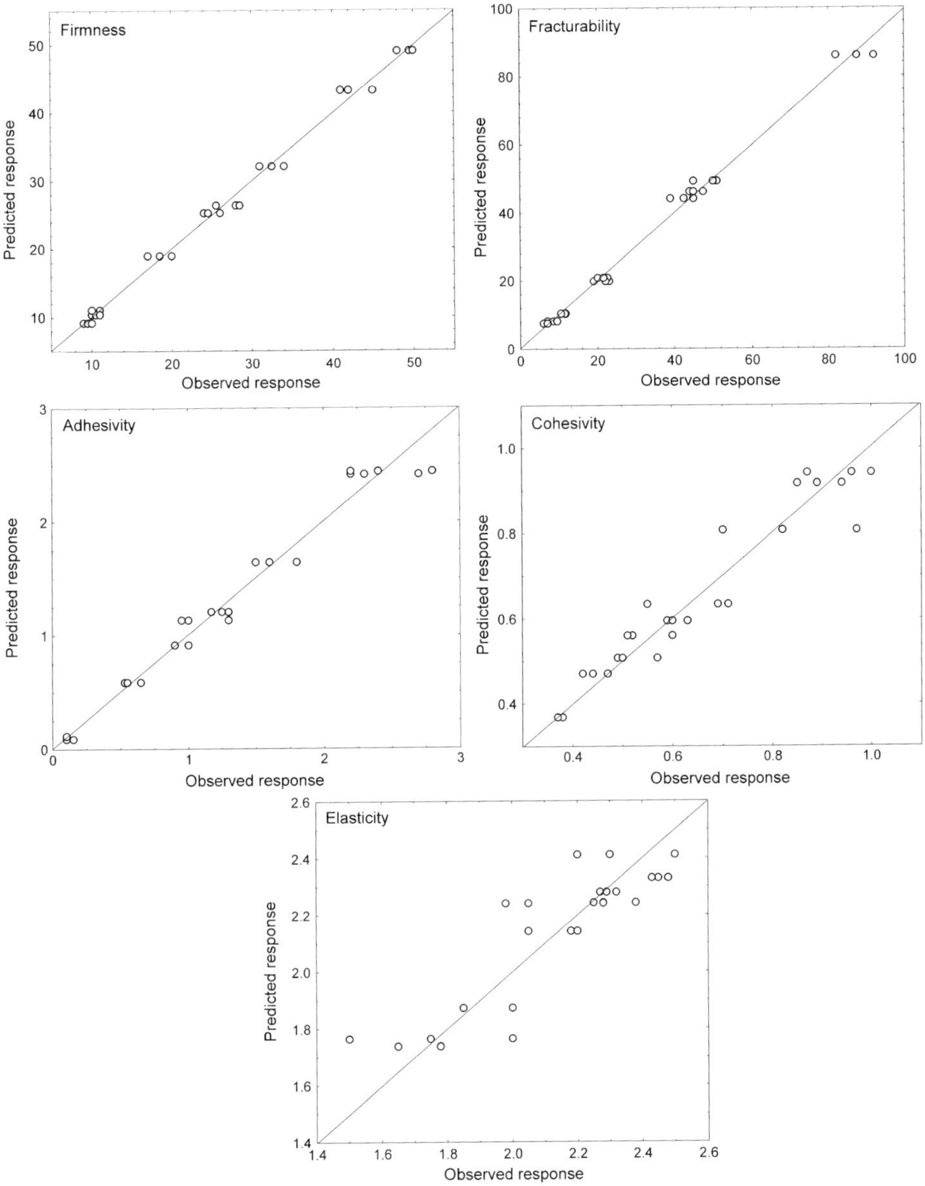

Fig. 7A.10. Comparison of the predicted and observed responses for the five pudding characteristics.

7A.6. Designing new ceramic materials

Silicon carbide (SiC) is an important structural ceramic material used for making abrasives, refractory materials, automotive engine components

Table 7A.9
Oxide component proportions, pseudocomponent values, density (d) and mass loss (ML) values for duplicate runs of the simplex centroid mixture design used for studying SiC-based ceramics

Run	$\%SiO_2$	$\%Al_2O_3$	$\%Y_2O_3$	x_1	x_2	x_3	d[a]	ML[b]
1	60	20	20	1.00	0.00	0.00	89.8726	7.7235
2	20	60	20	0.00	1.00	0.00	94.8675	3.1728
3	20	20	60	0.00	0.00	1.00	94.8015	7.9008
4	40	40	20	0.50	0.50	0.00	92.2523	5.6242
5	40	20	40	0.50	0.00	0.50	94.1623	9.2139
6	20	40	40	0.00	0.50	0.50	94.8109	4.7092
7	33	33	33	0.33	0.33	0.33	95.2047	6.5610
8	60	20	20	1.00	0.00	0.00	90.8765	7.7866
9	20	60	20	0.00	1.00	0.00	96.6749	3.6454
10	20	20	60	0.00	0.00	1.00	94.5585	7.8855
11	40	40	20	0.50	0.50	0.00	92.0386	5.3348
12	40	20	40	0.50	0.00	0.50	94.1119	8.6163
13	20	40	40	0.00	0.50	0.50	95.4626	4.4978
14	33	33	33	0.33	0.33	0.33	95.6606	6.3458

[a] Percentage of the theoretical density.
[b] Percent mass loss.

and cutting tools. The properties of SiC-based ceramics can be modified by sintering them with oxide additives. A simplex centroid mixture design with duplicate runs was performed to determine how mixtures of the oxides SiO_2, Al_2O_3 and Y_2O_3 affect the density, stability, hardness and other properties of silicon-based ceramics (Marchi, 2003). Table 7A.9 contains the real proportions of the three oxide additives, the corresponding pseudocomponent values and the density and mass loss values for ceramic samples sintered at 1950 °C for 1 h.

The ANOVA results for the quadratic and special cubic models for the density are given in Table 7A.10. The calculated $MS_{\text{lof}}/MS_{\text{pe}}$ ratio of 8.34 for the quadratic model is larger than the $F_{1,7,\ 95\%}$ value of 5.59, indicating significant lack of fit at this confidence level. The special cubic model cannot be tested for lack of fit because there are no degrees of freedom left for this. Assuming though that it is adequate (this seems reasonable, since the lack of fit of the quadratic model is no longer significant at the 99% confidence level) and using the square root of the pure error mean square as an estimate of experimental error, the special cubic model for the density can be expressed as

$$d = \underset{(\pm 0.42)}{90.38}\,x_1 + \underset{(\pm 0.42)}{95.77}\,x_2 + \underset{(\pm 0.42)}{94.68}\,x_3 - \underset{(\pm 2.07)}{3.71}\,x_1x_2$$

$$+ \underset{(\pm 2.07)}{6.44}\,x_1x_3 - \underset{(\pm 2.07)}{0.36}\,x_2x_3 + \underset{(\pm 14.59)}{42.13}\,x_1x_2x_3.$$

Table 7A.10
ANOVA results for the quadratic and special cubic models for the density of the sintered ceramic samples

Source of variation	*SS*	df.	*MS*	*F*
Regression	44.08 (*47.05*)	5 (*6*)	8.82 (*7.84*)	12.78 (*21.78*)
Residual	5.49 (*2.51*)	8 (*7*)	0.69 (*0.36*)	
Lack of fit	2.99 (*0.00*)	1 (*0*)	2.99 (***)	8.34 (***)
Pure error	2.51	7	0.36	
Total	49.57	13		

Notes: The values for the special cubic model are within parentheses.

The three linear coefficients[75] are similar, but it can be shown that the coefficient for the pseudocomponent proportion x_1 is significantly lower than the other two, at the 95% level. One way of verifying this is calculating the contrast between the value of b_1 and the average of b_2 and b_3. This is given by $C_1 = b_1 - (b_1+b_3)/2 = -4.85$, which has a standard error of 0.52, as you can discover applying the rules for the variance of a linear combination, discussed in Chapter 2. Combining this information with $t_{7,\,95\%} = 2.365$, we find that the confidence interval for the value of C_1 is [–3.62, –6.08]. Since this interval does not include zero, we conclude that the blending coefficient for pseudocomponent 1 is indeed smaller. A similar calculation for the contrast $b_2 - b_3$ shows that these two coefficients are statistically indistinguishable.[76]

Since the special cubic model could not be tested for lack of fit, confirmatory experiments were carried out to test the model's validity. Duplicate results of 92.2641 and 92.1490 were obtained using 50%, 25% and 25% SiO_2, Al_2O_3 and Y_2O_3, respectively. Converting these proportions into the respective x_i values and feeding them into the model equation, we obtain a predicted density value of 92.33, with a prediction confidence interval of [90.78, 93.88], in excellent agreement with the observed values and confirming the validity of the special cubic model.

The ANOVA results for fitting linear and quadratic models to the mass loss data are shown in Table 7A.11. The linear model has a large MS_R/MS_r ratio but it also has a significant lack of fit, since $MS_{\text{lof}}/MS_{\text{pe}} = 17.01$ is larger than $F_{4,\,7,\,95\%} = 4.12$. The quadratic model shows no significant lack of fit (its $MS_{\text{lof}}/MS_{\text{pe}}$ ratio is less than one) and its MS_R/MS_r value, 187.32, is much larger. Its equation is given by

$$ML = \underset{(\pm 0.16)}{7.75}\, x_1 + \underset{(\pm 0.16)}{3.40}\, x_2 + \underset{(\pm 0.16)}{7.88}\, x_3 - \underset{(\pm 0.72)}{0.24}\, x_1x_2 + \underset{(\pm 0.72)}{4.54}\, x_1x_3 - \underset{(\pm 0.72)}{4.02}\, x_2x_3,$$

[75] They are also called *blending* coefficients.

[76] How about trying your hand at verifying this assertion?

Table 7A.11
ANOVA results for the linear and quadratic models for the mass loss of the sintered ceramic samples

Source of Variation	*SS*	df.	*MS*	*F*
Regression	43.11 (*46.78*)	2 (*5*)	21.55 (*9.36*)	58.15 (*187.3*)
Residual	4.08 (*0.40*)	11 (*8*)	0.37 (*0.05*)	
Lack of fit	3.70 (*0.02*)	4 (*1*)	0.92 (*0.02*)	17.04 (*0.367*)
Pure error	0.380	7	0.054	
Total	47.18	13		

Note: The values for the quadratic model are within parentheses.

where the square root of the residual mean square has been used as an estimate of the experimental error.

Further analysis shows that the blending coefficient of 2, the pseudocomponent rich in Al_2O_3, is significantly different from the other two, whereas these latter appear to have equivalent blending properties for density. Also there is a significant synergistic effect between the pseudocomponents rich in SiO_2 and Al_2O_3 (x_1x_3) and an antagonistic one for those rich in Al_2O_3 and Y_2O_3 (x_2x_3). No significant interaction effect is found for the pseudocomponents rich in SiO_2 and Al_2O_3 (x_1x_2). Evidence of the x_1x_3 interaction effect can be seen in the contour curves of the triangular diagram in Fig. 7A.11. Relatively high-valued contour curves focus on the central region of the $x_1 - x_3$ axis (pseudocomponents rich in SiO_2 and Y_2O_3, respectively), clearly showing the synergistic effect between these two pseudocomponents: the positive x_1x_3 cross term in the quadratic model makes large contributions to the mass loss in this region. The antagonistic effect represented by the significant negative cross term between x_2 and x_3 is harder to see, because the linear coefficients for x_2 and x_3 are very different (3.40 and 7.88, respectively).

7A.7. Improving selectivity in high-performance liquid chromatography

Separating the components of a mixture by reversed-phase high performance liquid chromatography (RP-HPLC) requires a non-polar stationary phase (SP) and a polar mobile phase (MP). In this analytical method, the silica used to fill the chromatographic columns – the SP – is usually coated with a polymer identified by the number of carbon atoms in its alkyl chain: C1, C4, C8 or C18, for example. For the MP, water and an organic solvent are mixed in the proportion required to give an ideal chromatographic strength. The separation process takes place because of the different interactions of each mixture component with the stationary

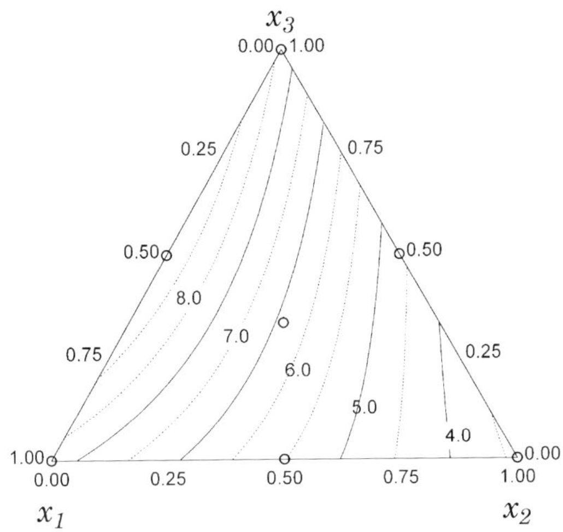

Fig. 7A.11. Contour curves for mass loss as a function of the x_1(60:20:20), x_2 (20:60:20) and x_3 (20:20:60) pseudocomponents (% SiO2 : % Al_2O_3 : %Y_2O_3).

and MPs. Liquid chromatography thus differs from gas chromatography in the sense that its MP is not inert. Besides having the function of leaching the chemical components of the sample through the column, it also interacts with them.

As part of their research in optimizing chromatographic methods, M.C. Breitkreitz and co-workers employed a mixture design to search for the MP composition that would give the best selectivity for a 10-component mixture separation (Breitkreitz et al., 2005). The organic solvents acetonitrile (ACN), methyl alcohol (MeOH) and tetrahydrofuran (THF) were used as water modifiers. Initially, the solvent strength of ACN in water was varied to produce a desirable separation power for all the solutes. Once this composition had been experimentally determined, the water contents of the other two solvents were predicted based on empirical relationships. In this manner, the "pure" components used to specify the design were, in fact, binary mixtures ACN:H_2O, MeOH:H_2O and THF:H_2O.

For the first step of the study, a simplex centroid design was chosen, augmented with three mixtures and with a replicate performed at each point. The experimental runs were carried out in random order. Previous screening experiments had indicated that the best separations would be found in the experimental region defined by the $0.3 \leq \text{ACN} \leq 1.0$; $0 \leq \text{MeOH} \leq 0.7$ and $0 \leq \text{THF} \leq 0.7$ intervals. This region is shown in Fig. 7A.12, together with the points of the design. The compositions of the MP for each point and the responses obtained in the respective runs are given in Table 7A.12. The response is the sum of the chromatographic peak-valley ratios for the 10 mixture components. Since the experimental

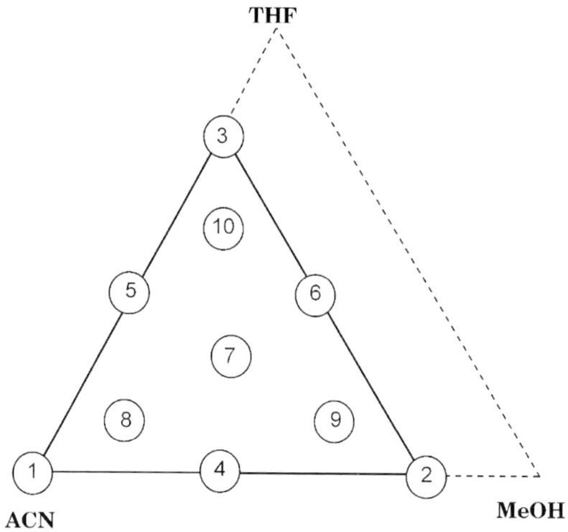

Fig. 7A.12. Simplex centroid mixture design and the experimentally constrained region. The points represent the compositions shown in Table 7A.12.

Table 7A.12
Solvent proportions used in the first design of the HPLC study and the separation responses obtained

	True proportions			Pseudocomponents				
Run	ACN	MeOH	THF	ACN	MeOH	THF	Response	
1	1	0	0	1	0	0	8.035	7.925
2	0.3	0.7	0	0	1	0	7.366	7.460
3	0.3	0	0.7	0	0	1	7.368	7.487
4	0.65	0.35	0	0.5	0.5	0	8.000	8.000
5	0.65	0	0.35	0.5	0	0.5	8.893	8.790
6	0.3	0.35	0.35	0	0.5	0.5	7.600	7.557
7	0.533	0.233	0.233	0.333	0.333	0.333	8.848	8.115
8	0.767	0.117	0.117	0.666	0.166	0.166	8.768	8.475
9	0.416	0.466	0.117	0.166	0.666	0.166	7.963	7.267
10	0.416	0.117	0.466	0.166	0.166	0.666	7.363	8.152

region is constrained, each MP composition can be converted into pseudo-components using the following expressions derived from Eq. (7.23):

$$x_{\text{ACN}} = \frac{c_{\text{ACN}} - 0.3}{0.7}; \quad x_{\text{MeOH}} = \frac{c_{\text{MeOH}}}{0.7}; \quad x_{\text{THF}} = \frac{c_{\text{THF}}}{0.7}.$$

Linear, quadratic and special cubic models were fitted to the data. The linear model shows lack fit, whereas the quadratic and special cubic do not. Since the cubic term is not statistically significant (95% confidence level throughout, as usual), the quadratic model should be preferred. However, since the interactions terms x_1x_2 and x_2x_3 of the full quadratic model are not significant either, we can refit the data to a simpler model where only the interaction x_1x_3 is retained. This has the additional advantage of leaving more degrees of freedom to test the lack of fit. As expected, the reduced model, given by

$$\hat{y} = \underset{(\pm 0.19)}{8.17}\, x_1 + \underset{(\pm 0.16)}{7.54}\, x_2 + \underset{(\pm 0.19)}{7.37}\, x_3 + \underset{(\pm 0.96)}{4.56}\, x_1x_3,$$

yields the best ANOVA results of all, which are presented in Table 7A.13. Note that the pure error sum of squares, 0.889, is quite large, representing almost 30% of the *total* variance, which is 0.297. The plot of the predicted against the observed responses (Fig. 7A.13) shows that

Table 7A.13
ANOVA results for the reduced quadratic model fitted to the data of Table 7A.12

Source of variation	*SS*	df.	*MS*	*F*
Regression	4.143	3	1.381	14.76
Residual	1.497	16	0.094	
Lack of fit	0.608	6	0.101	1.14
Pure error	0.889	10	0.089	
Total	5.640	19	0.297	

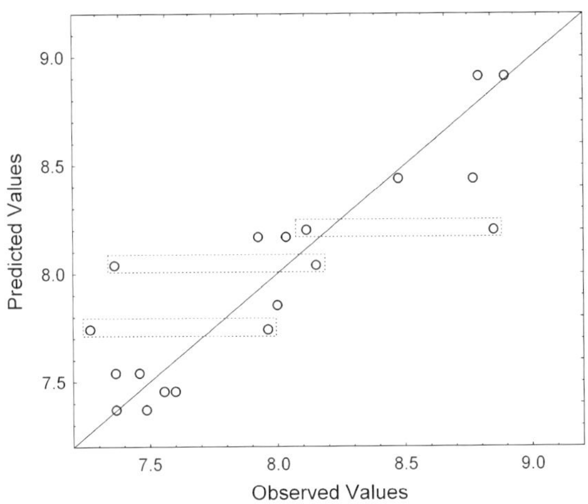

Fig. 7A.13. Predictions of the reduced quadratic model. The boxes highlight replicate response with very high variation.

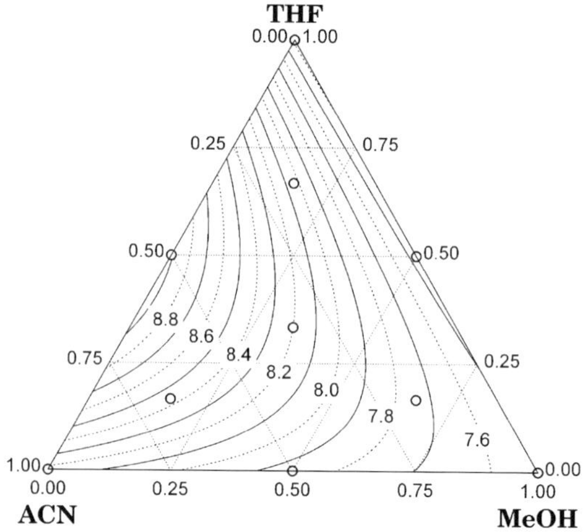

Fig. 7A.14. Contour curves of the response surface for the reduced quadratic model fitted to the data of Table 7A.12.

this is due to abnormally high variation between some replicates. The more drastic cases are highlighted by dashed boxes. The range spanned by each of these pairs of replicates is more than half the range of all the responses obtained in the design. This explains why the predictions look so poor. The contour curves of the response surface for the reduced quadratic model can be seen in Fig. 7A.14.

Despite these shortcomings, we can see that the best response values are situated near points 5 and 8. For a more detailed investigation, the experimental region was then constrained a little bit more, to increase the number of runs in the most favorable region. A new design was conducted, defined by the constraints $0.5 \leq \text{ACN} \leq 1.0$, $0 \leq \text{MeOH} \leq 0.5$ and $0 \leq \text{THF} \leq 0.5$, which amounts to shrinking the pseudocomponent triangle toward the ACN vertex. Since the response surface has already been studied and we know that the quadratic model describes it as well as possible, a seven-run central composite design was chosen for the new region, with four replicates performed at the center point. With this simpler design, the number of experiments was reduced, time saved, and 1 degree of freedom was still left to test lack of fit. Table 7A.14 shows the MP compositions for each point of this design and the responses obtained.

The best model fitted to the new data is again the quadratic model retaining only the interaction x_1x_3, and its equation is practically the same obtained for the first design:

$$\hat{y} = \underset{(\pm 0.20)}{7.53}\, x_1 + \underset{(\pm 0.17)}{7.92}\, x_2 + \underset{(\pm 0.20)}{7.81}\, x_3 + \underset{(\pm 0.91)}{4.58}\, x_1 x_3.$$

Table 7A.14
Solvent proportions used in the second HPLC mixture design and the responses obtained

	True proportions			Pseudocomponents			
Run	ACN	MeOH	THF	ACN	MeOH	THF	Response
1	1	0	0	1	0	0	7.385
2	0.5	0.5	0	0	1	0	7.705
3	0.5	0	0.5	0	0	1	7.751
4	0.75	0.25	0	0.5	0.5	0	8.000
5	0.75	0	0.25	0.5	0	0.5	8.785
6	0.5	0.25	0.25	0	0.5	0.5	7.965
7	0.666	0.1665	0.1665	0.333	0.333	0.333	7.982
7′	0.666	0.1665	0.1665	0.333	0.333	0.333	8.351
7″	0.666	0.1665	0.1665	0.333	0.333	0.333	8.425
7‴	0.666	0.1665	0.1665	0.333	0.333	0.333	8.349

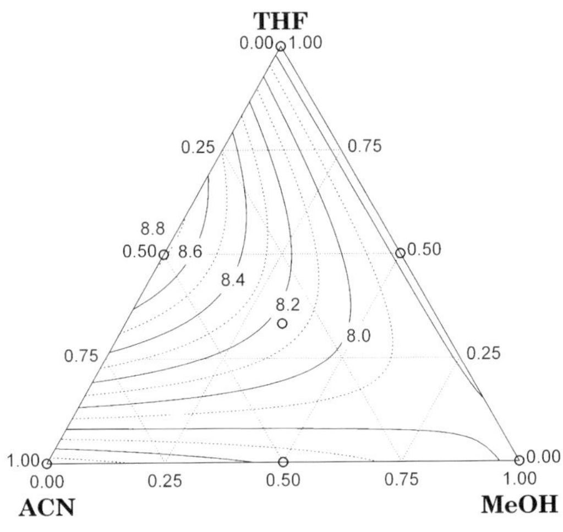

Fig. 7A.15. Contour curves of the response surface for the reduced quadratic model fitted to the data of Table 7A.12.

This model presents no lack of fit and its MS_R/MS_r ratio is 8.96, indicating statistical significance when compared with $F_{3,6} = 4.76$, at the 95% confidence level. Note that the number of degrees of freedom have changed with respect to the first ANOVA, because less replicates were made in the second design.

The contour curves for the second model, shown in Fig. 7A.15, bear a striking resemblance to those of Fig. 7A.14, and point to an optimum composition close to the point ($x_1 = x_3 = 0.50$, $x_2 = 0$), where a response of about 8.8 is expected. In true proportions, Table 7A.12 tells us that this corresponds to a mixture made of three parts ACN and one part THF. In the first design, the closest to this composition are the mixtures of runs 5 (65% ACN : 0% MeOH ; 35% THF) and 8 (76.7% ACN : 11.7% MeOH : 11.7% THF). The average value of the four responses obtained in these runs is 8.73. This is in excellent agreement with the value predicted by the second model and indicates the presence of a plateau in the neighborhood of these points.

CHAPTER 8
Simplex optimization

In the optimization methods discussed in the previous chapters, the system response was expressed as a mathematical function of the factors being optimized. Experimental results were used to obtain numerical values for model parameters. There exists another class of methods that permit us to optimize a system without needing to know, or even postulate, any mathematical relation between the response and the independent variables. In this chapter, we study one of these methods, the sequential simplex, that is widely used in engineering and even enjoys much popularity among analytical chemists. The simplex methods work well in the presence of experimental error and are capable of optimizing systems controlled by a large number of independent variables. Besides this, they do not require the use of significance tests (such as the t- and F-tests), which is a distinct advantage to research workers who have little time to study statistical methods.

The simplex methods, as the very name implies, are based on very simple algorithms that can be very easily implemented on analytical instruments, transforming the optimization of their performance into an automatic procedure. On the other hand, simplex optimization is always sequential since we can only go to the next step after we know the result of the immediately preceding step. Whereas when we are determining a response surface we can perform several experiments at the same time to complete a factorial design, the simplex methods only permit us to do one experiment at a time (that is why they are called *sequential*). This characteristic makes simplex use most convenient for rapid response instruments that are often encountered in analytical chemistry laboratories.

The simplex methods have other limitations that we should take into account when we choose an optimization method. In the first place, they work only with quantitative variables. Qualitative factors, such as catalyst type or the presence/absence of a certain component cannot be a part of simplex optimization, owing to the specific characteristics of its algorithm. Secondly, and for the same reasons, the researcher does not control the levels of the experimental factors. The simplex dictates the factor levels to be used and if the factors are not easily manipulated to attain these conditions, the system cannot be optimized by this method. Finally, we can only optimize one response at a time. If there are several responses of interest, we need to select one in order to carry out the simplex optimization. This last restriction may not be too serious, since sometimes we can circumvent this problem using a linear combination of experimental responses as the response value for the simplex algorithm. These can be weighted according to their relative importance or the desirability function of the Derringer and Suich method can be employed.

The 10th edition of *Webster's Collegiate Dictionary* defines simplex as a spatial configuration of n-dimensions determined by $n+1$ points in a space of dimension equal to or greater than n. In the simplex optimization methods this configuration is a polygon (or its multidimensional equivalent) of $p+1$ vertices, where p is the number of independent variables that we wish to adjust. With two variables, therefore, the simplex is a triangle. With three, a tetrahedron. With four or more, a hyperpolyhedron. The number of factors defines the number of dimensions in which the simplex moves.

There exist several simplex methods. In this chapter, we will discuss three of them, in increasing order of complexity: the basic simplex, the modified simplex and the super-modified simplex. The more sophisticated methods are able to adapt themselves better to the response surface studied. However, their construction requires a larger number of experiments. In spite of this, the modified and super-modified simplexes normally are able to come closer to the maximum (or minimum if this were of interest) with a *total* number of experiments that is smaller than would be necessary for the basic simplex. In this chapter, we will see examples with only two or three variables, so that we can graphically visualize the simplex evolution for instructive purposes. However, the efficiency of the simplex, in comparison with univariate optimization methods, increases with the number of factors.

8.1. The basic simplex

Yes, the basic simplex is the simplest of all. This simplex is always a regular geometrical figure, whose form and size do not vary during the optimization process. For this reason it may not be very efficient (Spendley et al., 1962; Deming, 1981). With two factors, the simplex is an equilateral triangle. With three, it is a regular tetrahedron. Fig. 8.1 illustrates graphically the movement of the basic simplex in a two-dimensional problem, in which the factors are the reaction time and the reagent concentration, and the response is the reaction yield, all represented in arbitrary units. To facilitate our arguments, we are assuming the response surface to be almost planar and ascending from the lower left corner to the upper right one of the figure, as indicated by the contour curves. This is only an illustrative device, however. In a real problem, we probably would have no previous information about the form of the response surface. If we did, we could optimize much faster using the

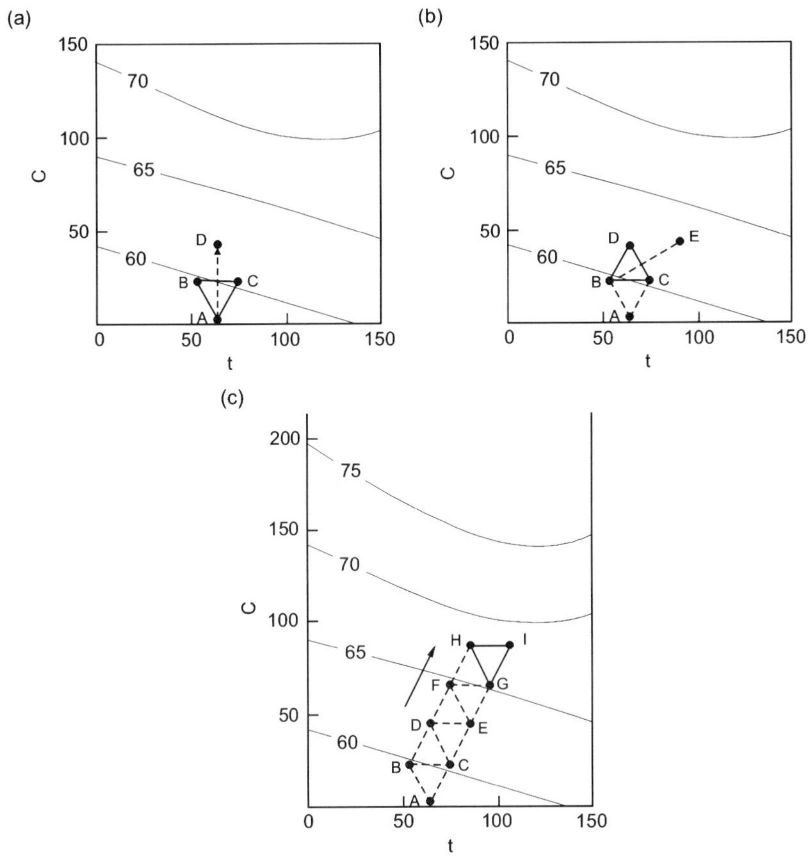

Fig. 8.1. Displacement of the bidimensional basic simplex. (a) The initial simplex is the **ABC** triangle. The reflection from the worst vertex through the **BC** line generates the new simplex, **BCD**. (b) The rejection of the **B** vertex produces the **CDE** simplex. (c) The successive displacements result in a movement such that the simplex rises on the response surface.

response surface methodology instead of the simplex. Here, the contour lines help us understand how the simplex moves in different situations.

The basic idea of the methods discussed in this chapter is to displace the simplex over the response surface in such a way as to *avoid* regions with unsatisfactory responses. In the present example, since we obviously would like to obtain a maximum yield, we should stay away from points that correspond to low yields. This is done using five rules.

Rule 1 – *The first simplex is determined performing a number of experiments equal to the number of factors plus one.* The size, position and orientation of the initial simplex are chosen by the researcher, taking into account his experience and available information about the system under study (Burton and Nickless, 1987). In Fig. 8.1a, the first simplex is defined by the **A**, **B** and **C** vertexes. Performing experiments at the conditions indicated by these vertexes and comparing the results, we verify that they correspond, respectively, to the worst, second worst and the best of the three observed responses. You can easily verify this by observing the location of the simplex relative to the contour curves of the response surface. This classification is necessary so that we can define the location of the second simplex, done according to rule 2.

Rule 2 – *The new simplex is formed rejecting the vertex corresponding to the worst response* (in Fig. 8.1a, vertex **A**) *and substituting it by its mirror image through the hyper-face defined by the other vertexes* (**B** and **C** here). In our example, where the hyper-face is simply the side **BC**, this produces the **D** vertex (Fig. 8.1a), and the new simplex will be the **BCD** triangle. In the language of descriptive geometry, this movement is called reflection of the **ABC** triangle through the **BC** side.

The worst response of the new simplex (**BCD**) occurs at the **B** vertex, whose reflection results in the **CDE** simplex (Fig. 8.1b). Doing this several times, we get a sort of zig-zag displacement with its resultant almost perpendicular to the contour curves of the response surface, approximately corresponding to the path of maximum inclination (Fig. 8.1c).

For a planar surface, as in Fig. 8.1, the simplex also being planar, adapts itself perfectly to the shape of the surface and its displacement takes place without any problems. If the surface has significant curvature, however, the application of rule 2 can lead to a paralization of the simplex. The **TUV** simplex in Fig. 8.2a illustrates this problem. It is obtained from the **STU** simplex by rejection of the **S** vertex that is substituted by the **V** vertex. The **V** vertex, however, happens to have the worst response of the three responses of the new simplex, **TUV**. By rule 2 it should be discarded, to obtain the following simplex. This brings us back to the **S** vertex, making the new simplex identical to the starting one, the **STU** triangle. Under these conditions, if we continue using rule 2 we will maintain the simplex stuck in the same place on the response surface, oscillating between the **S** and **V** vertexes. To escape this situation, we apply rule 3.

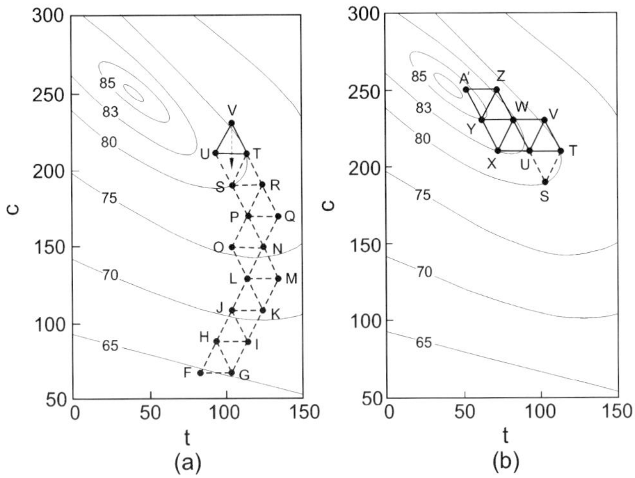

Fig. 8.2. (a) The oscillating simplex. Application of rule 2 causes the **S** vertex to be substituted by **V**, and vice versa. (b) Displacement of the simplex after applying rule 3 to the **TUV** simplex.

Rule 3 – *When the reflected vertex has the worst response of the new simplex, we reject the second worst vertex.* Applying this rule to the **TUV** simplex, we discard the **T** vertex (instead of the **V** vertex) and obtain the **UVW** simplex (Fig. 8.2b). From this simplex, returning to rule 2, we obtain the **UWX**, **WXY** and **WYZ** simplexes.[77] To leave the **WYZ** simplex, nevertheless, we have to use rule 3 again, since the **Z** vertex, generated by the reflection of the **X** vertex, is the worst of the three. Therefore, we reject the **W** vertex and obtain the **YZA′** simplex.

Since replicate runs are not performed with the simplex method, we do not have an estimate of experimental error, and for this reason we cannot evaluate the precision of the response measurements. We run the risk, therefore, to maintain wrong response values in the simplex calculations, whether they are too high or too low. In a study where the objective is maximization, to include a low-false response, that is, a response that is lower than its true value, does not have serious consequences. The following response measurements will probably be higher, and the simplex algorithm will automatically eliminate the erroneous value. The danger lies in determinations that are false highs. The next measured responses, perhaps, will not be higher than the false value and it will be inadvertently retained in the simplex. Our simplex will then start gyrating around the false maximum. If the study aims at minimization, clearly we have the opposite problem, and the risk of the simplex being

[77]The purest prefer the plural form *simplices*.

stuck at a low-false response, that can result in fictitious minimums. To avoid this problem we should use rule 4.

Rule 4 – *If the same vertex is maintained in p+1 simplexes* (like the **U** and **W** vertexes in Fig. 8.2b), *we should make another measurement of the response corresponding to this vertex before constructing the next simplex.* If the first response value has been too high because of experimental error, it is improbable that this phenomenon will repeat itself in the second determination if the error is random. The new response will probably be lower, and the vertex will be eliminated from the simplex. If, on the contrary, the response remains high, then it is probable that we are really close to the maximum point, and the vertex will be deservedly retained in the simplex. In the simplex movement shown in Fig. 8.2b we should, according to this rule, perform new runs at the **U** and **W** points to confirm if the responses at these vertexes are as high as those measured in the first determination.

Sometimes the simplex method specifies experimental conditions for the next run that are impossible or extremely difficult to execute. For example, reflection of the simplex can take it to a region with negative concentrations. Or perhaps, in the optimization of an analytical method, to a reaction time that is too long to be of practical use. Rule 5 tells us how to proceed in these cases.

Rule 5 – *If the new vertex passes beyond acceptable limits for any one of the variables that is being adjusted, we should attribute an undesirable value for the response at this vertex.* The application of rules 2 and 3 will then force the simplex to return to a region with acceptable values for the variable in question. Later, we will see a real example of this kind of behavior.

Fig. 8.3 shows what winds up happening when the basic simplex comes sufficiently close to the pursued value. Having arrived in the neighborhood of the maximum, which is the desired value in this example, the simplex starts to describe a circular movement around the highest observed response (point **A**′, in Fig. 8.3), and is not able to move from there. At this point there is nothing more for us to do, since the basic simplex cannot reduce its size. The optimization process should be stopped, and the precision with which the optimized conditions are determined depends on the size and location of the initial simplex. In our example, the maximum response is a little more than 86. The maximum value reached by the simplex is quite close: about 85.3, at the experimental conditions defined by the **A**′ vertex, $t \cong 50$ and $c \cong 247$.

Exercise 8.1. In Fig. 8.3, which are the simplexes obtained by application of rule 2? Which are the results of rule 3? Did we need to apply rule 4 to any vertex?

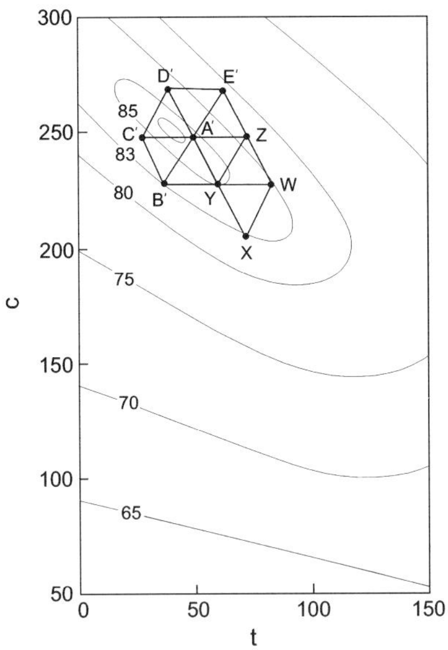

Fig. 8.3. Circular movement of the basic simplex in the vicinity of the maximum. Vertex **A**′ is retained in all the simplexes.

8.2. The modified simplex

In the modified algorithm (Nelder and Mead, 1965), the simplex can change its size and form, and consequently adapt itself more efficiently to the response surface. This flexibility permits a more precise determination of the optimum point, because the simplex can "shrink" in its proximity. Besides this desirable characteristic, the modified method, compared to the basic simplex, can reduce the number of runs necessary to find the optimum, because it can "stretch" itself when it is far from the desired point, usually on a planar portion of the response surface. For this reason it approaches the experimental region of interest more rapidly.

The possible movements of the modified simplex are illustrated in Fig. 8.4. The initial simplex is the **BNW** triangle. This notation already classifies the vertexes according to their corresponding responses. **B** is the vertex with the *best* response and **W** is the one with the *worst*. The **N** vertex corresponds to the *next to worst* result. The first movement, starting from the initial vertex, is identical to the one for the basic simplex: a reflection of the worst vertex through the centroid of the rest of the vertexes. In Fig. 8.4, this corresponds to reflecting point **W** through the midpoint of the **BN** line segment identified as $\bar{\mathbf{P}}$. The result is the **R** point and the new simplex becomes **BNR**. To decide if this simplex will be maintained, however, we need to compare the observed response at **R** with the responses for the generating simplex, **BNW**. Three cases are

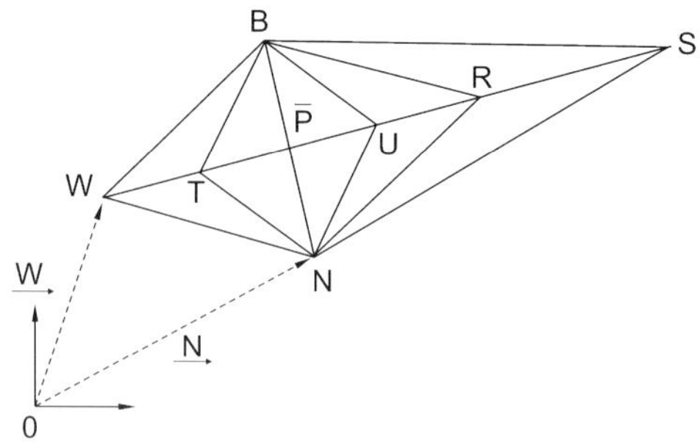

Fig. 8.4. Possible movements for the modified simplex. The **BNR** simplex is obtained from the initial simplex, **BNW**, by means of a simple reflection. The other three correspond to the following movements: expansion (**BNS**), contraction (**BNU**) and contraction with direction change (**BNT**).

possible. To discuss them, we are going to identify the response values with the same letters used to label the vertexes. We also assume, to simplify the discussion that we are in search of a maximum.

First case – **R**>**B**. The new response is better than all those in the preceding simplex. This suggests assuming that the simplex is on the right path, and that we should continue our investigations in that direction. We then perform a new determination at point **S**, located on the **WP̄R** line in such a way that the **P̄S** distance is double the **P̄R** distance. Depending on the value of the response at point **S** we have two possibilities:

1a – **S**>**R**. The response is even better. The expansion was worth it, and the new simplex becomes **BNS**.

1b – **S**<**R**. The result is worse with the expansion. We should maintain the first unexpanded simplex, **BNR**.

Second case – **N**<**R**<**B**. The observed response after reflection is worse than the best response of the initial simplex, but is better than the next-to-worst response. In this case, it is not worth performing another experiment in an attempt to expand. Neither expansion nor contraction is indicated and the **BNR** simplex is maintained.

Third case – **R**<**N**. The new response is less than the next-to-worst response of the starting simplex. We conclude that the direction of movement is not satisfactory, and that we need to correct the trajectory. We again have two possibilities:

3a – **R**<**W**. The worst happened: the new response is less than any of those previously observed. We should retreat. The new simplex becomes **BNT**, with **T** being the midpoint of **WP̄** line segment. In this case, not only a contraction occurs but also a change in direction of the displacement.

3b – **W**<**R**<**N**. Although the new response is bad, it is still better than the worst of the preceding results. We should retreat, but with moderation. The simplex is contracted to the intermediate position between $\bar{\mathbf{P}}$ and $\mathbf{R}$. The new simplex becomes **BNU**, where **U** is the midpoint of the $\bar{\mathbf{P}}\mathbf{R}$ segment.

Using elementary analytical geometry, we can easily calculate the locations of the various vertexes as the simplex moves. It is necessary to consider the coordinates of each point as components of a vector and only apply the rules of vector algebra. Thus, for example, the vector located at point $\bar{\mathbf{P}}$ is the average of the **B** and **N** vectors:

$$\bar{\mathbf{P}} = \frac{\mathbf{B} + \mathbf{N}}{2}. \tag{8.1a}$$

In our example the simplex is a triangle, and for this reason the $\bar{\mathbf{P}}$ vector contains the averages of the coordinates of only two vertexes. If we were optimizing a system with three variables, the simplex would be a tetrahedron, and the $\bar{\mathbf{P}}$ point would be given by the average of the three vertexes situated on the face opposite to the worst vertex. Higher dimensional simplexes involve the same kind of calculation but with more vertexes corresponding to the hyper-face opposite the worst vertex.

A simple reflection (i.e., without expansion or contraction) generates point **R**, given by

$$\mathbf{R} = \bar{\mathbf{P}} + \left(\bar{\mathbf{P}} - \mathbf{W}\right). \tag{8.2a}$$

In case the reflection is carried out using the second worst vertex, that is, the result of applying rule 3, one only needs to substitute the **W** vertex by **N**

$$\mathbf{R} = \bar{\mathbf{P}} + \left(\bar{\mathbf{P}} - \mathbf{N}\right). \tag{8.2b}$$

In this case, the $\bar{\mathbf{P}}$ point must change, now being given by

$$\bar{\mathbf{P}} = \frac{\mathbf{B} + \mathbf{W}}{2}. \tag{8.1b}$$

With vectorial Equations (8.1) and (8.2) we can determine the coordinates of all the points spanned by the basic simplex.

Exercise 8.2. Use "the parallelogram rule" of vectorial geometry, to demonstrate that Eq. (8.2a) is true.

Exercise 8.3. The coordinates of the **A**, **B** and **C** vertexes in Fig. 8.1a are given, respectively, by $(t, C) = (68, 0)$, $(56, 21)$ and $(80, 21)$. (a) What are the coordinates of point **D**? (b) What are the coordinates of point **E**?

In the modified simplex, we have expansion and contraction displacements as we have just seen. For these movements the $(\bar{\mathbf{P}} - \mathbf{W})$ vector (or, if it is the case, $(\bar{\mathbf{P}} - \mathbf{N})$) is multiplied by a factor that increases or decreases its contribution. In the expansion we have

$$\mathbf{S} = \bar{\mathbf{P}} + 2(\bar{\mathbf{P}} - \mathbf{W}). \tag{8.3}$$

For contraction, we have

$$\mathbf{U} = \bar{\mathbf{P}} + \frac{1}{2}(\bar{\mathbf{P}} - \mathbf{W}). \tag{8.4}$$

If besides contraction, the simplex needs to change its direction, we would have a subtraction of vectors rather than an addition:

$$\mathbf{T} = \bar{\mathbf{P}} - \frac{1}{2}(\bar{\mathbf{P}} - \mathbf{W}). \tag{8.5}$$

By using different coefficients instead of 2 and 1/2 we can vary the size of the expansion or contraction of the simplex, but these values are the ones most commonly used. All of these equations are easily solved using a linear algebra program, or even using most spread-sheets available on the commercial market.

In Figs. 8.5 and 8.6, we use the same response surface as the one in Figs. 8.1 and 8.2 to illustrate the application of the modified simplex. The corresponding numerical values of concentration and time for the various vertexes are given in Table 8.1. Note that the initial **ABC** simplex is the same as in the example of the basic simplex, permitting us to compare the efficiencies of the two algorithms.

The first movement is a reflection that takes us to point **D**′. Since the response at this point is better than all the responses in the initial simplex, we attempt an expansion until point **D**, whose response is even better. The new simplex is, therefore, the **BCD** triangle whose worst vertex is point **B**. This, on reflection, takes us to point **E**′, and from there, by expansion, to point **E**, forming **CDE** simplex.

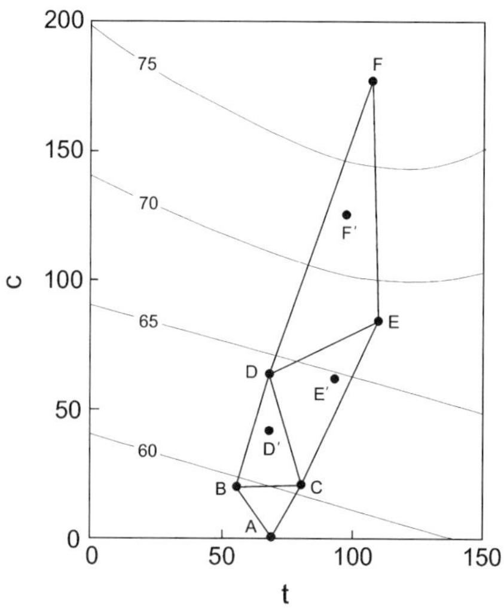

Fig. 8.5. Displacement of the modified bidimensional simplex on a planar portion of the response surface.

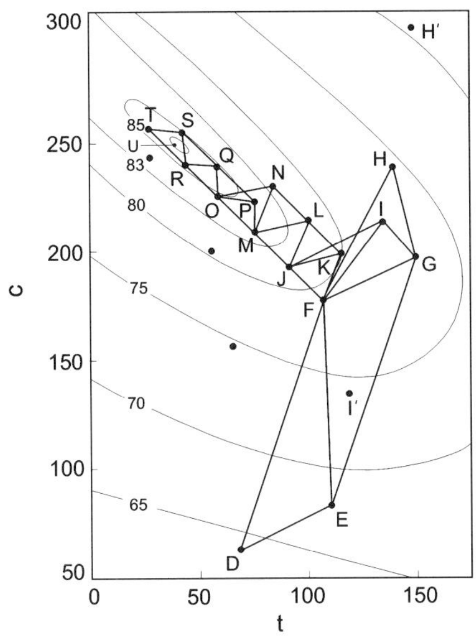

Fig. 8.6. Displacement of the modified bidimensional simplex on a raising ridge. The points not included in the simplexes are vertexes that were rejected by the algorithm rules.

Exercise 8.4. Use the coordinates for the **BCD** simplex given in Table 8.1 and calculate the values of the time and concentration at point **E** shown in Fig. 8.5.

Both Fig. 8.1c as well as Fig. 8.5 present the situation of the simplex after nine experiments. In Fig. 8.1c, which shows the application of the basic simplex, the response of the ninth point (the **I** vertex) is ≅ 68. Now in Fig. 8.5, in which the modified algorithm was applied, the final response (vertex **F**) raises to around 79, because the modified simplex was accelerated on the rising portion of the surface, since the new responses were proving to be better than the preceding ones. In this region, however, the response surface starts curving, and the simplex, if it continues in the same direction, would wind up bypassing the maximum. Fig. 8.6 shows what really happens. The displacement **DEF** → **EFG** is trivial. From this last simplex a reflection takes us to point **H**′, in the upper right of the figure. The response at this point is larger than the response at point **E**, but less than the one at point **G**, which is the second worst vertex of the **EFG** simplex. The recommended procedure in this case is a simple contraction (rule 3b) that produces the **FGH** simplex. A simple reflection then takes us to point **I**′ (lower right), whose response, however, is worst than the worst response in the **FGH** simplex. By rule 3a, we should perform not only a contraction but also a change in direction. The result is point **I**, and the new simplex becomes **FGI**.

Table 8.1
Vertex coordinates of the bidimensional modified simplex whose movement is illustrated in Figs. 8.5 and 8.6

Vertex	Simplex	*t*	Concentration	Movement
A		68	0	Initialization
B		56	21	Initialization
C	**ABC**	80	21	Initialization
D	**BCD**	68	63	Expansion
E	**CDE**	110	84	Expansion
F	**DEF**	107	178.5	Expansion
G	**EFG**	149	199.5	Reflection
H	**FGH**	137	241.5	Contraction
I	**FGI**	132.5	215.25	Contraction and direction change
J	**FIJ**	90.5	194.26	Reflection
K	**JFK**	115.63	200.82	Contraction and direction change
L	**JKL**	99.13	216.58	Reflection
M	**JLM**	74.01	210.02	Reflection
N	**LMN**	82.63	232.34	Reflection
O	**MNO**	57.54	225.78	Reflection
P	**MOP**	74.19	225.12	Contraction and direction change
Q	**OPQ**	57.69	240.88	Reflection
R	**OQR**	41.08	241.54	Reflection
S	**QRS**	41.19	256.64	Reflection
T	**RST**	24.50	257.30	Reflection
U	**RTU**	36.97	253.03	Contraction and direction change

Exercise 8.5. Use the data for the **FIJ** simplex (Table 8.1) to calculate the time and concentration values corresponding to the **K** vertex in Fig. 8.6.

Exercise 8.6. Which is the simplex immediately following the **RST** simplex, in Fig. 8.6? Which rule was applied in this case?

The last vertex reached in Fig. 8.6 (point **U**) is very close to the maximum of the response surface. In case it is necessary to obtain better precision, one can perform more experiments that could decrease the size of the simplex even more and produce a still smaller region in the vicinity of the maximum. It is obvious that the simplex could not reduce itself to a single point, and for this reason it is improbable that the mathematical maximum will be obtained. To avoid this search from going on indefinitely, the investigation is usually stopped when the differences in the responses of the vertexes of the simplex are the same size as the expected experimental error.

To conclude our discussion of the modified simplex we present an example with real data, obtained in an attempt to optimize the analysis of molybdenum by a method based on the catalysis of the reaction of KI and H_2O_2 (Eiras, 1991). The investigation of this same system using fractional factorial designs was discussed in Chapter 4. In this example, the research workers studied a different region of the response surface. The coordinates of the vertexes of all the simplexes constructed in this investigation and the values of the corresponding analytical responses are presented in Table 8.2. Since three factors are studied (the H_2SO_4, KI and H_2O_2 concentrations), the simplex is a tetrahedron. Its displacements are too complicated to follow graphically in two dimensions, but the concentration values for each vertex are easily calculated using Eqs. (8.1)–(8.5).

The first simplex is defined by the 1–4 vertexes. Of these, the one presenting the lowest response is vertex 1. Since we are trying to increase the signal, this is also the worst response, and therefore vertex 1 should be eliminated from the simplex. We should reflect it through the central point on the opposite face that is given by

$$\bar{\mathbf{P}} = 1/3(\mathbf{B} + \mathbf{N} + \mathbf{I}), \tag{8.6}$$

where **B** and **N** have their usual meanings and **I** represents the remaining vertex. Substituting the appropriate numerical values we can write

$$\bar{\mathbf{P}} = 1/3\left\{\begin{bmatrix}0.3200\\0.0400\\0.0040\end{bmatrix} + \begin{bmatrix}0.3200\\0.0300\\0.0040\end{bmatrix} + \begin{bmatrix}0.3200\\0.0300\\0.0050\end{bmatrix}\right\} = \begin{bmatrix}0.3200\\0.0333\\0.0043\end{bmatrix}.$$

The new vertex, **R**, can be obtained reflecting vertex 1 (which in this case is the **W** vertex) through point $\bar{\mathbf{P}}$. Applying Eq. (8.2a) we have

$$\mathbf{R} = \begin{bmatrix}0.3200\\0.0333\\0.0043\end{bmatrix} + \left\{\begin{bmatrix}0.3200\\0.0333\\0.0043\end{bmatrix} - \begin{bmatrix}0.4200\\0.0300\\0.0040\end{bmatrix}\right\} = \begin{bmatrix}0.2200\\0.0366\\0.0046\end{bmatrix}.$$

Exercise 8.7. Calculate the H_2SO_4, KI and H_2O_2 concentrations corresponding to vertex 8, that is the result of an expansion of the simplex formed by the 2, 3, 5 and 6 vertexes in Table 8.2.

Exercise 8.8. Calculate the H_2SO_4 concentration at vertex 15 that is obtained by reflection from the simplex formed by the 8, 9, 11 and 14 vertexes in Table 8.2.

Table 8.2
Application of the modified simplex to the optimization of Mo(VI) determination as a function of the H_2O_2, KI and H_2SO_4 concentrations. The observed response, that is to be maximized, is represented by ΔA

Vertex	Displacement[a]	Retained vertexes	$C_{H_2SO_4}$	C_{KI}	$C_{H_2O_2}$	ΔA
1			0.4200	0.0300	0.0040	0.183
2			0.3200	0.0400	0.0040	0.314
3			0.3200	0.0300	0.0050	0.236
4[b]			0.3200	0.0300	0.0040	0.198
5	R	2,3,4	0.2200	0.0366	0.0046	0.253
6	R	2,3,5	0.2533	0.0411	0.0051	0.307
7	R	2,5,6	0.2089	0.0485	0.0042	0.352
8	E	2,5,6	0.1533	0.0578	0.0038	0.372
9	R	2,6,8	0.2644	0.0559	0.0039	0.353
10	R	2,8,9	0.2385	0.0614	0.0027	0.341
11	R	8,9,10	0.1175	0.0767	0.0029	0.457
12	E	8,9,10	0.0162	0.0951	0.0024	0.370
13	R	8,9,11	0.1183	0.0655	0.0044	0.523
14	E	8,9,11	0.0582	0.0676	0.0052	0.528
15	R	8,11,14	−0.0451[c]	0.0788	0.0041	—
16	C	8,11,14	0.1870	0.0617	0.0040	0.426
17	R	11,14,16	0.0885	0.0795	0.0042	0.542
18	E	11,14,16	0.0561	0.0903	0.0044	0.595
19	R	11,14,18	−0.0325[c]	0.0948	0.0044	—
20	CDC	11,14,18	0.1321	0.0699	0.0040	0.421
21[d]		—	0.0868	0.0835	0.0037	0.479
22		—	0.0572	0.0790	0.0048	0.517
23		—	0.1216	0.0760	0.0042	0.481
24	R	18,22,23	0.0698	0.0800	0.0052	0.516
25	R	18,22,24	0.0005	0.0902	0.0054	0.116
26	CDC	18,22,24	0.0913	0.0796	0.0045	0.531
27	R	18,22,26	0.0666	0.0859	0.0039	0.550
28	R	18,26,27	0.0855	0.0915	0.0037	0.527
29	C	18,26,27	0.0784	0.0884	0.0040	0.560
30	R	18,27,29	0.0427	0.0968	0.0037	0.503

[a] Simplex movements: R = Reflection; E = Expansion; C = Contraction; CDC = Contraction with direction change.

[b] 01–04: Initial simplex vertexes.

[c] Since the acid concentration is negative, the response for this vertex was considered the worst result.

[d] 21–23: Massive contraction maintaining vertex 18.

The result of Exercise 8.8 is an example of the acceptable limit violation for the experimental conditions. It is clear that a negative sulfuric acid concentration has no physical meaning, and therefore it is not possible to carry out the experiment specified by vertex 15. That being the case, we arbitrarily attribute an undesirable value for the response at this vertex. This causes the simplex to move away from this point and return to the permitted experimental region.

Exercise 8.9. Calculate the H_2SO_4, KI and H_2O_2 concentrations corresponding to vertex 16, starting from the simplex formed by vertexes 8, 11, 14 and 15 in Table 8.2.

One of the simplexes in Table 8.2 was obtained in a way we have not yet discussed. Vertex 20 was the result of a contraction and a change of direction. This, in turn, was the consequence of an unsuccessful reflection that had produced the experimentally unacceptable point (vertex 19). It happens, however, that vertex 20 presents a worst response than any of the responses in the generating simplex. In other words, neither reflection nor contraction with direction change was successful. In this situation, the recommended procedure is a drastic contraction of the simplex in the direction of the vertex having the best response. This movement, called a *massive contraction*, is illustrated in Fig. 8.7 for a triangular simplex. In this figure, the **T** vertex that is obtained from the **BNW** simplex by means of a contraction and direction change presents a response that is lower than the worst response of the preceding **BNW** simplex, which is point **W**. This suggests that to obtain an optimized response we should investigate the region of point **B** that can be done by means of a massive contraction. In this kind of movement only the vertex corresponding to the best response (**B**) is maintained. The rest (**N** and **W**, in our example) are substituted by points located at the midpoints of the sides that are connected to the **B** vertex. The new simplex then becomes the **BX′Y′** triangle. In Table 8.2, since the simplex is a tetrahedron, it is necessary to eliminate three vertexes. Vertex 18, which has the best response of the four, is maintained, and vertexes 11, 14 and 16 are substituted. The 18, 21, 22 and 23 vertexes now form the new simplex and the optimization continues from it.

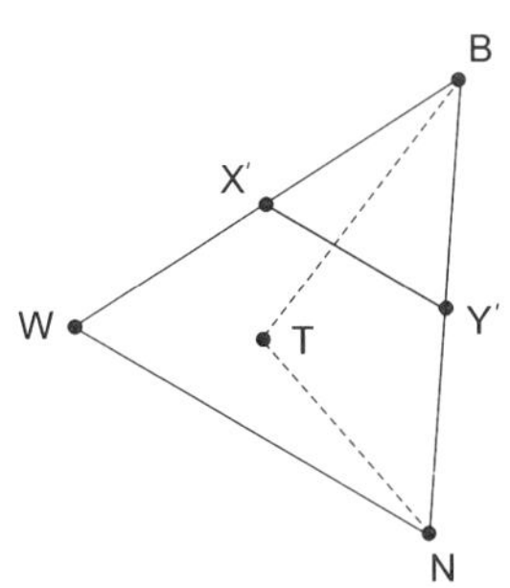

Fig. 8.7. Massive contraction of a triangular simplex. The response at point **T** is worse than the response at point **W**. The new simplex is **BX′Y′**.

Exercise 8.10. Calculate the coordinates of the 21, 22 and 23 vertexes, resulting from the massive contraction of the simplex formed by the 11, 14, 16 and 18 vertexes in Table 8.2.

Continuing with the optimization process of Table 8.2 we can note that vertex 18 is maintained in all of the simplexes constructed after the massive contraction. In other words, the simplex remains anchored at this point, and makes a circular movement around it, indicating that we have arrived in the region of the maximum. At this point, to determine the maximum with more precision it is better to abandon the simplex method and map the region around vertex 18, applying the response surface methodology we discussed in Chapter 6.

8.3. The supermodified simplex

In the modified algorithm the initial simplex, **BNW**, can be submitted to five different operations: reflection, expansion, contraction, contraction with direction change and massive contraction. In the supermodified simplex (Routh et al., 1977), the selection of options is amplified.

Equations (8.2)–(8.5), that govern the movement of the modified simplex, can be considered special cases of only one equation,

$$\mathbf{Y} = \bar{\mathbf{P}} + \alpha(\bar{\mathbf{P}} - \mathbf{W}), \tag{8.7}$$

where **Y** represents the new vertex, whose location depends on the value of the α parameter. When α is equal to 1, 2, 0.5 and –0.5, the **Y** vertex corresponds, respectively, to the **R**, **S**, **U** and **T** vertexes of Eqs. (8.2)–(8.5) and of Fig. 8.4. In the last section, we remarked that other values of α could be used, although these are the most common. The ideal value, that is, the one that takes us to the desired point faster, depends on several factors, such as the form of the surface, the size of the simplex and the importance of experimental error. In the supermodified simplex, this is taken into consideration and an α value, more appropriate for the surface studied, is determined. In this way, the movement of the simplex can become more efficient. In compensation, the determination of each new simplex requires the performance of more experiments than in the case of the modified algorithm.

To discuss the supermodified algorithm it is convenient to make a small modification in Eq. (8.7), substituting the α parameter by (β–1). With this and a small rearrangement the equation becomes

$$\mathbf{Y} = \beta\bar{\mathbf{P}} + (1 - \beta)\mathbf{W}. \tag{8.8}$$

Letting $\beta = 2$, 3, 1.5 and 0.5, respectively, we obtain the **R**, **S**, **U** and **T** points, that are positioned on an axis, as shown in Fig. 8.8. To discuss the application of the supermodified simplex let us imagine that the responses

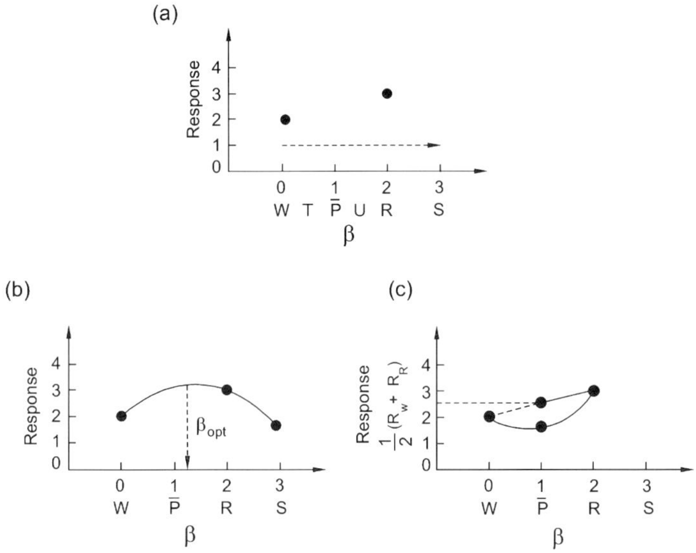

Fig. 8.8. The supermodified simplex. (a) The responses observed at **W** and **R** would indicate an expansion of the modified simplex. (b) With this response surface expansion would not be the best movement. A β value of about 1.3 would produce a larger response. (c) For the concave surface the **R** vertex is maintained.

determined for the **W** and **R** vertexes are as they appear in Fig. 8.8a. In the modified method the displacement indicated in this case would be an expansion, that would lead to the performance of an experiment at point **S**. Suppose now that the response surface in this region is described by the curve shown in Fig. 8.8b, that contains a maximum between the **W** and **R** points. It is obvious, in this case, that point **S** (i.e., a β value of 3) would not be a good choice. A smaller β value, around 1.3, would produce a better response. These wise considerations, unfortunately, can only be made *a posteriori*, after the response at point **S** has been measured, or if we know beforehand the response surface, in which case we do not have to perform any experiments.

To avoid situations such as those presented in Fig. 8.8b the supermodified simplex calls for the realization of a new experiment not at point **S**, but at an intermediate point, $\bar{\mathbf{P}}$. The responses at the three points, **W**, $\bar{\mathbf{P}}$ and **R** (corresponding to $\beta =$ 0, 1 and 2) are then fit to a second-order polynomial in β. Taking the derivative of this function we then obtain the optimum β value, that is given by the expression

$$\beta_{\text{opt}} = \frac{R_{\text{W}} - R_{\bar{\text{P}}}}{R_{\text{R}} - 2R_{\bar{\text{P}}} + R_{\text{W}}} + 0.5, \qquad (8.9)$$

where R_{W}, $R_{\bar{\text{P}}}$ and R_{R} are the responses determined at the **W**, $\bar{\mathbf{P}}$ and **R** points, respectively. The new simplex is formed by the **B**, **N** and **Z** vertexes where

$$\mathbf{Z} = \beta_{\text{opt}}\bar{\mathbf{P}} + (1 - \beta_{\text{opt}})\mathbf{W}. \quad (8.10)$$

Exercise 8.11. Fit the equation $R = a\beta^2 + b\beta + c$ to the points whose coordinates are $(0, R_W)$, $(1, R_{\bar{P}})$ and $(2, R_R)$. Take the derivative of this equation relative to β, set this derivative to 0 and show that

$$\beta_{\text{opt}} = \frac{R_R - 4R_{\bar{P}} + 3R_W}{2R_R - 4R_{\bar{P}} + 2R_W}.$$

Show that this expression is equivalent to Eq. (8.9).

Applying Eq. (8.9) to the curve shown in Fig. 8.8b, where $R_W = 2.0$, $R_{\bar{P}} = 3.2$ and $R_R = 3.0$, we obtain $\beta_{\text{opt}} = 1.36$ and therefore

$$\mathbf{Z} = 1.36\bar{\mathbf{P}} - 0.36\mathbf{W}.$$

The curvature of the response surface can be determined by comparing the observed response at point $\bar{\mathbf{P}}$ with the average of the responses at the **W** and **R** points. If by chance $R_{\bar{P}} < 1/2(R_W + R_R)$, the surface is concave, as in Fig. 8.8c. Obviously, in this kind of situation, it is not of interest to continue investigating the region between **W** and **R**, and the **R** point is maintained as the new vertex.

The number of possible values for β_{opt} is infinite. Some ranges of values, however, are inconvenient. In case the β_{opt} value falls in one of these ranges, such as illustrated in Fig. 8.9, it is discarded and the new vertex comes to be defined by the following rules.

1. β_{opt} values less than -1 or larger than 3 represent larger extrapolations of the simplex than we would obtain with the modified algorithm, and this is considered excessive. In this case, the expansion (or contraction) determined by the modified simplex is adopted.
2. β_{opt} values close to 0 should be avoided, because they result in a new simplex that is very similar to the original one. Normally, a safety margin,

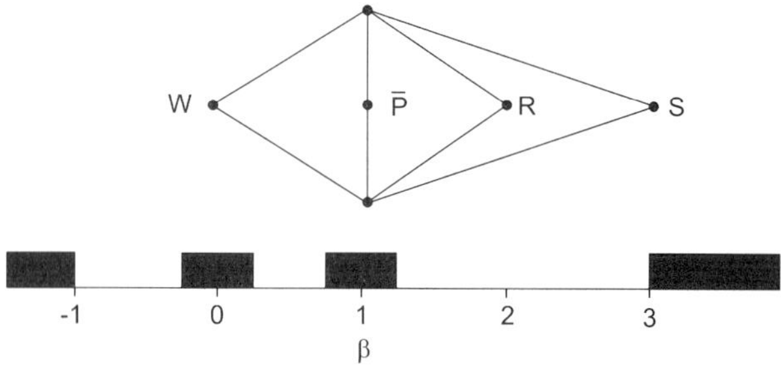

Fig. 8.9. Prohibited intervals for the β values, shown in black.

s_β, is defined and in case β_{opt} falls in the $(-s_\beta,\ s_\beta)$ interval its value is substituted by s_β or $-s_\beta$. Typically, the value of s_β is between 0 and 0.5.

3. If the β_{opt} value is exactly equal to 1 the simplex loses a dimension. The new point, **Z**, in Eq. (8.10) coincides with point $\bar{\mathbf{P}}$, and the reflection cannot be carried out. β_{opt} values close to 1, therefore, also are to be avoided. If $(1 - s_\beta) \leq \beta_{opt} \leq (1 + s_\beta)$, the β_{opt} value is substituted by the closest limit, $(1-s_\beta)$ or $(1+s_\beta)$.

In the cases of violations of established experimental limits for the variables being optimized, the choice of a new supermodified simplex is more complicated than for the other algorithms, because the response at the **R** vertex (that could be impossible to determine) is necessary to calculate β_{opt}. The reader can discover how to treat this problem in Morgan et al. (1990).

The biggest disadvantage of the supermodified simplex is that the definition of each new simplex requires an extra experiment, made at point $\bar{\mathbf{P}}$. Some researchers have preferred to avoid this additional effort, using the average of the responses at all the vertexes of the simplex (except **W**) as an estimate of the response at $\bar{\mathbf{P}}$. This variant of the supermodified algorithm has been tested on several response surfaces (Brown, 1990).

References

Uncited References

Bruns, et al. (1996); DESCARTES (1952); Pearson Wishart (1943); RUBO (1991)

References

AGUIAR, P. F. de; BOURGUIGNON, B.; KHOTS, M. S.; MASSART, D. L. and PHAN-THAN-LUU, R. D-optimal designs. *Chemom. Intell. Lab. Systems* **30**, 1995, pp. 199–210.

ANDRADE, J. C.; EIRAS, S. P. and BRUNS, R. E. Study of the Mo(VI) catalytic response in the oxidation of iodide by hydrogen peroxide using a monosegmented continuous flow system. *Anal. Chim. Acta* **255**, 1991, pp. 149–155.

ANDRADE, V. S.; NETO, B. B.; SOUZA, W. and CAMPOS-TAKAKI, G. M. A factorial design analysis of chitin production by *Cunninghamella elegans*. *Can. J. Microbiol.* **46**, 2000, pp. 5–12.

ANSCOMBE, F. J. Graphs in statistical analysis. *Am. Stat.* **27**, 1973, pp. 17–21.

AZEREDO, H. M. C. and JARDINE, J. G. Desidratação osmótica de abacaxi aplicada à tecnologia de métodos combinados. *Ciênc. Tecnol. Aliment.* **20**, 2000, pp. 78–82.

AZEVEDO, A. L. M. S.; NETO, B. B.; SCARMINIO, I. S.; OLIVEIRA, A. E. and BRUNS, R. E. A chemometric analysis of ab initio vibrational frequencies and infrared intensities of methyl fluoride. *J. Comp. Chem.* **17**, 1996, pp. 167–177.

BARNETT, V. and LEWIS, T. *Outliers in statistical data, 2nd edition*. New York, Wiley, 1984.

BHOTE, K. R. *Qualidade de classe mundial*. Rio de Janeiro, Qualitymark, 1996.

BOX, G. E. P. The exploration and exploitation of response surfaces: some considerations and examples. *Biometrics* **10**, 1954, pp. 16–60.

BOX, G. E. P. Evolutionary operation: a method for increasing industrial productivity. *Appl. Stat.* **6**, 1957, pp. 3–23.

BOX, G. E. P. and BEHNKEN, D. W. Some new three-level designs for the study of quantitative variables. *Technometrics* **2**, 1960, pp. 455–475.

BOX, G. E. P. and DRAPER, N. R. *Evolutionary operation: a statistical method for process improvement*, New York, Wiley, 1969.

BOX, G. E. P. and DRAPER, N. R. *Empirical model-building and response surfaces*, New York, Wiley, 1987.

BOX, G. E. P. and HUNTER, J. S. Multifactor experimental designs for exploring response surfaces. *Ann. Math. Stat.* **28**, 1957, pp. 195–242.

BOX, G. E. P.; HUNTER, W. G. and HUNTER, J. S. *Statistics for experimenters: an introduction to design, data analysis and model building*. New York, Wiley, 1978.

BOX, G. E. P. and WETZ, J. Criteria for judging adequacy of estimation by an approximate response function. *University of Wisconsin Technical Report* **9**, 1973.

BOX, G. E. P. and WILSON, K. B. On the experimental attainment of optimum conditions. *J. Roy. Stat. Soc.* **B13**, 1951, pp. 1–38.

BOX, G. E. P. and YOULE, P. V. The exploration and exploitation of response surfaces: an example of the link between the fitted surface and the basic mechanism of the system. *Biometrics* **11**, 1955, pp. 287–323.

BRAGA NETO, J. A. Desenvolvimento de produto alimentar assistido por computador: uma sistematização interativa pela aplicação combinada de métodos para planejamento, modelagem, análise e otimização na formulaç ão de pudim. Doctoral thesis, Departamento de Tecnologia de Alimentos e Medicamentos, Universidade Estadual de Londrina, 1998.

BREITKREITZ, M. C.; JARDIM, I. C. S. F. and BRUNS, R. E. Otimização da seletividade da FM empregada em CLAE-FR através de planejamento de misturas. *28ª Reunião Anual da Sociedade Brasileira de Química*, Poços de Caldas, MG, 2005, QA146.

BREYFOGLE, F. W. III. *Implementing six sigma: smarter solutions using statistical methods*. New York, Wiley, 1998.

BRUNHARA-SALUM, S. S. Otimização do processo de extração de colesterol do óleo de manteiga utilizando extrato de *Quilaia saponaria*. Master's dissertation, Faculdade de Engenharia Química, Universidade Estadual de Campinas, 1997.

BROWN, S. D. Chemometrics. *Anal. Chem.* **62**, 1990, p. 84R.

BRUNS, R. E.; ANDRADE, J. C.; REIS, C. and NAKAI, D. Is statistical lack of fit a reliable criterion for chemical complexity? *Chemom. Intell. Lab. Systems* **19**, 1996, pp. 159–166.

BURTON, K. W. C. and NICKLESS, G. Optimization via simplex. Part 1. Background, definitions and a simple application. *Chemom. Intell. Lab. Systems* **1**, 1987, pp. 135–149.

CAVALCANTE, R. M. Estudo de bioequivalência de comprimidos de lamivudina. Master's dissertation, Departamento de Ciências Farmacêuticas, Universidade Federal de Pernambuco, 1999.

CESTARI, A. R.; BRUNS, R. E. and AIROLDI, C. A fractional factorial design applied to organofunctionalized silicas for adsorption optimization. *Colloids Surfaces A Physicochem. Eng. Aspects* **117**, 1996, pp. 7–13.

CORNELL, J. A. How to apply response surface methodology, in *The ASQC basic references in quality control: statistical techniques*. SHAPIRO, S. S. and MYKYTKA, E. F. (eds.), Milwaukee, WI, American Society for Quality Control, EUA, 1990a.

CORNELL, J. A. *Experiments with mixtures: designs, models and the analysis of mixture data, 2nd edition*. New York, Wiley, 1990b.

CORNELL, J. A. How to run mixture experiments for product quality, in *The ASQC basic references in quality control: statistical techniques*. SHAPIRO, S. S. e. and MYKYTKA, E. F. (eds.), Milwaukee, WI, American Society for Quality Control, EUA, 1990c.

COSTA, R. A.; HATTORI, R. S.; REDONDO, E. G.; BRUNS, R. E. and SCARMINIO, I. S. Optimization of EPDM compounds for resistant insulators to electrical tracking. *Proceedings of the 3rd International Conference on Properties and Applications of Dielectric Materials*, 1991, pp. 300–304.

COSTA FERREIRA, S. L. C.; SANTOS, W. N. L.; QUINTELLA, C. M.; NETO, B. B. and BOSQUE-SENDRA, J. M. Doehlert matrix: a chemometric tool for analytical chemistry – review. *Talanta* **63**, 2004, pp. 1061–1067.

DEMING, S. N. Experimental designs: response surfaces, in *Chemometrics, mathematics and statistics in chemistry*. KOWALSKI, B. R. (ed.), Dordrecht, Reidel, 1981.

DERRINGER, G. and SUICH, R. Simultaneous optimization of several response variables. *J. Qual. Technol.* **12**, 1980, pp. 214–219.

DESCARTES, R. *Discours de la méthode de bien conduire sa raison et chercher la vérité dans les sciences; plus la dioptrique, les météores et la geométrie, qui sont des essais de cette méthode.* Leyden, 1637. There are several English translations. The text quoted here was translated by ELIZABETH S. HALDANE and G. R. T. ROSS (Chicago, Encyclopaedia Britannica, 1952).

DOEHLERT, D. H. Uniform shell designs. *Appl. Stat.* **19**, 1970, pp. 231–279.

DOEHLERT, D. H. and KLEE, V. L. Experimental designs through level reduction of a D-dimensional cuboctahedron. *Discrete Math.* **2**, 1972, pp. 309–334.

DUDEWICZ, E. J. and MISHRA, S. N. *Modern mathematical statistics*. New York, Wiley, 1985.

EIRAS, S. P. Determinação catalítica de molibdênio em plantas usando análise em fluxo contínuo monossegmentado com detecção espectrofotométrica. Doctoral thesis, Instituto de Química, Universidade Estadual de Campinas, 1991.

GALTON, F. Regression towards mediocrity in hereditary stature. *J. Anthropol. Inst.* **15**, 1886, pp. 246–263.

GAUCHI, J. -P. Plans d'expériences optimaux pour modèles linéaires; Plans d'expériences optimaux pour modèles de régression non linéaire, in *Plans d'expériences – applications à l'entreprise*. DROESBEKE, J. -J.; FINE, J. and SAPORTA, G. (eds.), Paris, Technip, 1997, pp. 326–438.

GEYER, A. L. M.; MOREIRA, J. C.; FAIGLE, J. F.; BRUNS, R. E. and CURTIUS, A. J. Local and temporal variations in essential elements and agar of the marine algae *Pterocladia capillacea*. *Hydrobiologia* **194**, 1990, pp. 143–148.

GOUPY, J. *Plans d'expériences pour surfaces de réponse*. Paris, Dunod, 1999.

GOUPY, J. *Plans d'éxpériences: les mélanges*. Paris, Dunod, 2000.

ISHIKAWA, K. *What is total quality control?* Englewood Cliffs, NJ, Prentice-Hall, 1985.

KALIL, S. J.; MAUGERY, F. and RODRIGUES, M. I. Response surface analysis and simulation as a tool for bioprocess design and optimization. *Process Biochem.* **35**, 2000, pp. 539–550.

LOPES, L. Separação e pré-concentração de cádmio, chumbo, cobre, níquel e zinco usando extração em fase sólida com ditizona e naftaleno coprecipitados, Master's dissertation, Instituto de Química, Universidade Federal de Bahia, 1999.

MARCHI, J. Sinterização via fase líquida de cerâmicas à base de carbeto de silício com aditivos óxidos utilizando planejamento de experimentos. Ph. D. Thesis, Instituto de Pesquisas Energéticas e Nucleares, Universidade de São Paulo, 2003.

MASSART, D. L.; VANDEGINSTE, B. G. M.; BUYDENS, L. M. C.; JONG, S. de; LEWI, P. J. and SMEYERS-VERBEKE, J. *Handbook of chemometrics and qualimetrics: part A*, Amsterdam, Elsevier, 1977.

MATHIAS, A.; SCARMINIO, I. S. and BRUNS, R. E. Planejamento fatorial 2^3 aplicado á decomposição do oxalato de cálcio monohidratado e da celulose usando termogravimetria. *22ª Reunião Anual da Sociedade Brasileira de Química*, Poços de Caldas, MG, 1999, QA109.

MENDES, A. S.; de CARVALHO, J. E.; DUARTE, M. C. T.; DURÁN, N. and BRUNS, R. E. Factorial design and response surface optimization of crude violacein for *Chromobacterium violaceum* production. *Biotech. Lett.* **23**, 2001, pp. 1963–1969.

MONTGOMERY, D. C. *Introduction to statistical quality control, 3rd edition*. New York, Wiley, 1997.

MONTGOMERY, D. C. *Design and analysis of experiments, 6th edition*. New York, Wiley, 2004.

MONTGOMERY, D. C. and RUNGER, G. C. *Applied statistics and probability for engineers, 2nd edition*. New York, Wiley, 1999.

MORGAN, E.; BURTON, K. W. C. and NICKLESS, G. Optimization using the super-modified simplex method. *Chemom. Intell. Lab. Systems* **8**, 1990, pp. 97–107.

MYERS, R. H. and MONTGOMERY, D. C. *Response surface methodology: process and product optimization using designed experiments, 2nd edition*. New York, Wiley, 2002.

NELDER, J. A. and MEAD, R. A simplex method for function minimization. *Comput. J.* **7**, 1965, pp. 308–312.

NEVES, C. F. C. Desenvolvimento de uma instalação de separação de gases por adsorção e modelagem do processo. Doctoral thesis, Faculdade de Engenharia Química, Universidade Estadual de Campinas, 2000.

OAKLAND, J. S. and FOLLOWELL, R. F. *Statistical process control*. Oxford, Butterworth-Heinemann, 1990.

PATERLINI, W. C. Fotodegradação de herbicidas utilizando processo foto-Fenton, Master's dissertation, Instituto de Química, Universidade Estadual Paulista, Araraquara, SP, 2002.

PIMENTEL, M. F. and NETO, B. B. Calibração: uma revisão para químicos analíticos. *Química Nova* **19**, 1996, pp. 268–275.

PLACKETT, R. L. and BURMAN, J. P. The design of optimum multifactorial experiments. *Biometrika* **33**, 1946, pp. 305–325, 328–332.

REIS, C.; ANDRADE, J. C.; BRUNS, R. E. and MORAN, R. C. C. P. Application of the split-plot experimental design for the optimization of a catalytic procedure for the determination of Cr(VI). *Anal. Chim. Acta* **369**, 1998, pp. 269–279.

RIBEIRO, R. L. V.; GRESPAN, C. B.; COLLINS, C. H.; COLLINS, K. E. and BRUNS, R. E. Optimization through factorial planning of the use of ethanol:water as a mobile phase for reversed phase HPLC. *J. High Resol. Chromatogr.* **22**, 1999, pp. 52–54.

ROCHA, R. F.; ROSATTO, S. S.; BRUNS, R. E. and KUBOTA, L. T. Factorial design optimization of redox properties of methylene blue adsorbed on a modified silica gel surface. *J. Electroanal. Chem.* **433**, 1997, pp. 73–76.

RONCONI, C. M. and PEREIRA, E. C. Electrocatalytic properties of Ti/TiO_2 electrodes prepared by the Pechini method. *J. Appl. Electrochem.* **31**, 2001, pp. 319–323.

RORABACHER, D. B. Statistical treatment for rejection of deviant values: critical values of Dixon's "Q" parameter and related subrange ratios at the 95% confidence level. *Anal. Chem.* **63**, 1991, pp. 139–146.

ROUTH, M. W.; SWARTZ, P. A. and DENTON, M. B. Performance of the super modified simplex. *Anal. Chem.* **49**, 1977, pp. 1422–1428.

RUBO, A. N. Obtenção de filmes poliméricos por calandragem a partir de poliisobutileno, polietileno e cera parafínica. Master's dissertation, Instituto de Química, Universidade Estadual de Campinas, 1991.

SATO, L. M. Planejamento estatístico de experimentos aplicado ao processo de produção de protetores auditivos de espuma de poliuretano. Master's dissertation, Instituto de Química, Universidade Estadual de Campinas, 2002.

SCARMINIO, I. S.; CAMARA, C. A. P. da; ANDRADE, S. R. C. and POPPI, R. J. Desenvolvimento de programas computacionais para análise da composiç ão química de misturas. *VII Encontro Nacional de Química Analítica*, Rio de Janeiro, R. J., Livro de resumos, 86, 1993.

SILVA, G. G.; MARZANA, B. E. and BRUNS, R. E. A statistically designed study of ternary electrolyte polymer material ($PEO/LiClO_4$/ethylene carbonate). *J. Mater. Sci.* **35**, 2000, pp. 4721–4728.

SIMONI, J. A. *Química geral – manual de laboratório.* Instituto de Química, Universidade Estadual de Campinas, 1998.

SIQUEIRA, D. F.; BRUNS, R. E. and NUNES, S. P. Optimization of polymer blends using statistical mixture analysis. *Polym. Networks Blends* **3**, 1993, pp. 63–69.

SKOOG, D. A.; WEST, D. M. and HOLLER, F. J. *Fundamentals of analytical chemistry.* Philadelphia, Saunders College Publishing, 1996.

SPENDLEY, W.; HEXT, G. R. and HIMSWORTH, F. R. Sequential application of simplex designs in optimization and evolutionary operation. *Technometrics* **4**, 1962, pp. 441–461.

STATSOFT, INC. *STATISTICA (data analysis software system), version 6.* www.statsoft.com, 2004.

STUDENT (GOSSET, W. S.) The probable error of a mean. *Biometrika* **6**, 1908, pp. 1–25.

VASCONCELOS, A. F.; BARBOSA, A. M.; DEKKER, R. F. H.; SCARMINIO, I. S. and RESENDE, M. I. Optimization of laccase production by *Botryosphaeria* sp. in the presence of veratryl alcohol by the response-surface method. *Process Biochem.* **35**, 2000, pp. 1131–1138.

WU, C. F. J. and HAMADA, M. *Experiments: planning, analysis and parameter design optimization*. New York, Wiley, 2000.

Answers to exercises

Chapter 2

2.1 We have already done this. Now it is your turn.

2.2 Our experiments indicate that it is around 5000.

2.3 Infinite. The possible values correspond to an (indeterminate) interval on the axis of real numbers.

2.4 You have two options: (a) sum the frequencies of the corresponding intervals and (b) sum the number of beans in these intervals and divide by the total number, 140. The results are 54.3% and 54.28%. The difference is due to rounding errors.

2.5 We suggest intervals with widths of 0.1%.

2.6 $\bar{x} = 0.1887$; $s = 0.0423$.

2.7 $x = \mu + z\sigma = 3.80\% + 2.5 \times 0.15\% = 4.18\%$.

2.8 (a) 73.24%; (b) (0.1313, 0.2735); (c) 52.6%, if our guess is 5.000 beans kg^{-1}, that is, $\mu = 0.2\,\text{g}$, and we interpolate between the closest values in Table A.1.

2.9 (a) 0; (b) 50%; (c) 15.87%; (d) 84.13%; (e) 100%. For the form of the curve, see Fig. 3.7.

2.10 (2955, 5098).

2.11 $r(x,y) = 0$, because for each positive value of x there is a value of the same size, but with negative sign, that has the same ordinate value. The products of deviations arising from these pairs (x, y) have the same size, but opposite signs. When they are added to give the correlation coefficient, they cancel out two by two. (b) Any even function, that is, one satisfying $f(x) = f(-x)$. For example: $y = \cos x$, $y = x^4$.

2.12 $\mathrm{Cov}(x, y) = 0.00167$; $r(x, y) = 0.9864$.

2.13 Applying Eq. (2.11), we have (a) $s_y^2 = s_1^2 + s_2^2 + 2(1)(-1)s_1 s_2(1) = s_1^2 + s_2^2 - 2s_1 s_2$, (b) $s_y^2 = s_1^2 + s_2^2 + 2(1)(-1)s_1 s_2(0) = s_1^2 + s_2^2$. Since the variances are equal to 1, we have (a) $s_y^2 = 0$ and (b) $s_y^2 = 2$.

2.14 From 4798 to 5092, with 95% confidence.

2.15 From 3858 to 7429.

2.16 (4796, 5094). These values are practically identical to those of Exercise 2.14, because the number of degrees of freedom is very large.

2.17 The values of the last line of Table A.2 are the values of the standard normal distribution (z) corresponding to the tail areas on the right 0.4, 0.25, 0.1, 0.05, 0.025, 0.01, 0.005, 0.0025, 0.001 and 0.0005.

2.18 This is not true. What happens is that the transformation from the mass of a bean to the number of beans per kilogram is not linear. The mass enters into the denominator, in a fraction with constant numerator. Therefore, the same interval for mass variation, when centered on a smaller value, results in a larger variation in terms of number of beans.

2.19 Using Eq. (2.18), we obtain $0.00064 < \sigma^2 < 0.01757$, and from this $0.0253 < \sigma < 0.1326$.

2.20 The 95% confidence interval is $3.527\% < \mu < 4.307\%$. Since this interval does not include the 4.45% value, we reject the null hypothesis and conclude that the concentration of acid in the lot is inferior to the required minimum.

2.21 The confidence interval is $13.65\% < \mu < 14.20\%$, and does not include the true value, 14.3%. This evidence indicates that the four determinations do not come from a distribution with a mean of 14.3%. In other words, we reject the null hypothesis at this confidence level and conclude that the new method does not have the necessary accuracy.

2.22 $\left(\frac{z\sigma}{L}\right)^2 = \left(\frac{1.96 \times 0.5}{0.2}\right)^2 = 24.01$. Therefore, we need at least 25 determinations.

2.23 The interval is defined by $\bar{x} \pm \left[t/\sqrt{n}\right]s$. Since we want it to be equal to $\bar{x} \pm 0.5s$, we must search in Table A.2, in the column corresponding to 95% confidence, a number of beans such that $t/\sqrt{n} = 0.5$. The whole number closest to satisfying this requirement is $n = 18 (\nu = 17)$.

2.24 The 95% confidence interval for the difference between the two averages is $(-3.74, -1.00)$, and does not include zero. We can reject the null hypothesis and conclude that there does exist a systematic difference between the results obtained with and without correction for the baseline, the corrected results being systematically smaller.

2.25 Use Eq. (2.29), substituting Δ by the reference value and $\bar{d}$ and s_d by the sample average and standard deviation. Then compare the $\hat{t}$ value with the tabulated value at the desired confidence level. If the calculated value is larger than the tabulated one, we reject the null hypothesis and conclude that the sample is incompatible with the reference value.

2.26 $\hat{t} = \dfrac{|-2.37 - 0|}{\dfrac{1.021}{\sqrt{3}}} = 4.02$. This value corresponds to a confidence level of about 94%.

Chapter 3

3.1 Think.

3.2 (2 temperature levels) $\times$ (2 catalyst levels) $\times$ (3 pressure levels) = 12 runs.

3.3 $\mathbf{CT} = \frac{1}{2}\{(\bar{y}_4 - \bar{y}_2) - (\bar{y}_3 - \bar{y}_1)\} = \frac{1}{2}(\bar{y}_4 - \bar{y}_2 - \bar{y}_3 + \bar{y}_1) = \frac{1}{2}\{(\bar{y}_4 - \bar{y}_3) - (\bar{y}_2 - \bar{y}_1)\} = \mathbf{TC}$.

3.4 With $n - 1 = 1$ and $\bar{x} = (x_1 + x_2)/2$, we can write

$$s^2 = \frac{\Sigma(x_i - \bar{x})^2}{n-1} = \left[x_1 - \frac{(x_1 + x_2)}{2}\right]^2 + \left[x_2 - \frac{(x_1 + x_2)}{2}\right]^2$$
$$= \left[\frac{(x_1 - x_2)}{2}\right]^2 + \left[\frac{(x_2 - x_1)}{2}\right]^2$$
$$= \frac{d^2}{4} + \frac{d^2}{4} = \frac{d^2}{2},$$

where $d^2 = (x_1 - x_2)^2 = (x_2 - x_1)^2$. For a pooled estimate we have

$$s^2 = \frac{(v_1 s_1^2 + v_2 s_2^2 + \cdots + v_n s_n^2)}{(v_1 + v_2 + \cdots + v_n)}.$$

With $v_1 = v_2 = \ldots = v_n = 1$, the expression reduces to

$$s^2 = \left(\frac{(d_1^2/2) + (d_2^2/2) + \cdots + (d_n^2/2)}{n}\right) = \sum_i \frac{d_i^2}{2n}.$$

3.5 Any effect is always given by a difference $(\bar{y}_+ - \bar{y}_-)$, where each average has $n/2$ observations. We can then write

$$\sigma^2_{\text{effect}} = \sigma^2(\bar{y}_+ - \bar{y}_-) = \sigma^2(\bar{y}_+) + \sigma^2(\bar{y}_-) = \frac{\sigma^2}{(n/2)} + \frac{\sigma^2}{(n/2)} = \frac{4\sigma^2}{n},$$

which is four times larger than the variance of the mean, σ^2/n. Taking the square root, we have

$$\sigma_{\text{effect}} = \frac{2\sigma}{\sqrt{n}} = 2\sigma_{\bar{y}}.$$

3.6 Applying Eq. (3.5), we obtain $s^2 = 7.99$ and therefore $s = 2.83$, with a total of $\upsilon_T = 8$ degrees of freedom.

3.7 $\mathbf{G} = -1.60$, $\mathbf{A} = -2.11$ and $\mathbf{GA} = 0.52$. The standard error of an effect is 0.22, which means that the **GA** interaction is not significant at the 95% level. The main effects show that the curing time is lowered by 1.6 min when the finer granulation is used (150–200 mesh) and also decreases by 2.11 min when the residual water increases by 7.5%.

3.8

$$\mathbf{A}^{-1}\mathbf{e} = \begin{bmatrix} 1 & -\frac{1}{2} & -\frac{1}{2} & \frac{1}{2} \\ 1 & \frac{1}{2} & -\frac{1}{2} & -\frac{1}{2} \\ 1 & -\frac{1}{2} & \frac{1}{2} & -\frac{1}{2} \\ 1 & \frac{1}{2} & \frac{1}{2} & \frac{1}{2} \end{bmatrix} \times \begin{bmatrix} 67.75 \\ 22.50 \\ -13.50 \\ -8.50 \end{bmatrix} = \begin{bmatrix} 59 \\ 90 \\ 54 \\ 68 \end{bmatrix};$$

$$\mathbf{Xb} = \begin{bmatrix} 1 & -1 & -1 & 1 \\ 1 & 1 & -1 & -1 \\ 1 & -1 & 1 & -1 \\ 1 & 1 & 1 & 1 \end{bmatrix} \times \begin{bmatrix} 67.75 \\ 11.25 \\ -6.75 \\ -4.25 \end{bmatrix} = \begin{bmatrix} 59 \\ 90 \\ 54 \\ 68 \end{bmatrix}$$

The two columns of $\mathbf{A}^{-1}$ and $\mathbf{X}$ are identical. The others differ by a factor of $\frac{1}{2}$, which is also responsible for the difference between vectors **e** and **b**.

3.9 Level (−) of factor **3**: Effect $\mathbf{12}(-) = \frac{1}{2}(54.0-86.5-48.0+63.0) = -8.75$
Level (+) of factor **3**: Effect $\mathbf{12}(+) = \frac{1}{2}(63.0-93.5-58.5+72.0) = -8.50$
Interaction of factor **3** with interaction **12**: $\frac{1}{2}\ [\mathbf{12}(+)-\mathbf{12}(-)] = \frac{1}{2}(-8.50+8.75) = 0.125$
Level (−) of factor **1**: Effect $\mathbf{23}(-) = \frac{1}{2}(54.0-48.0-63.0+58.5) = 0.75$
Level (+) of factor **1**: Effect $\mathbf{23}(+) = \frac{1}{2}(86.5-63.0-93.5+72.0) = 1.0$
Interaction of factor **1** with interaction **23**: $\frac{1}{2}\ [\mathbf{23}(+)-\mathbf{23}(-)] = \frac{1}{2}(1.0-0.75) = 0.125$.

The final values are identical to the **123** interaction value given in the text.

3.10 $V_{\text{effect}} = V(\bar{y}_+ - \bar{y}_-) = V(\bar{y}_+) + V(\bar{y}_-) = \dfrac{s^2}{n/2} + \dfrac{s^2}{n/2}$

For a 2^3 factorial without replicates $n = 8$ and therefore $V_{\text{effect}} = (s^2/4) + (s^2/4) = s^2/2$.

3.11 $\mathbf{t} = 3.87$, $\mathbf{C} = 12.36$, $\mathbf{P} = -2.17$, $\mathbf{tC} = 2.54$, $\mathbf{t}\boldsymbol{P} = -0.02$, $\mathbf{CP} = 0.47$ and $\mathbf{tCP} = 0.07$. The standard error of an effect is 0.325. One can simply compare the response values with the design matrix to see that the pattern of oscillation of the responses is associated with the pattern of signs in the concentration column.

3.12 The same conclusions drawn from Fig. 3.3.

3.13 A contrast between two tetrahedrons.

3.14

$$\hat{\mathbf{y}} = \begin{bmatrix} 1 & -1 & -1 & -1 & 1 & 1 & 1 & -1 \\ 1 & 1 & -1 & -1 & -1 & -1 & 1 & 1 \\ 1 & -1 & 1 & -1 & -1 & 1 & -1 & 1 \\ 1 & 1 & 1 & -1 & 1 & -1 & -1 & -1 \\ 1 & -1 & -1 & 1 & 1 & -1 & -1 & 1 \\ 1 & 1 & -1 & 1 & -1 & 1 & -1 & -1 \\ 1 & -1 & 1 & 1 & -1 & -1 & 1 & -1 \\ 1 & 1 & 1 & 1 & 1 & 1 & 1 & 1 \end{bmatrix} \times \begin{bmatrix} 67.3 \\ 11.4 \\ -6.9 \\ 4.4 \\ -4.3 \\ 0 \\ 0 \\ 0 \end{bmatrix} = \begin{bmatrix} 54.1 \\ 85.5 \\ 48.9 \\ 63.1 \\ 62.9 \\ 94.3 \\ 57.7 \\ 71.9 \end{bmatrix};$$

$$\mathbf{y} - \hat{\mathbf{y}} = \begin{bmatrix} 54.0 \\ 86.5 \\ 48.0 \\ 63.0 \\ 63.0 \\ 93.5 \\ 58.5 \\ 72.0 \end{bmatrix} - \begin{bmatrix} 54.1 \\ 85.5 \\ 48.9 \\ 63.1 \\ 62.9 \\ 94.3 \\ 57.7 \\ 71.9 \end{bmatrix} = \begin{bmatrix} -0.1 \\ 1.0 \\ -0.9 \\ -0.1 \\ 0.1 \\ -0.8 \\ 0.8 \\ 0.1 \end{bmatrix}$$

The residuals are much smaller than the observed values, indicating that the simplified model explains most of the variation in the observations. This point will be taken up again in Chapter 5.

3.15
$$\hat{y}(x_1x_2x_3x_4) = b_0 + b_1x_1 + b_2x_2 + b_3x_3 + b_4x_4 + b_{12}x_1x_2 + b_{13}x_1x_3 + b_{14}x_1x_4 + b_{23}x_2x_3 + b_{24}x_2x_4 + b_{34}x_3x_4 + b_{123}x_1x_2x_3 + b_{124}x_1x_2x_4 + b_{134}x_1x_3x_4 + b_{234}x_2x_3x_4 + b_{1234}x_1x_2x_3x_4$$

3.16 At the 95% confidence level the t-value with 5 degrees of freedom is 2.571. The limiting value for a significant effect will therefore be $0.54 \times 2.571 = 1.39$. Only the **1**, **2**, **3**, and **12** effects are significant at this confidence level.

3.17 (a) 50%; (b) 84.1% and (c) 97.5%.

3.18
$$(V_{\text{effect}})_{\upsilon=11} = \frac{3(-0.625)^2 + (-0.125)^2 + 3(0.325)^2 + 4(0.8755)^2}{11} = 0.425$$
$$\frac{(V_{\text{effect}})_{\upsilon=11}}{(V_{\text{effect}})_{\upsilon=5}} = \frac{0.425}{0.291} = 1.459 < F_{11,5} = 4.71, \text{ at the 95\% confidence level.}$$

Therefore, the two estimates can be seen as coming from the same population.

3.19 We can interpret it as the effect of changing the batch of starting material.

3.20 The interaction is given by $\mathbf{23} = \frac{1}{4}(y_1 + y_2 + y_7 + y_8) - \frac{1}{4}(y_3 + y_4 + y_5 + y_6)$. Let η_i stand for the population value of the response y_i. In the first block we have $y_i = \eta_i$, for $i = 1$, 4, 6 and 7. In the second block, $y_i = \eta_i + h$, for $i = 2$, 3, 5 and 8. In the calculation of the **23** interaction the systematic difference h is cancelled, since it appears twice with positive sign (for $i = 2$ and 8), and another two times with negative sign ($i = 3$ and 5).

Chapter 4

4.1 For example: $\mathbf{2} = 2 = \frac{1}{8}(-52 - 61 + 124 + \ldots + 289 + 286) = 109.38$.

4.2 $\mathbf{134} = \frac{1}{4}(-52 - 86 + \ldots + 286) = 114.75 = l_2$

1234 = Twice the average of all the response values.
Table 4.5 shows the relations that hold for each contrast. It is not a good idea to interpret these values as estimates of the **134** and **1234** effects, because they are mixed with hierarchically superior effects.

4.3 16.

4.4 $\mathbf{2} = \frac{1}{8}(-y_1 - y_2 + y_3 + \ldots + y_{16})$

$\mathbf{134} = \frac{1}{8}(-y_1 + y_2 - y_3 + \ldots + y_{16})$

$\mathbf{2} + \mathbf{134} = \frac{1}{8}(-2y_1 + 2y_4 + \ldots + 2y_{16}) = \frac{1}{4}(-y_1 + y_4 + \ldots + y_{16}) = l_2.$

4.5 Because the divisor is 8 for l_1, but only four for the contrasts.

4.6 $l_1^* = -2.50$; $l_2^* = 104.0$; $l_3^* = 57.0$; $l_4^* = 64.5$.

Effect $\mathbf{i} = \frac{1}{2}(l_i + l_i^*)$, for $i = 1$, 3 and 4. For example: $3 = 54.38 = \frac{1}{2}(l_3 + l_3^*)$.

4.7 $\mathbf{1234} = l_1 - l_1^* = 138.87 - 147.75 = -8.88$.

4.8 The confounding patterns are the same as those of Table 4.5, obtained from $\mathbf{I} = \mathbf{1234}$. The following values are calculated for the contrasts:

$$l_{\mathrm{I}} = 58.13, \quad l_1 = 8.75, \quad l_2 = -11.25 \quad l_3 = 48.75,$$
$$l_4 = 48.75, \quad l_{12} = l_{34} = -18.75, \quad l_{13} = l_{24} = 6.25 \quad l_{14} = l_{23} = -8.75.$$

The contrasts for solvent (**3**) and catalyst (**4**) have large positive effects (coincidentally equal). The largest yields are obtained when these two factors are at their higher levels. The −18.75 value probably can be attributed to the interaction between these factors.

4.9 The main effects are confounded with fifth-order interactions. The binary interactions are confounded with fourth-order interactions.

4.10 When t, [KI] and $[H_2O_2]$ are at their higher levels, either (a) the response is not influenced by changes in the $[H_2SO_4]$ and flow rate levels or (b) the variations produced by changes in these levels cancel one another.

4.11 The first four factors have the same signs as the full 2^4 factorial. When the first four factors are placed in standard order the signs of the fifth are, from top to bottom, $(-++--++-+--++--+)$. The relations between the contrasts and the effects are as follows:

$$l_1 = \mathbf{1} + \mathbf{245}, \quad l_2 = \mathbf{2} + \mathbf{145}, \quad l_3 = \mathbf{3} + \mathbf{12345},$$
$$l_4 = \mathbf{4} + \mathbf{125}, \quad l_5 = \mathbf{3} + \mathbf{124}, \quad l_{12} = \mathbf{12} + \mathbf{45},$$
$$l_{13} = \mathbf{13} + \mathbf{2345}, \quad l_{14} = \mathbf{14} + \mathbf{25}, \quad l_{15} = \mathbf{15} + \mathbf{24},$$
$$l_{23} = \mathbf{23} + \mathbf{1345}, \quad l_{24} = \mathbf{15} + \mathbf{24}, \quad l_{25} = \mathbf{25} + \mathbf{14},$$
$$l_{34} = \mathbf{34} + \mathbf{1235}, \quad l_{35} = \mathbf{35} + \mathbf{1234}, \quad l_{45} = \mathbf{12} + \mathbf{45},$$

Main effect **3** is contaminated by an interaction of fifth order, instead of one of fourth order. In compensation, the other four effects are probably estimated with less accuracy, since they are confounded with third-order interactions rather than the fourth-order ones that appear in resolution V factorials. The binary interactions involving factor **3** are contaminated only by fourth order interactions. This factorial, therefore, favors estimation of factor **3** at the expense of the rest. Unless there is a good reason for using this factorial, we do not recommend it.

4.12 (a) Simply multiply factor **1** by the identity, $\mathbf{I} = \mathbf{1234} = \mathbf{125} = \mathbf{345}$:
$\mathbf{1} \times \mathbf{I} = \mathbf{234} = \mathbf{25} = \mathbf{1345} \rightarrow l_1 = \mathbf{1} + \mathbf{25} + \mathbf{234} + \mathbf{1345}$
(b) The same thing, but with factor **5**:
$\mathbf{5} \times \mathbf{I} = \mathbf{12345} = \mathbf{12} = \mathbf{34} \rightarrow l_5 = \mathbf{5} + \mathbf{12} + \mathbf{34} + \mathbf{12345}$.

4.13 $l_5 = \frac{1}{4}(56 - 66 + 51 - 52 - 54 + 70 - 42 + 64) = 6.75$.

4.14 With some effort, you can discover that $l_5 = \mathbf{5} + \mathbf{13} + \mathbf{27} + \mathbf{46} + \mathbf{126} + \mathbf{147} + \mathbf{234} + \mathbf{367} + \mathbf{1245} + \mathbf{1567} + \mathbf{2356} + \mathbf{3457} + \mathbf{12357} + \mathbf{13456} + \mathbf{24567} + \mathbf{123467}$

4.15 The only difference is the side of the tennis court. In Table 4.10 run no. 4 involves serving with a slice, with high frequency, during the day, on a concrete surface, from the right side, with a shirt on and with a medium-head racquet. In Table 4.12, the same serve is made from the left side.

4.16 Multiplying the first three generating relations of the design in Table 4.10 by factor **8**, which corresponds to the identity, does not affect their signs. However, they all now have an even number of factors (four), so that changing all their signs does not change the sign of the product.

4.17 $l_1 = \mathbf{1} + \mathbf{248} + \mathbf{358} + \mathbf{237} + \mathbf{346} + \mathbf{256} + \mathbf{678} + \mathbf{457}$.

Chapter 5

5.1 Equation (5.9) is

$$b_1 = \frac{\sum(X_i - \bar{X})(y_i - \bar{y})}{\sum(X_i - \bar{X})^2}.$$

Manipulating the numerator and the denominator we have
For the denominator:

$$\begin{aligned}\sum(X_i - \bar{X})^2 &= \sum(X_i^2 - 2X_i\bar{X} + \bar{X}^2) \\ &= \sum X_i^2 - 2n\bar{X}^2 + n\bar{X}^2 = \sum X_i^2 - nX^2 \\ &= \sum X_i^2 - n\left[\left(\frac{1}{n}\right)\sum X_i\right]^2 = \sum X_i^2 - \frac{1}{n}\left(\sum X_i\right)^2\end{aligned}$$

For the numerator:

$$\begin{aligned}\sum(X_i-\bar{X})(y_i-\bar{y}) &= \sum(X_iy_i-\bar{y}X_i-\bar{X}y_i+\bar{X}\bar{y})\\ &= \sum X_iy_i-\bar{y}\sum X_i-\bar{X}\sum y_i+\sum\bar{X}\bar{y}\\ &= \sum X_iy_i-\left(\frac{1}{n}\right)\sum y_i\sum X_i-\left(\frac{1}{n}\right)\sum X_i\sum y_i+n\bar{X}\bar{y}\\ &= \sum X_iy_i-\left(\frac{2}{n}\right)\sum y_i\sum X_i+\left(\frac{1}{n}\right)\sum X_i\sum y_i\\ &= \sum X_i\sum y_i-\frac{1}{n}(\sum y_i)(\sum X_i)\end{aligned}$$

Putting these expressions back in the fraction, we arrive at Eq. (5.8):

$$b_1=\frac{\sum X_iy_i-(1/n)(\sum y_i)(\sum X_i)}{\sum X_i^2-(1/n)(X_i)^2}.$$

5.2

$$\mathbf{X'X}=\begin{bmatrix}1 & 1 & \dots & 1\\ X_1 & X_2 & \dots & X_n\end{bmatrix}\times\begin{bmatrix}1 & X_1\\ 1 & X_2\\ \vdots & \vdots\\ 1 & X_n\end{bmatrix}=\begin{bmatrix}n & \sum X_i\\ \sum X_i & \sum X_i^2\end{bmatrix}$$

$$\mathbf{X}^t\mathbf{y}=\begin{bmatrix}1 & 1 & \dots & 1\\ X_1 & X_2 & \dots & X_n\end{bmatrix}\times\begin{bmatrix}y_1\\ y_2\\ \vdots\\ y_n\end{bmatrix}=\begin{bmatrix}\sum y_i\\ \sum X_iy_i\end{bmatrix}$$

5.3 $s^2=\dfrac{\sum(y_i-\bar{y})^2}{n-1}=\dfrac{S_{yy}}{(n-1)}$. Therefore, $s=\sqrt{\dfrac{S_{yy}}{n-1}}$.

5.4 (a) Nonlinear; (b) linear; (c) linear; (d) linear; (e) nonlinear, but by taking the logarithm of both sides we arrive at a linear model.

5.5 $\mathbf{b}=(\mathbf{C'C})^{-1}\mathbf{C'A}=\begin{bmatrix}0.000897\\ 0.1838\end{bmatrix}$.

5.6

$$\mathbf{X'X}=\begin{bmatrix}4 & 0 & 0 & 0\\ 0 & 4 & 0 & 0\\ 0 & 0 & 4 & 0\\ 0 & 0 & 0 & 4\end{bmatrix};\quad(\mathbf{X'X})^{-1}=\begin{bmatrix}\frac{1}{4} & 0 & 0 & 0\\ 0 & \frac{1}{4} & 0 & 0\\ 0 & 0 & \frac{1}{4} & 0\\ 0 & 0 & 0 & \frac{1}{4}\end{bmatrix};$$

$$(\mathbf{X}'\mathbf{X})^{-1}\mathbf{X}'\mathbf{y} = \begin{bmatrix} 67.75 \\ 11.22 \\ -6.75 \\ -4.25 \end{bmatrix}.$$

The results are the numerical values of the coefficients of Eq. (3.11).

5.7 $\sum(\hat{y}_i - \bar{y})(y_i - \hat{y}) = \sum[\bar{y} + b_1(X_i - \bar{X}) - \bar{y}][y_i - \hat{y}_i] = b_1 \sum(X_i - \bar{X})$
$(y_i - \hat{y}) = b_1 \sum(X_i - \bar{X})[y_i - \bar{y} - b_1(X_i - \bar{X})] =$
$b_1\{\sum(X_i - \bar{X})(y_i - \bar{y}) - b_1 \sum(X_i - \bar{X})^2\} = 0,$
because the terms in curly brackets cancel out by Eq. (5.9).

5.8
$$b_1 = \frac{\sum(X_i - \bar{X})(y_i - \bar{y})}{\sum(X_i - \bar{X})^2} = \frac{\sum(X_i - \bar{X})y_i - \bar{y}\sum(X_i - \bar{X})}{\sum(X_i - \bar{X})^2}.$$

The second term in the numerator is zero, because it contains a sum of residuals about the average. Remembering that by definition the denominator is S_{xx}, we can write

$$b_1 = \frac{\sum(X_i - \bar{X})y_i}{S_{xx}}.$$

5.9
$$\mathrm{Cov}(\bar{y}, b_1) = \sigma^2 \sum a_i c_i = \frac{\sigma^2 \sum(1/n)[X_i - \bar{X}]}{S_{xx}} = \frac{\sigma^2 \sum(X_i - \bar{X})}{nS_{xx}} = 0,$$

for the same reason as in the preceding exercise.

5.10

Source of variation	Sum of squares	Degree of freedom	Mean square
Regression	0.295146	1	0.295146
Residual	0.000279	10	0.000028
Lack of fit	0.000170	4	0.000043
Pure error	0.000109	6	0.000018
Total	0.295425	11	

% explained variance = 99.906
Maximum explainable % variance = 99.963
$MS_{\mathrm{lof}}/MS_{\mathrm{pe}} = 2.39$. At the 95% confidence level. $F_{4,6} = 4.53$
Conclusion: there is no evidence of lack of fit.

5.11 $S^2 = 0.000028$

Standard error of b_0:0.00348; standard error of b_1:0.00179.
With these values and $t_{10} = 2.228$ (95% confidence level) only the value of b_1 is significant.

Chapter 6

6.1 Effect x_1 (concentration): -10.5
Effect x_2 (stirring speed): 8.5
Interaction effect, x_1x_2: -0.5
The main effects are twice the b_1 and b_2 coefficient values in the regression, because while b_1 and b_2 represent changes in the response caused by unit variations in x_1 and x_2, in the effects the changes correspond to two units, from $x_i = -1$ to $+1$. The interaction effect is much smaller than the main effects. This is to be expected, since the response surface is adequately represented by a linear model.

6.2 $\hat{V}_{\bar{y}} = 0.333, \quad s_{\bar{y}} = 0.58;$

$\hat{V}_{\text{effect}} = 2.33, \quad s_{\text{effect}} = 1.53.$

The standard error of the average is equal to the standard error of b_0, because $b_0 = \bar{y}$. The standard error of the effects is twice the standard errors of b_1 and b_2 because, as we have already seen in the preceding exercise, the effect values are twice the coefficient values.

6.3 $MS_R/MS_r = 66.4$ and $F_{2,4} = 6.69$ at the 95% confidence level. The regression is highly significant. Again, we could have expected this, since all of the coefficients in Eq. (6.3) are significant.

6.4 $t_2 = 4.303$, with 95% confidence.
$S(b_0) = 0.58 \times 4.303 = 2.496$
$S(b_1) = s(b_2) = 0.76 \times 4.303 = 3.27$
The 95% confidence intervals are $\beta_0 : (65.55, 70.45)$; $\beta_1 : (-1.98, -8.52)$; $\beta_2 : (0.98, 7.52)$.
Since none of these intervals includes zero, all of the coefficients are significant at the chosen confidence level.

6.5 $\mathbf{G} = 115\text{gL}^{-1}$, $\mathbf{A} = 11.4\text{gL}^{-1}$, $\mathbf{T} = 0.19\text{mgL}^{-1}$ and $\hat{y} = 78.4\%$.

6.6 $t_4 = 2.776$, at the 95% confidence level. The intervals are as follows:

For $\beta_0 : 85.71 \pm 2.776 \times 0.49$ *or* $(84.35, 87.07)$
For $\beta_1 : 1.25 \pm 2.776 \times 0.65$ *or* $(-0.55, 3.05)$
For $\beta_2 : 2.25 \pm 2.776 \times 0.65$ *or* $(-0.45, -4.05)$

Since the b_1 interval includes the zero value, this coefficient is not statistically significant at the 95% confidence level. The other two are significant.

6.7 $MS_R/MS_r = 52.2 > F_{5,5}$ (95% confidence level).

6.8 (a) Starting from Eq. (6.8) we have

$$\frac{\partial \hat{y}}{\partial x_1} = 1.51 - 5.62x_1 + 1.75x_2 = 0$$

$$\frac{\partial \hat{y}}{\partial x_2} = -2.36 + 1.75x_1 - 5.62x_2 = 0.$$

Solving this system of equations, we arrive at $x_1 = 0.152$, which corresponds to a concentration of 35.8%, and $x_2 = -0.374$ (i.e., a stirring speed of 121.3 rpm).

(b) With Eq. (6.6), the result is

$$\frac{\partial \hat{y}}{\partial x_1} = +1.25 \neq 0 \text{ and } \frac{\partial \hat{y}}{\partial x_2} = 2.25 \neq 0.$$

That is, this function has no extreme points. Why is that so?

6.9 Since there is practically no lack of fit for the quadratic model, we can take the square root of the residual mean square as an estimate of pure error:

$$\text{Error} = \sqrt{MS_r} = \sqrt{0.028} = 0.167\,\text{MPa}$$

This is an estimate with 96 degrees of freedom.

6.10 Using $t_{79} = 1.99$ (95% confidence level) and the standard errors given in Eq. (6.10) we calculate confidence intervals showing that the significant coefficients are the constant term, the linear terms in concentration and particle size, the squared term for concentration and the cross term of concentration and particle size. No term involving the reaction time is significant.

Chapter 7

7.1 We hope so.

7.2 (a) $\hat{y} = b_A^* x_A + b_B^* x_B$ substituting $b_A^* = 14$ and $b_B^* = 6$, we have $\hat{y} = 14x_A + 6x_B$.

(b) $\hat{y} = 14 \times 0.5 + 6 \times 0.5 = 10\,\text{km L}^{-1}$.

(c) $\hat{y} = 14 \times 0.7 + 6 \times 0.3 = 11.6\,\text{km L}^{-1}$.

7.3 (a) $b_A^* = 14.0$; $b_B^* = 6.0$; $b_{AB}^* = 4 \times 12 - 2(14 + 6) = 8.0$.

(b) The model predicts the yield to be $\hat{y} = \left(\frac{2}{3}\right)14 + \left(\frac{1}{3}\right)6 + \left(\frac{2}{3}\right)\left(\frac{1}{3}\right)8 = 13.1\,\text{km L}^{-1}$, which is in good agreement with the observed value.

7.4 Coefficients: $b_A^* = 1.44$; $b_B^* = 1.17$; $b_{AB}^* = -0.82$. The standard error of an observation is $s = 0.035$. With this we calculate the standard errors of the coefficients as $\Delta b_A^* = \Delta b_B^* = 0.025$ and $\Delta b_{AB}^* = 0.107$.

7.5 The response surface is horizontal. The response does not depend on the composition.

7.6 At the 95% confidence level, with $t_9 = 2.262$, the limiting values for significance are $2.262 \times 0.17 = 0.38$ for the absolute values of the linear terms and $2.262 \times 0.75 = 1.70$, for the squared terms. The terms with $b_1^* = 3.10$ and $b_{13}^* = 9.62$ are highly significant. The linear term in $x_2(b_2^* = 0.45)$ is significant by a small margin.

7.7 Substituting x_3 into Eq. (7.16) by $(1-x_1)$ we arrive at $\hat{y} = 12.62x_1 - 9.62x_1^2$. Taking the derivative and setting it equal to zero, we have $x_1 = 0.66$. From this, $x_3 = 0.34$ and therefore

$$\hat{y}_{\max} = 3.10 \times 0.66 + 9.62 \times 0.66 \times 0.34 = 4.20\,\text{cm}.$$

7.8 (a) Al_2O_3. Just compare the tracking values with the compositions.
(b) $\hat{y}_{\text{mass loss}} = 2.84x_1 + 5.24x_2 + 3.80x_3 - 11.44x_1x_2 - 4.56x_1x_3 - 4.56x_2 x_3$; $\hat{y}_{\text{tracking}} = 94.3x_1 + 9.0x_2 + 11.5x_3 + 294.0x_1x_2 + 200.0x_1x_3 + 1.3x_2x_3$;

7.9 (a) $b_{123}^* = 6.0$. (b) Standard error of b_{123}^* : 5.41. Since this error is almost equal to the coefficient value, the cubic term can be considered practically non-significant.

7.10 From Eqs. (7.16) and (7.20) we obtain $\hat{y}_{\text{quad}} = 2.10$cm and $\hat{y}_{\text{sp cubic}} = 3.30$cm. The estimate from the special cubic model is much closer to the observed value, 3.50 cm.

7.11

$$\mathbf{b} = \mathbf{X}^{-1}\mathbf{y}, \quad \text{where } \mathbf{X} = \begin{bmatrix} 1 & 0 & 0 & 0 & 0 & 0 & 0 \\ 0 & 1 & 0 & 0 & 0 & 0 & 0 \\ 0 & 0 & 1 & 0 & 0 & 0 & 0 \\ \frac{1}{2} & \frac{1}{2} & 0 & \frac{1}{4} & 0 & 0 & 0 \\ \frac{1}{2} & 0 & \frac{1}{2} & 0 & \frac{1}{4} & 0 & 0 \\ 0 & \frac{1}{2} & \frac{1}{2} & 0 & 0 & \frac{1}{4} & 0 \\ \frac{1}{3} & \frac{1}{3} & \frac{1}{3} & \frac{1}{9} & \frac{1}{9} & \frac{1}{9} & \frac{1}{27} \end{bmatrix}$$

and $\mathbf{y}' = \begin{bmatrix} 3.10 & 0.45 & 0.35 & 1.70 & 4.13 & 0.27 & 3.50 \end{bmatrix}$.

7.12 (0.333, 0.333, 0.333);
(0.567, 0.217, 0.217);
(0.217, 0.567, 0.217);
(0.217, 0.217, 0.567);

Chapter 8

8.1 Rule 2: **WYZ**, **YA′B′**, **A′C′D′**.
Rule 3: **YZA′**, **A′B′C′**, **A′D′E′**, **ZA′E′**;
Rule 4: To the **Y** and **A′** vertices.

8.3 Applying Eq. (8.2a), we have

(a) $\mathbf{D} = \frac{1}{2}(\mathbf{B}+\mathbf{C}) + \left[\frac{1}{2}(\mathbf{B}+\mathbf{C}) - \mathbf{A}\right] = \mathbf{B}+\mathbf{C}-\mathbf{A} = (68, 42)$;

(b) $\mathbf{E} = \frac{1}{2}(\mathbf{C}+\mathbf{D}) + \left[\frac{1}{2}(\mathbf{C}+\mathbf{D}) - \mathbf{B}\right] = \mathbf{C}+\mathbf{D}-\mathbf{B} = (92, 42)$.

8.4 Using Eq. (8.3), we have $\mathbf{E} = \frac{1}{2}(\mathbf{C}+\mathbf{D}) + 2\left[\frac{1}{2}(\mathbf{C}+\mathbf{D}) - \mathbf{B}\right] = \frac{3}{2}(\mathbf{C}+\mathbf{D}) - 2\mathbf{B}$. Substituting the coordinates from Table 8.1, we arrive at $t = 110$ and $C = 84$.

8.5 Point **K** is produced from the **FIJ** simplex by contraction with change of direction (Eq. (8.5)). We have therefore $\mathbf{K} = \frac{1}{2}(\mathbf{J}+\mathbf{F}) - \frac{1}{2}\left[\frac{1}{2}(\mathbf{J}+\mathbf{F}) - \mathbf{I}\right] = \frac{1}{4}(\mathbf{J}+\mathbf{F}) + \frac{\mathbf{I}}{2}$. With the appropriate numerical values, we arrive at $t = 115.63$ and $C = 200.82$.

8.6 The **RTU** simplex, produced by contraction and change of direction.

8.7 The starting simplex is (**2, 3, 5, 6**), with $\mathbf{W} = \mathbf{3}$ and $\bar{\mathbf{P}} = \frac{1}{3}(\mathbf{2}+\mathbf{5}+\mathbf{6})$. Reflection of vertex **3** produces vertex **7**. Point **8** is obtained by expansion, through Eq. (8.3). The results are in the table.

8.8 $\mathbf{W} = \mathbf{9}$, and $\bar{\mathbf{P}} = \frac{1}{3}(\mathbf{8}+\mathbf{11}+\mathbf{14})$. We only have to apply Eq. (8.2a) to arrive at $C_{H_2SO_4} = -0.045M$.

8.9 $C_{H_2SO_4} = 0.1870M$, $C_{KI} = 0.0617M$ and $C_{H_2O_2} = 0.0039M$. Use Eq. (8.4), with $\bar{\mathbf{P}} = \dfrac{\mathbf{8}+\mathbf{11}+\mathbf{14}}{3}$ and $\mathbf{W} = \mathbf{15}$.

8.10 $\mathbf{21} = \frac{1}{2}(\mathbf{18}+\mathbf{11}) = (0.0868,\ 0.0835,\ 0.0037)$;

$\mathbf{22} = \frac{1}{2}(\mathbf{18}+\mathbf{14}) = (0.0572,\ 0.0790,\ 0.0048)$;

$\mathbf{23} = \frac{1}{2}(\mathbf{18}+\mathbf{16}) = (0.1216,\ 0.0760,\ 0.0042)$.

SUBJECT INDEX